The Impact of Plant Molecular Genetics

Cover photograph depicts flowering sugarcane (*Saccharum* spp.)
Photograph courtesy of Bruno W. S. Sobral.

Cover design by David Gardner, Dorchester, MA.

The Impact of Plant Molecular Genetics

Bruno W. S. Sobral
Editor

Springer Science+Business Media, LLC
1996

Bruno W. S. Sobral
CAMBIA Americas
11099 North Torrey Pines Road, suite 295
La Jolla, CA 92037
USA

Library of Congress Cataloging-in-Publication Data

The impact of plant molecular genetics / Bruno W. S. Sobral, editor.
p. cm.
Includes bibliographical references and index.

1. Plant genetic engineering. 2. Plant molecular genetics.
I. Sobral, Bruno W. S. (Bruno Walther Santos), 1958-
SB123.57.I57 1996
631.5′23--dc20 95-38473
CIP

Printed on acid-free paper

ISBN 978-1-4615-9857-2 ISBN 978-1-4615-9855-8 (eBook)
DOI 10.1007/978-1-4615-9855-8

Typeset by University Graphics, York, PA

9 8 7 6 5 4 3 2 1

I dedicate this book to the memories of Manoel Sobral, Isabel Perreira Barros, and Joaquim dos Santos, and to the future, which lies in children like Paulo Lacerda Sobral, Juliana Lacerda Sobral, and Bruna Lacerda Sobral.

Contents

Preface

The impact of molecular genetics on plant breeding and, consequently, agriculture, is potentially enormous. Understanding and directing this potential impact is crucial because of the urgent issues that we face concerning sustainable agriculture for a growing world population as well as conservation of the world's rapidly dwindling plant genetic resources.

This book is largely devoted to the applications of genetic markers that have been developed by the application of molecular genetics to practical problems. These are known as DNA markers. They have gained a certain notoriety in forensics, but can be used in a variety of practical situations.

We are going through a period of accelerated breakthroughs in molecular genetics. Therefore, the authors of each chapter were encouraged to speculate about both current bottlenecks and the future of their subfields of research. We can certainly apply molecular genetic tools and approaches to help resolve crucial genetic resource problems that face humanity. However, little has been discussed with respect to when or how we should use such tools, nor to who specifically should use them; therefore, social and economic analyses are important in the planning stages of projects that are aimed at practical results.

To cover such vast areas of research and discovery and to keep the book down to a manageable size, I have had to make tough decisions regarding contributions. To focus, I have chosen key contributors in areas that are most relevant to agriculture and biodiversity, concentrating mostly, but not solely, on tropical agriculture. My motivation is two-fold: (1) personal, because of my own childhood in the tropics and (2) social, because it is my belief that the potential impact of plant molecular genetics will be greatest and most rewarding in tackling difficult questions concerning the use and preservation of the world's plant genetic resources, most of which are in the tropics. I say potential because much of what actually transpires will depend heavily upon the choices made now, concerning which approaches and targets should be selected for study.

Part I of this book focusses on genetics and breeding. Although there are many books dealing with various aspects of plant biotechnology, this one is perhaps unique in that its goal is to identify and discuss key areas in which the use of molecular genetic techniques, primarily DNA-based genetic markers, has allowed breakthrough discoveries or approaches. For example, genetics of many polyploid species were not studied extensively until DNA markers were applied to these plants. Chapter 1 focuses on sugarcane as an example of such a polyploid. Around 1987, to the best of my knowledge, the first Southern hybridizations were done on sugarcane DNA using heterologous maize probes; the results showed, about 100 years after Mendel's work, that sugarcane actually has genes! Until that time, despite relatively large gains afforded by almost 100 years of sugarcane breeding, it was not known what was the basic chromosome complement and the mode of transmission genetics of this tropical grass; therefore breeders needed to use a holistic approach based on phenotype only. However, after less than six years of investment in sugarcane molecular genetics, it is now known that the basic species of *Saccharum* are polysomic polyploids of very high ploidy. In addition, a worldwide cytoplasmic monoculture exists for this crop. It is also suggested that some of the major traits of agronomic importance may be controlled by only a few loci. Marker-assisted selection therefore may be possible in the near future. These advances imply changes in sugarcane breeding strategies.

DNA markers have enabled rapid acquisition of very large genetic data sets, thus requiring further refinement of various applied statistical fields. These large data sets have prompted the development of more and more sophisticated and powerful computer algorithms to analyze and handle data. In addition, studies involving quantitative trait loci (QTL) are growing in magnitude and significance. Such studies aim to reduce the complexity of multigenic traits using statistical procedures, so that they can then be treated as Mendelian characters from the perspective of marker-assisted breeding. As these studies grow, changes in the culture of breeders are required. Biological and statistical validation for the results of QTL studies are also required. Chapter 2 focuses on the problem of deciding how to validate a QTL, and mentions the possibility of map-independent QTL detection as an option for largely uncharacterized crops in which good breeding records and populations are readily available. The question of linking phenotype to genotype, a crucial one in plant breeding, ecology and evolution, is also posed by Keith Crandall in Chapter 8. Crandall expands on the work of Alan Templeton in this field, taking a phylogenetic approach to unraveling this link.

Perhaps no area of genetics has received as great an impact by the application of DNA markers as forest tree genetics. This is because forest tree breeding, due to the generation time of the species, has been a very slow process. As a result, many of our forest tree resources have suffered little domestication and improvement. The excellent Forestry Biotechnology group directed by Ron Sederoff at North Carolina State University has changed many of our perceptions of forest tree genetics, and continues to do so. In Chapter 3, David O'Malley, one member of this team, discusses the conceptual advances that are being made in forestry because of the application of DNA markers. Although David focuses on forest trees, the idea is general for all plants with long generation times. An increase in the output of intensive forestry, perhaps allied with the use of new, renewable sources of fiber, such as sugarcane, will be required if we are to preserve much of the remaining native forests worldwide, as it seems that the human appetite for fiber is unlikely to diminish in the short term.

Temperate agriculture and the temperate world's remaining plant resources of economic importance are already in the hands of large corporations; as a whole, these private interests have sculpted the public efforts in a manner that removes potential conflicts from public and private plant biotechnology interests. So, I believe that most crops important to temperate agriculture will largely "take care of themselves". However, in the tropics, especially in crops that do not have a large commercial infrastructure but which are planted for subsistence, it seems that collaborative, participatory, de-centralized international efforts focusing on development and application of user-friendly, high-technology solutions to critical genetic questions could reap enormous benefits. In order to reap these benefits, funding mechanisms must become sensitive to this approach. For example, global projects aimed at unraveling phylogenetic relations within families of plants that contain tropical crop species (or incipient tropical crop species), together with coordinated efforts to generate and study comparative genetic maps of these families would be of great importance to the vast majority of tropical plant genetic resources that have not been developed and which may harbor useful genes and products that will be critical now or in the future. Chapter 4 focuses on current experience in comparative genomic analysis, another key area of research that was made possible only through the application of DNA markers. Comparative mapping allows groups working on phylogenetically related species to benefit from the work of others, facilitating even greater advances in such research. Information from comparative mapping programs could be provided on the Internet and accessed worldwide by breeders and conservationists. The resultant flow of information between fields will also spur on research in other areas.

Although some areas of plant molecular biology, such as transformation with exogenous DNA, fall outside of the focus of this book, there is a role to be played by such technology, especially when it is allied with genetic studies. In Chapter 5, Richard Jefferson, the father of GUS (β-glucuronidase), the key genetic marker that has allowed plant transformation to attain the status of a highly useful technology, delineates a crucial role for transformation, allied with genetic studies, in generation of apomictic plants for agricultural crops. In apomicts, the genotype of the mother plant is passed on to the offspring without recombination, thereby allowing the farmer to simply use selected seed from the field in the next generation, while still reaping the benefits thus far enjoyed only by crops in which hybrid seeds could be made. The crucial social difference is that the farmers can replicate the genotype independently of seed companies, much like what was done in the past or with inbred crops. Genetic studies are underway to characterize the genetic basis of apomixis in various species in which apomixis occurs naturally, with the ultimate goal of transferring thc trait into other major crops.

In current plant taxonomies, below the rank of families and genera of plant groups that include crops and their relatives is the genetic variation within the cultivated species themselves, which is largely uncharacterized for most tropical crops. Furthermore, little is known about the processes, both genetic and human (agricultural practices), that lead to genetic diversification and maintenance and amplification of genetic diversity in tropical crops. This is particularly true for crops that have facultative vegetative reproduction, many of which were domesticated in the tropics by indigenous peoples. Yet it is precisely the genetic variation at the level of species that will likely provide most of the raw materials that farmers and breeders will use to develop new varieties with desirable characteristics. By now it is clear to most people that this genetic diversity, while largely uncharacterized, is also greatly threatened by accelerated worldwide economic and cultural change. In Chapter 6 we get a look at a mechanism, called intragenic recombination, that may be very important in generating new alleles in maize, and that may have revolutionary implications in our understanding of diversification of grass genes.

Systematic relations at these lower taxonomic levels, such as species, are unclear for many crops and their relatives. However, it is at the species level that most conservation efforts need to operate. Crucial questions pertaining to the use of DNA markers for the purpose of characterization of species diversity is discussed in Chapter 7. In fact, through the pioneering applications of DNA markers to various crops, we have found that even our definition of species is called into question! The concept of species is excellently presented and discussed by Jerrold Davis in Chapter 10 and is relevant to all our conservation and breeding efforts.

At the level of genes and biochemical pathways, we are on the verge of being able to link knowledge from biochemistry and genetics at a much faster rate than previously possible. One perhaps unexpected yet very powerful research direction that has emerged from the application of DNA markers to plants is the integration of previously loosely linked research programs in genetics, phylogenetics, and development. Elizabeth Kellogg presents this subject superbly in Chapter 9. Further acceleration in the rate of integration of genetics, biochemistry and developmental biology will be largely due to new technologies, such as RNA arbitrarily primed PCR (also known as differential display), which allow an unprecedented amount of information to be obtained about temporal and spatial gene expression in a biochemical context (especially if pathways are known and genes from those pathways are cloned). By linking specific biochemical pathways and key regulatory enzymes in those pathways to responses in gene expression caused by specific stimuli (environmental or otherwise) or particular combinations of regulatory alleles, and then feeding the data through neural networks, our understanding of basic genetic questions, such as heterosis, or more practical questions, such as the genetic and biochemical basis of adaptive traits, will be understood in much greater detail than has been previously possible. Such an approach may also facilitate discovery of potentially useful natural products from largely uncharacterized plants; this approach should be faster and more fruitful than the method of fractionation of plant extracts.

Agriculture depends not only on the plants that we grow for food and fiber.

It also depends heavily on the interaction of those plants with their environment, including the community of microorganisms that attack or benefit plants in an agricultural setting. Part III provides two chapters that exemplify the impact that DNA markers have had on our understanding and capacity to detect microorganisms relevant to agriculture. First, there is the chapter by Rhonda Honeycutt and Michael McClelland on how to detect pathogens of plants. Next there is the practical application of molecular genetics to ecology of beneficial microorganisms, a dream developed by Kate Wilson.

As mentioned previously, we have perhaps reached a stage at which our capacity to generate genetic data using DNA markers has outstripped our capacity to analyze such data meaningfully. However, because of computer technology, we can see bioinformatics as a way of avoiding the data poisoning that rapidly occurs in laboratories working on data-intensive problems. Part IV develops a couple of examples of the directions this field is taking. In Chapter 13, Carol Bult and Chris Fields give us an idea of the resources that are out there already to help those looking for ways to handle data analysis and avoid data poisoning. In Chapter 14, the team at Perkin-Elmer (ABI Division) gives us a peek at the way the hardware is developing to accommodate the analysis of very large numbers of samples with large numbers of markers, in a reasonable time. An outstanding issue is how to make cost per data point diminish to the level that those working in plant sciences can afford, since the main development in this field is driven by the human genetics/diagnostics field, where cost per data point is irrelevant but accuracy is paramount.

In Part V, a couple of examples of how DNA markers can be used in practical situations are shown. First, in Chapter 15, the group working at Pioneer Hi-Bred discusses their viewpoint. Pioneer can be seen to represent one of the most capable and forward-thinking of the private companies that work with DNA markers for plant breeding. Chapter 16, written by Mauricio Lopes, shows how molecular biology and classical breeding can be used together to generate new varieties in maize. Mauricio represents the Brazilian national program in maize, and therefore presents a synthesis more closely related to the public breeding sector.

Finally, in Part VI, I have included contributions that allow us to think about social and economic implications of the powerful technologies that are being developed and applied to agricultural genetics worldwide. Chapter 17 discusses the economic impacts of technologies on international forestry, giving the perspective of the United Nations Food and Agriculture Organization (FAO). In Chapter 18, Hans Rosling, a visionary doctor who has worked in Africa on physiological problems related to cassava, a major starch crop for the third world, shows great insight into the fundamental question of when to apply our powerful technological toolbox, suggesting that in some cases it may not be needed. In my view, one needs a jeep and not a cadillac if one wants to drive to the farm. I thank Hans for bringing these issues to the forefront and forcing us to think out what we plan to do carefully and in an interdisciplinary manner. Anything else would be senseless in the face of the limited resources available for genetic research in many socially important crops, such as cassava. In the final Chapter, Carlos Moreira Filho gives us an analysis of the economic opportunities and adjustments that biotechnology is creating by using the vast developing agriculture of Brazil as the subject of study.

In summary, I believe we are living in an exciting and dangerous time in the realm of plant molecular genetics. Exciting because we have powerful new tools and approaches with which to better understand the basic biology of plants in much greater detail than ever before–and the possibility of using our knowledge for enhancing our sustainability on this planet. Dangerous because we have little time to harvest the fruits from new approaches, tools, and knowledge, since population growth and accelerated cultural change, primarily in the tropics, are potentially a serious threat to the survival of our precious genetic resources into the next century, when they will surely be needed!

Bruno Walther Santos Sobral, Encinitas CA USA 9-25-95

List of Contributors

Stephen Bates, Perkin Elmer, Applied Biosystems Division, Foster City, CA, USA

William Beavis, Pioneer Hi-Bred International, Inc, Research and Product Development, Agronomic Traits Group, 7300 NW 62nd Ave, P.O. Box 1004, Johnston, IA 50131-1004, USA

Jeffrey Bennetzen, Department of Biological Sciences, Purdue University, West Lafayette, IN, 47907-1392, USA

Ross Bicknell, Crop and Research Institute, Private Bag, 4704, Christchurch, New Zealand

William Bridges, Clemson University, Department of Experimental Statistics, F148-Poole Agricultural Center, Clemson, SC, 29634-0367, USA

Carol Bult, Department of Gene Discovery and Comparative Genomics, The Institute for Genomic Research, 932 Clopper Road, Gaithersburg, MD 20878, USA

Laura Civardi, Department of Biochemistry and Biophysics, Iowa State University, Ames Iowa, 50011, USA

Keith A. Crandall, Department of Zoology, The University of Texas, Austin, TX, 78712-1064, USA

Jerrold I Davis, L.H. Bailey Hortorium, Cornell University, 462 Mann Library, Ithaca, NY, 14583, USA

Guilherme L.S. Dias, Department of Economy, University of São Paulo, 05508-900, Brazil

Christopher A. Fields, National Center for Genome Resources, 1800 Old Pecos Trail, Santa Fe, NM 87505, USA

Carlos A. Moreira-Filho, Instituto de Ciências Biomédicas da USP, Avenida Professor Lineu Prestes 2415, 05508-900, São Paulo, Brazil

Francesca T. Grifo, Center for Biodiversity and Conservation, American Museum of Natural History, Central Park West at 79th Street, New York, NY 10024, USA

William J. Hahn, Laboratory of Molecular Systematics, MRC 534, Smithsonian Institution, Washington, DC 20560, USA

Rhonda Honeycutt, CAMBIA Americas, 11099 North Torrey Pines Road, Suite 295, La Jolla, CA 92037, USA

An-Ping-Hsia, Department of Zoology and Genetics, Iowa State University, Ames Iowa, 50011, USA

Richard A. Jefferson, CAMBIA, GPO Box 3200, Canberra City, ACT 2601, Australia

Elizabeth A. Kellogg, Harvard University Herbaria, 22 Divinity Avenue, Cambridge, MA 02138, USA

David A. Knorr, Perkin Elmer, Applied Biosystems Division, Foster city, CA, USA

Brian A. Larkins, University of Arizona, Department of Plant Sciences, Tucson, AZ 85721, USA

Maurício A. Lopes, National Maize and Sorghum Research Center, CNPMS/ EMBRAPA Caixa Postal 151, CEP 35701-970, Sete Lagoas-MG, Brazil

Michael McClleland, Sidney Kimmel Cancer Center, 11099 North Torrey Pines Road, Suite 290, La Jolla, CA 92037, USA

Basil J. Nikolau, Department of Biochemistry and Biophysics, Iowa State University, Ames, Iowa 50011, USA

David M. O'Malley, Forest Biotechnology Group, 6113 Jordan Hall, North Carolina State University, Raleigh, NC 27695-8008, USA

Hans Rosling, Unit for International Child Health, Uppsala University, S-751 85 Uppsala, Sweden

Patrick S. Schnable, Departments of Agronomy and Zoology and Genetics, Iowa State University, Ames, Iowa 50011, USA

Jorge A.G. da Silva, Copersucar Technology Center, Piracicaba, São Paulo, Brazil

Marcos E. da Silva, University of São Paulo, Department of Economy, 05508-900, São Paulo, Brazil

Stephen Smith, Pioneer Hi-Bred International, Inc., Research and Product Development, Agronomic Traits Group, 7300 NW 62nd Ave, P.O. Box 1004, Johnston, IA 50131-1004, USA

Bruno W.S. Sobral, California Institute of Biological Research and CAMBIA Americas, 11099 North Torrey Pines Road, Suite 295, La Jolla, CA 92037, USA

Victor M. Villalobos, CINVESTAV, Unidad Irapuato, P.O. Box 629, Irapuato 36500, Irapuato, Gto. Mexico

Jennifer W. Weller, Perkin Elmer, Applied Biosystems Division, Foster City, CA, USA

Kate J. Wilson, CAMBIA, GPO Gox 3200, Canberra, ACT 2601, Australia

Adriana P.A. Xavier, University of São Paulo, Biomedical Sciences Institute, 05508-900, São Paulo, Brazil

Yiji Xia, Department of Zoology and Genetics, Iowa State University, Ames, Iowa 50011, USA

Xiaojie Xu, Department of Zoology and Genetics, Iowa State University, Ames, IA 50011, USA

Lei Zhang, Department of Zoology and Genetics, Iowa State University, Ames, IA 50011, USA

Janet S. Ziegle, Perkin Elmer, Applied Biosystems Division, Foster City, CA, USA

Part I

GENETICS AND BREEDING

1

Genetics of Polyploids

JORGE A. G. DA SILVA AND BRUNO W.S. SOBRAL

Introduction and Terminology

Large gains in the knowledge of plant genetics have been made during the last decade due to the development and application of DNA markers. The use of DNA markers has allowed the construction of linkage maps, assessment of genetic variability, and gene tagging in a variety of species (Sobral and Honeycutt, 1994). Densely populated DNA marker linkage maps have enabled map-based gene cloning and marker-assisted selection in some plants. Furthermore, DNA markers have contributed information to the fields of ecology, population genetics, and evolution (see the relevant chapters in Parts I and II).

While the great majority of plants that have benefited from the application of DNA marker technology are diploids, there are also heteroploids. Heteroploidy is defined as "deviation from the normal chromosome number in a cell, tissue, or whole organism" (Sharp, 1934). Polyploids are a type of heteroploid that have three or more genomes. Polyploids have been classified in a variety of ways, and polyploid terminology has not been consistently used. The genetic complexity inherent to most polyploids has, until recently, precluded genetic studies. However, polyploids are biologically interesting because of these complexities as well as because they are economically important in many cases as major crops.

As an evolutionary process, polyploidization is a drastic change that has been differently adopted by different taxa. Polyploidy is almost nonexistent in mammals, rare and concentrated in certain groups of insects, amphibians, fishes, and reptiles. However, it is common in plants, both pteridophytes and angiosperms (Lewis, 1980). It has been suggested that a major cause for differences in occurrence of polyploidy may be sex determination mechanisms (Mac Key, 1987).

There are two major types of polyploids: aneuploids, in which the chromosome number is not a whole number of the basic chromosome number (known

The Impact of Plant Molecular Genetics
BWS Sobral, Editor

as x); and euploids, in which the entire chromosome set has been repeated (Sharp, 1934). Table 1 exemplifies use of this terminology. As usual, n is the gametic chromosome number, and $2n$ is the somatic chromosome number.

For the purposes of our discussion, we will further divide polyploids into two other types based on what is known about their origin, phylogenetic relations, and evolution. One type of polyploid will be called simple, and it has extent diploid relatives and well characterized genetic, cytogenetic, and phylogenetic relationships, such as is the case of hexaploid wheat or polyploid Brassicas. These examples are of simple *allo*polyploids. Allopolyploids have evolved or been domesticated via interspecific or intergeneric hybridization and therefore contain genetically different sets of chromosomes from distinct species (Kihara and Ono, 1926). *Auto*polyploids are derived by multiplication of the haploid chromosome complement (see Table 1).

Another type of polyploid will be called complex, in which diploid relatives are extinct (or unknown), phylogenetic and genetic relations also are largely unknown, and cytology is either difficult or unstudied. Such more complex and frequently less well characterized polyploids may be allopolyploids or autopolyploids, with respect to their origin.

Stebbins (1947) broadened and expanded on the terminology of Clausen et al (1945), which classified polyploids into autopolyploids, segmental allopolyploids, allopolyploids, and autoallopolyploids. The terminology and its use has become confusing and controversial since its creation. Stebbins (1980) was aware of problems in defining polyploid terminology and considered that information from a wide range of sources needed to be considered in studying

Table 1. Terminology Used in Polyploids*

Term	Representation	Somatic ($2n$) chromosomal complement**
Aneuploids		
nullisomic	$2x - 2$	(12)(12), $2n = 4$
monosomic	$2x - 1$	(123)(12), $2n = 5$
double monosomic	$2x - 1 - 1$	(12)(13), $2n = 4$
trisomic	$2x + 1$	(123)(123)(1), $2n = 7$
double trisomic	$2x + 1 + 1$	(123)(123)(1)(2), $2n = 8$
tetrasomic	$2x + 2$	(123)(123)(1)(1), $2n = 8$
Euploids		
monoploid	x	(123), $n = x = 3$
diploid	$2x$	(123)(123), $2n = 6$
triploid	$3x$	(123)(123)(123), $2n = 9$
autotetraploid	$4x$	(123)(123)(123)(123), $2n = 12$
allotetraploid	$2x + 2x'$	(123)(123)(1′2′3′), $2n = 12$
autooctoploid	$8x$	(123)(123)(123)(123)(123) (123)(123)(123), $2n = 24$
allooctoploid	$4x + 4x'$	(123)(123)(123)(123)(1′2′3′) (1′2′3′)(1′2′3′)(1′2′3′), $2n = 24$

* Adapted from Allard, 1966.

** 1, 2, and 3 are nonhomologous chromosomes of a hypothetical species with $x = n = 3$ chromsomes.

polyploid origin. Jackson and Casey (1980) noted that factors that affect chromosome pairing are of major importance in classifying and understanding the evolution of polyploids, with which we strongly agree. Therefore, with respect to our discussion, we need to introduce the terms *polysomic* and *disomic* inheritance. We will take the evolutionary perspective and think of autopolyploidy as synonymous with polysomic inheritance and allopolyploidy as synonymous with disomic inheritance because it is this chromosomal behavior that causes the major differences between these types of polyploids. These differences are characterized by divergent expectations for polysomic and disomic polyploids (Mac Key, 1987; Uhl, 1992). For example, disomy allows allelic interactions to be fixed. It also may increase segregational load, enhance divergent genetic evolution, and break down polyploidy (leading to diploidization). On the other hand, polysomic polyploidy typically does not function in combination with autogamy and tends to exaggerate the effects of allogamy, thus extending conditions for heterozygosity and for heterosis.

At meiosis diploid organisms pair homologous chromosomes two-by-two, yielding what are known as bivalents. Diploids display disomic inheritance of markers and disomic segregation ratios. Because polyploids can have multiple copies of chromosomes containing the same (or similar) genetic information, pairing at meiosis can include various distinct combinations of bivalents composed of homologous chromosomes or multivalents in which three or more chromosomes are paired. Polyploids, therefore, can give polysomic segregation ratios, and multivalent formation allows the possibility of additional complexity because of potential double reduction. Furthermore, if the polyploid has arisen through hybridization of different species, then, in addition to homologous pairing partners, there is the potential for pairing of genetically equivalent chromosomes that came from distinct species in a process called homoeologous pairing.

It generally has been thought that autopolyploids would display chromosome pairing difficulties, thus resulting in frequent multivalent formation (Soltis and Rieseberg, 1986). However, it has also been shown that pairing may be under simple genetic control (Wall et al, 1971). Jackson (1982) proposed that autopolyploids might have genes that favor bivalent pairing, just as allopolyploid wheat (Riley and Chapman, 1958; Sears and Okamoto, 1958). Bivalent pairing in autopolyploid species has been documented in a few cases (Soltis and Rieseberg, 1986; Crawford and Smith, 1984), although type of inheritance was not studied in some of those cases. Clearly, though, chromosome pairing behavior at meiosis has a direct impact on inheritance of traits in polyploids. This emphasizes the importance of integrating data from cytological, morphological, ecological, and genetic studies.

Complex polyploids constitute a large proportion of plants particularly within the grasses, for example, sugarcane (*Saccharum* spp.). These polyploid grasses frequently display large variation in chromosome number, as well as a broad range of potentially productive hybridizations, and they frequently are placed in polyploid complexes of daunting taxonomic and genetic difficulty. As an example, within the *Saccharum* complex (Mukherjee, 1957), there are between two and four genera in which chromosome numbers within a species may vary from $2n = 20$ to $2n > 200$ (Sreenivasan et al, 1987).

Until recently, our understanding of genetics of most polyploid complexes has suffered from a severe lack of morphological markers showing disomic inheritance. Because DNA-based genetic markers can now be acquired in increasingly large numbers, with increasing speed, and at decreasing cost per data point, on essentially any previously uncharacterized genome (Sobral and Honeycutt, 1993), we expect them to have a very large impact on the characterization of polyploid taxa of economic significance, provided there is adequate funding. Genetic and phylogenetic characterization of many tropical polyploid taxa is essential to direct and rationalize utilization and conservation of local biodiversity.

As we begin to reduce the complexity of polyploid genetics using DNA markers, we expect to uncover a wealth of information on population genetics and evolution of polyploids. Already, DNA markers have had an impact on our understanding of the genetics and evolution of polyploids, and we predict they will allow many hypotheses, previously only testable in theory, to be tested experimentally. For example, Rieseberg et al (1995) used DNA marker mapping data to demonstrate extensive genomic reorganization in the hybrid species *Helianthus anomalus*, which satisfied expectations of genetic models for speciation through hybrid recombination. They noted the importance of comparative linkage mapping (see chapter by J Bennetzen) for studying genomes during speciation because maps allow inference of the changes that accompany speciation. Maps also allow the genomic contributions of parental species to be determined on a chromosome by chromosome or finer level of resolution. Such maps have only become possible because of DNA markers.

Comparative maps also have been used to study the genomic organization and evolution of cotton, a disomic polyploid (Reinisch et al, 1994). This elegant study suggests that an earlier polyploidization event occurred within allotetraploid cotton, some 25 million years ago, thus lending support to the notion that many currently diploid species have undergone paleopolyploidy followed by diploidization. Additionally, DNA-marker-assisted studies suggest recurrent formation of polyploid species and a greater occurrence of natural autopolyploidy than previously imagined (Soltis and Soltis, 1993). Furthermore, largely because of DNA marker studies, autopolyploids are now viewed differently. Polysomy may be the source of genetic attributes such as enzyme multiplicity, increased heterozygosity, and increased allelic diversity, which may confer great potential success to autopolyploids in nature (Soltis and Soltis, 1993). Therefore, perhaps, it is not surprising that Da Silva et al (1993, 1995) and Al-Janabi et al (1993), using DNA marker linkage maps, showed that a wild relative of sugarcane, *Saccharum spontaneum* 'SES 208' ($2n = 64$), displays polysomic inheritance despite having bivalent pairing at meiosis, thus suggesting an autooctoploid behavior. This result may be seen as surprising by those who thought that bivalent pairing in polyploids was necessarily associated with allopolyploid evolution. However, the results of Timmis and Rees (1971) and Avivi (1976a, 1976b) had already suggested genetic control of low pairing in autotetraploids.

In the last few years the theory and practice of direct genetic mapping of even the most complex of polyploids has been developed, largely aided by the application of DNA markers. In this chapter we will summarize the current status of knowledge, describing the role DNA markers have played in our understanding of polyploid genetics, as well as the limitations of current methodologies, and we will discuss potential future directions for research. Most of the theory presented comes from our work with sugarcane and its relatives within the *Saccharum* complex, so sugarcane will be used as a model. We believe the greatest impact of DNA-marker-assisted genetic studies will be on previously intractable complex polyploid plants, like sugarcane, and perennial species. We also feel it fortunate that many such plants are crops that are socio-economically and ecologically important in the biodiversity-rich tropics.

Genetic Mapping in Polyploids

Introduction

Genetic mapping of polysomic polyploids is more difficult than it is in diploids because: (1) there is a multiplicity of segregating genotypes; (2) there are many possible modes of gamete formation; and (3) the frequencies of different gametic genotypes cannot be identified by a single backcross to a tester (Fisher, 1949). For example, in an polysomic tetraploid, five genotypes are possible at each locus: *AAAA* (quadruplex), *AAAa* (triplex), *AAaa* (duplex), *Aaaa* (simplex), and *aaaa* (nulliplex). In addition to the multiplicity of genotypes, two cytological events affect gametic output in polysomic polyploids: (1) frequency of multivalent formation, and (2) randomness of disjunction of multivalents. We will limit ourselves to species that exhibit bivalents only so that even under polysomic inheritance the key feature is that sister chromatids never appear in the same gamete. If we further impose that non-sister homologous chromatids pair with each other at equal frequencies, then we have defined random chromosome assortment (Allard, 1966), a situation that is typically used to define autopolyploid (polysomic) inheritance.

Random chromosome assortment is expected for polysomic polyploids that fail to show multivalent pairing. Random chromosome assortment or nonpreferential chromosome pairing is the other extreme of disomic inheritance or complete preferential pairing. However, intermediate scenarios can occur in which selective chromosome pairing occurs for some linkage groups but not others and occurs in varying levels of preference. Selective chromosome pairing suggests that the different homologues in the polyploid series are not equally homologous. It is believed that segmental allopolyploids or newly arisen amphidiploids are the most likely to show this intermediate type of pairing. Over time, bivalent pairing allied with selective or preferential chromosome pairing can diplodize autopolyploids (Allard, 1966).

As we have seen, polyploids can display various pairing configurations at meiosis. For our purposes, we will limit this discussion to polyploids in which only bivalents are formed at meiosis or, if multivalents and univalents are formed, they are rare. This simplifies genetic analysis and further highlights the importance of integrating cytological and molecular genetic analyses. Although we will not consider it further herein, multivalent formation complicates genetic analyses because it allows three distinct types of segregation, known as chromosome segregation (Muller, 1914), random chromatid segregation (Haldane, 1930), and maximum or complete equational segregation (Mather, 1935). In chromosome segregation, the chromatids in a particular multivalent belong to separate chromosomes in that multivalent. However, in the other types of segregation, gametes may be produced that are derived from sister chromatids, and chromatid segregation may occur (Bridges, 1916), yielding recombinant classes that could not be derived otherwise, known as double reduction products. Cytologically, double reduction requires (Mather, 1935, 1936; Little, 1945, 1958): (1) multivalent formation; (2) recombination between the gene and the centromere; (3) that chromatids involved in such a recombination pass to the same pole at Anaphase I; and (4) random separation of chromatids at Anaphase II.

Fortunately, it seems that many polyploids, particularly within the grasses, have developed a mechanism to control bivalent pairing (Jackson, 1982), which may be similar to the genetic mechanism that has been studied in wheat (Riley and Chapman, 1958; Sears and Okamoto, 1958). In wheat, it has been hypothesized that this mechanism is required to stabilize allopolyploids and allow meiosis to proceed normally, thereby making them disomic. However, perhaps unexpectedly, bivalent pairing has been shown in some euploid *Saccharum* species as well (Price, 1963), some of which seem to be largely polysomic polyploids (Al-Janabi et al, 1993, 1994a; Da Silva et al, 1993, 1995). Previously, excessive numbers of bivalents had been found in man-made autotetraploids of *Hordeum* and *Secale*, suggesting a genetic mechanism that might control pairing behavior even in these simple autopolyploids (Timmis and Rees, 1971; Avivi, 1976a, 1976b; Jackson, 1982). It has been suggested that chromosome size, location of initial pairing sites and chiasmata, and their frequency may be important for stable bivalent pairing (Mac Key, 1987). For example, acrocentric or very small chromosomes, such as observed in *Saccharum* (Janaki-Ammal, 1936; Price, 1963), are likely to make multivalent formation difficult while favoring the formation of open bivalents (Mac Key, 1987; Uhl, 1992).

Before the advent of DNA markers, several strategies had been used to facilitate linkage mapping in polyploids. Among them, the use of diploid relatives, aneuploid stocks, and haploids have been the most common in the literature (Sorrells, 1992). There are advantages and limitations to each approach, but they commonly require significant genetic characterization of the species or its relatives. However, DNA markers have made it possible to develop and apply novel approaches to genetically uncharacterized complex polyploids (Sobral and Honeycutt, 1993; Al-Janabi et al, 1994; Da Silva et al, 1993, 1995).

Genetic linkage maps have been constructed in cultivated diploid species by analyzing populations derived from crossing inbred lines. For complex polyploids, no equivalent populations exist. DNA-marker-based linkage maps have been constructed in simple polyploids by resorting to diploid relatives, as in potato (Bonierbale et al, 1988; Gebhardt et al, 1989) and wheat (Kam-Morgan and Gill, 1989). These studies introduced the usage of single-dose restriction fragment length polymorphisms (SDRFs) as a DNA-marker-based tool to study inheritance of what are essentially simplex alleles. However, PCR-based markers can also be used as single-dose (SD) markers in linkage studies, as shown by Sobral and Honeycutt (1993). The theoretical framework for linkage analysis using SD markers was described by Ritter et al (1990) and further refined for the case of polysomic polyploids by Wu et al (1992).

Generalization of the SD marker approach to linkage mapping of any biparental cross involving any two heterozygous individuals of any ploidy level was suggested by Sobral and Honeycutt (1993) and Al-Janabi et al (1993). This approach extends linkage mapping from inbred lines to virtually any cross between heterozygous individuals and has recently been named pseudo-testcross or double-pseudo-testcross by Grattapaglia and Sederoff (1994), depending on whether one or both parents in the cross are studied. The terminology is appropriate because the testcross mating configuration is not known *a priori*, rather it is inferred *a posteriori* from analysis of segregation. The beauty of the approach is that it can be immediately implemented on any species without any prior genetic information as long as enough progeny can be generated to allow estimation of recombination frequencies.

To apply SD markers to sugarcane, a haploid population derived by anther culture of *S. spontaneum* 'SES 208' ($2n = 64$), a wild accession from India, was available (Moore et al, 1989). Haploid populations are preferred for mapping mainly because of reduced development time and simplification of data analysis. Mapping efficiency is highest for a haploid population from a highly heterozygous genotype because virtually all SD markers can be mapped, and the progeny represents a sample of gametes from the parent genotype (Wu et al, 1992). This fact has not escaped the attention of pine tree geneticists who have exploited the natural haploidy in megagametophytic tissue of pines to quickly generate high-resolution genetic linkage maps of single trees (Grattapaglia et al, 1991).

In most species, however, haploid populations need to be constructed because naturally occurring haploid tissues, such as pollen, do not provide enough tissue for DNA extraction. Furthermore, haploids may not be readily available or easily produced for many if not most species. For example, the doubled haploid derived from SES 208, called ADP 85-0068 (Moore et al, 1989), was almost the only product from an intense effort to generate haploids in various species of *Saccharum* (Moore and Fitch, 1990). In addition, when considering the use of anther-culture-derived haploid lines for mapping, one must be concerned with segregation distortion that may have occurred during the in vitro phase. This may alter class frequency, thus reducing the probability of detection of SDRF or recombination. Deviation from the expected segregation ratio has been reported as occurring in doubled haploid barley lines, resulting from selection due to the anther culture process (Zivy et al, 1992). In the case of SES 208 haploids, RFLP segregation analysis showed a strong segregation distortion, and suggested aneuploids among the haploids. For this reason another mapping population was used (Da Silva et al, 1993).

The first step in mapping a complex polyploid is to assess the dosage or ploidy level (simplex, duplex, etc.) of each polymorphism. This is done by examining its segregation. Table 2 presents the expected segregation ratios for different dosages in disomic and polysomic octoploids.

The observed segregation ratio for presence:absence is tested against the expected one for each dosage level, using a χ^2 test. Once the dosage of each marker has been identified, the next step is to run linkage tests to determine linkage relationships. Finally, the homolog map can be coalesced into a chromosome map by using repulsion-phase linkages. Linkage analyses differ depending on the dosage of the markers involved on the test. Two types of linkage detection methods have been developed and applied to sugarcane: 1) using SD markers and 2) using Multi-dose (MD) markers. Following is a description of peculiarities of both methods.

Linkage Mapping Using Single-Dose (SD) Markers

SD markers (equivalent to simplex alleles in autopolyploids) segregate 1:1 for presence:absence in both auto and allopolyploids, by definition (Table 1). In diploids, segregation studies with RFLP markers take into account the whole band

Table 2. Gametic Segregation of a Di- and Polysomic Octoploid in the Absence of Double Reduction

Parent	Gametes					Ratio
	aaaa	*Aaaa*	*AAaa*	*AAAa*	*AAAA*	*A—:aaaa*
Disomic octoploid						
Aa aa aa aa	1	1	—	—	—	1:1
AA aa aa aa	—	1	—	—	—	—
Aa Aa aa aa	1	2	1	—	—	3:1
AA Aa aa aa	—	1	1	—	—	—
Aa Aa Aa aa	1	3	3	1	—	7:1
AA AA aa aa	—	—	1	—	—	—
AA Aa Aa aa	—	1	2	1	—	—
Aa Aa Aa Aa	1	4	6	4	1	15:1
Polysomic octoploid						
*Aaaaaaaa**	1	1	—	—	—	1:1
AAaaaaaa	3	8	3	—	—	11:3
AAAaaaaa	1	6	6	1	—	13:1
AAAAaaaa	1	16	36	16	1	69:1
AAAAAaaa	—	1	6	6	1	—
AAAAAAaa	—	—	3	8	3	—
AAAAAAAa	—	—	—	1	1	—

* Although expressed as octivalents for simplicity, pairing has been shown to be exclusively as bivalents (Al-Janabi et al, 1993).

profile since each fragment, generated by single-copy DNA probes, represents one allele. Because of the total correspondence between genotype and band phenotype, RFLP markers are codominantly inherited in diploids, and so heterozygotes can be detected by the presence of two different alleles at a locus. In polyploids, given the great number of RFLP fragments revealed by single-copy probes, one needs to consider the segregation of each fragment separately from the others. Because the presence of a fragment may represent the heterozygous and homozygous genotypes, the markers are dominant. In this respect, SDRFs and other SD markers are no different from PCR-based markers (except microsatellites) in diploids, which also are dominant.

We have used this method to generate an RFLP-, a PCR-based, and an integrated RFLP-PCR linkage map for SES 208 (Al-Janabi et al, 1993; Da Silva et al, 1993 and Da Silva et al, 1995). It is also being used to make linkage maps of the genomes of *S. officinarum* (Al-Janabi et al, 1994a; Kerher, 1994), the key genomic contributor to modern sugarcane varieties, and *S. robustum* (Al-Janabi et al, 1994a; Guimarães et al, 1995), hypothesized to be the progenitor species from which domestication occurred (Brandes, 1929).

One limitation of the SD method for polysomic polyploids is that it explores only linkages in coupling-phase. For detection of linkages in repulsion-phase, an extremely large mapping population is required (Wu et al, 1992). We used this peculiarity of SD markers to investigate the type of inheritance in *S. spontaneum* SES 208 (Al-Janabi et al, 1993; Da Silva et al, 1993, 1995) and in euploid *S. officinarum* and *S. robustum* (Al-Janabi et al, 1994a; Guimarães et al, 1995) (discussed later in this chapter).

Detection of only coupling-phase linkages results in a linkage map with $2n$ number of linkage groups (Al-Janabi et al, 1993; Da Silva et al, 1993, 1995). Information on pairing relationships is limited to the availability of highly polymorphic DNA probes. Herein, such DNA probes are defined as those that generate more than one SDRF which, once mapped, allows the allocation of linkage groups to homologous pairing, as illustrated in Figure 1 (Da Silva, 1993). However, their frequency is limited, particularly in species with high heterozygosity. Furthermore, the relationships revealed by this approach will include true pairing partners as well as duplicated loci on nonpairing chromosomes. Once a framework of SD markers is constructed, the next step is to map the MD markers, namely those with double and triple doses.

Linkage Mapping Using Multi-Dose (MD) Markers

Markers showing segregation ratios > 1:1 are likely to be in two or more doses. With the reminder that we are limited to bivalent pairing, in the case of a disomic octoploid, for instance, a double-dose (DD, or duplex) marker will segregate 3:1 for presence:absence (if the two alleles are on homoeologs that do not pair), compared to a segregation of 11:3 in a polysomic octoploid. A triple-dose (TD) marker will segregate 7:1 (if the three alleles are on homoeologs) in a disomic octoploid, compared to 13:1 in a polysomic octoploid (Table 2).

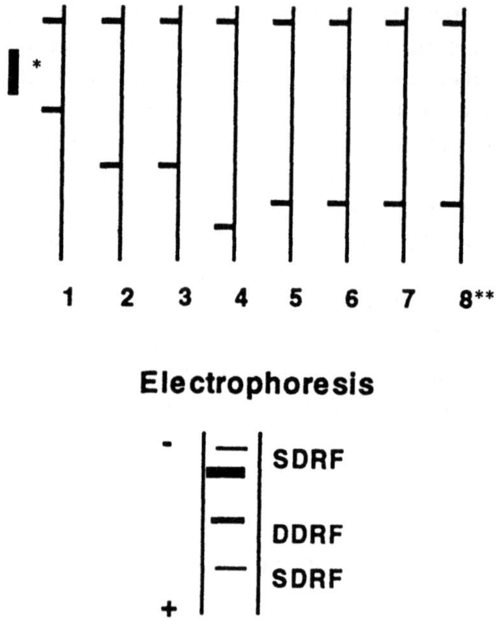

Figure 1. Detection of SDRF and DDRF on Southern blots of a polysomic octoploid. SDRF = Single Dose Restriction Fragments; DDRF = Double Dose Restriction Fragments.
* Vertical short bar represents a DNA probe detecting polymorphism.
** Long numbered vertical bars represent the eight homologous chromosomes of a polysomic octoploid species; horizontal bars represent restriction enzyme target sites. The RFLPs detected on chromosome 1 and 4 would appear on a Southern blot as two bands with different sizes, each segregating 1:1 for presence:absence in a polysomic octoploid progeny population. The RFLPs detected in homologous chromsomes 2 and 3 would appear as a single band on the blot, but segregate in 11:3 in the progeny of a polysomic octoploid. This hypothetical situation illustrates the assumption that multiple SDRFs detected by the same DNA probe, as well as the DDRFs, are located in homologous chomosomes.

Once DD and TD markers have been identified, two-point linkage tests are conducted using the expected frequencies for each of the four genotypic classes involved. These frequencies reflect the probability of each chromosome pairing configuration at Metaphase I. Da Silva (1993) derived the formulas for mapping DD and TD markers in polysomic segregation, in absence of double reduction (i.e., strict bivalent pairing) for any level of euploidy.

Once the type of chromosome pairing (i.e., random or preferential) and the ploidy are characterized for a given species, one can then use the expected ratios for DD and TD markers to test the observed gametic segregation. If the type of pairing and level of ploidy are uncertain, both segregation and linkage data can be used to address these questions, as explained below. Table 3 presents the frequencies for DD and TD markers in different euploidy levels.

Linkage analysis in diploids involves two gametic phases (or gametic se-

Table 3. Gene Frequencies (p) For the Presence of DD and TD markers in Different Polysomic Euploidy Levels

	Ploidy**						
Dosage	4	6	8	10	12	14	16
Double	$\frac{5}{6}$	$\frac{4}{5}$	$\frac{11}{14}$	$\frac{7}{9}$	$\frac{17}{22}$	$\frac{10}{13}$	$\frac{23}{30}$
Triple	$\frac{1}{1}$	$\frac{19}{20}$	$\frac{13}{14}$	$\frac{11}{12}$	$\frac{10}{11}$	$\frac{47}{52}$	$\frac{27}{30}$

* The frequency for absence of the marker equals $q = 1 - p$.
** basic chromosome number (4 = tetraploid, etc.)

ries), coupling and repulsion. In this case the first step is to test for linkage disequilibrium, usually with one degree of freedom χ^2 test in a two-way contingency table. If linkage disequilibrium is detected, the next step is to estimate the recombination fraction r. The two recombination classes are informative in the sense that both are used to estimate r.

In polysomic polyploids many more gametic phases are possible, thus complicating genetic analyses. In a polysomic octoploid, for instance, 30 gametic configurations are possible (Da Silva, 1993). The amount of linkage information that can be obtained from each gametic series is dictated by our ability to distinguish the ratio for the phenotypic classes under linkage from that same ratio under independent segregation. Under polysomic segregation, coupling phases provide more linkage information than repulsion phases.

Da Silva (1993) extended the linkage analysis in polysomic polyploids to the five most informative gametic series, deriving the formulas (Tables 4, 5, and 6) for linkage tests involving: (1) SD and DD markers in coupling (1&2-Asymmetrical Coupling); (2) SD and DD markers in repulsion (1&2-Asymmetrical Repulsion); (3) SD and TD markers in coupling (1&3-Asymmetrical Coupling); (4) DD markers only (Double Coupling); and (5) DD and TD markers in coupling (2&3-Asymmetrical Coupling). The mapping of DD markers in a SD-marker framework map requires two-point linkage tests between markers with different expected frequencies, resulting from the probability of each genotypic class.

PROBABILITIES OF GENOTYPIC AND PHENOTYPIC CLASSES

Figure 2 illustrates the linkage between a SD (A) and a DD marker (B) in coupling for a polysomic octoploid.

In this situation three events are possible:

(1) The AB-chromosome will pair with the B-chromosome (the two informative chromosomes) with probability 1/7, under nonpreferential pairing (polysomic segregation). Because they will migrate to different poles at Anaphase I, half of the gametes will bear the AB- and the other half the B-chromosome. The resulting frequencies of the genotypic classes AB and B are then $1/7 \times 1/2 = 1/14$ each.

Table 4. Formulas for Genotypic and Phenotypic Classes of Polysomic Polyploids for 1&2-Asymmetrical Coupling and Double Coupling Gametic Series

Class*	1&2-Asym. Coupling	Double Coupling
	Genotypic Classes	
AABB	—	$q_2 - 2q_2 + q_2r^2$
AAB	—	$2q_2r - 2q_2r^2$
ABB	$q_2 - q_2r$**	$2q_2r - 2q_2r^2$
AB	$p_2 - q_1$	$p_2 - q_2 - 4q_2r + 4q_2r^2$
AA	—	q_2r^2
A	q_2r	$2q_2r - 2q_2r^2$
BB	q_2r	q_2r^2
B	$p_2 - q_2$	$2q_2r - 2q_2r^2$
Nil	$q_2 - q_2r$	$q_2 - 2q_2r + q_2r^2$
	Phenotypic Classes	
AB	$p_1 - q_2r$	$p_2 - 2q_2r + q_2r^2$
A	q_2r	$2q_2r - q_2r^2$
B	$p_2 - q_1 + q_2r$	$2q_2r - q_2r^2$
Nil	$q_2 - q_2r$	$q_2 - 2q_2r + q_2r^2$
Sum	$p_2 + q_2 = 1$	$p_2 + q_2 = 1$

* For 1&2-Asymmetrical Coupling: A = SD marker; B = DD marker. For Double Coupling A and B = double dose.

** p_1, p_2, p_3 = frequencies for markers in SD, DD and TD (see text); r = recombination fraction $(0 \leq r \leq 0.5)$.

Table 5. Formulas for Genotypic and Phenotypic Classes of Polysomic Polyploids for 1&3-Asymmetrical Coupling and 2&3-Asymmetrical Coupling Gametic Series

Class	1&3-Asym. Coupling	2&3-Asymmetrical Coupling
	Genotypic Classes	
AABBB	—	$q_3 - 2q_3r + 2q_3r^2$
AABB	—	$q_2 - q_3 + 2q_3r - q_3r^2$
AAB	—	$q_2r - q_3r^2$
AABB	$q_2 - q_3r$	$2q_3r - 2q_3r^2$
ABB	$p_1 - 3q_3 - 2q_3r$	$p_2\backslash 2 - q_2\backslash 2 - 2q_3r + 2q_3r^2$
AB	$2q_3 + 2q_3r$	$p_2\backslash 2 - q_2\backslash 2 - 2q_3r + 2q_3r^2$
AA	—	q_3r^2
A	q_3r	$2q_3r - 2q_3r^2$
BBB	q_3r	q_3r^2
BB	$2q_3 + 2q_3r$	$q_2r - q_3r^2$
B	$p_3 - q_1 - 2q_3 - 2q_3r$	$q_2 - q_3 + 2q_3r - q_2r - q_3r^2$
Nil	$q_3 - q_3r$	$q_3 - 2q_3r + q_3r^2$
	Phenotypic Classes	
*AB**	$p_1 - q_3r$	$p_2 - 2q_3r + q_3r^2$
A	q_3r	$2q_3r - q_3r^2$
B	$p_3 - q_1 + q_3r$	$q_2 - q_3 + (q_2 - q_3)r - q_3r^2$
Nil	$q_3 - q_3r$	$q_3 - (q_2 - q_3)r + q_3r^2$
Sum	$p_3 + q_3 = 1$	$p_2 + q_2 = 1$

* For 1&3-Asymmetrical Coupling: A = SD marker; B = TD marker. For 2&3-Asymmetrical Coupling A = DD marker; B = TD marker.

** p_1, p_2, p_3 = frequencies for markers in single, double and triple doses (see text); r = recombination fraction $(0 \leq r \leq 0.5)$.

Table 6. Formulas for Phenotypic Classes for 1&2-Asymmetrical Repulsion in Polysomic Tetraploid and Octoploid Species

Class*	Polysomic Tetraploid	Polysomic Octoploid
AB	$\frac{2}{6} + \frac{1}{6}r$**	$\frac{5}{14} + \frac{1}{14}r$
A	$\frac{1}{6} - \frac{1}{6}r$	$\frac{2}{14} - \frac{1}{14}r$
B	$\frac{3}{6} - \frac{1}{6}r$	$\frac{6}{14} - \frac{1}{14}r$
Nil	$\frac{1}{6}r$	$\frac{1}{14} + \frac{1}{14}r$

* *A* = SD marker; *B* = DD marker.
** r = recombination fraction ($0 \leq r \leq 0.5$).

(2) The two informative chromosomes will not pair, which happens $1 - 1/7 = 6/7$ of the time, and go to the same pole at Anaphase I, which happens 1/2 of the time. Because these two conditions are independent, the event that satisfies them both has probability $6/7 \times 1/2 = 3/7$. Here, half of the gametes will bear both

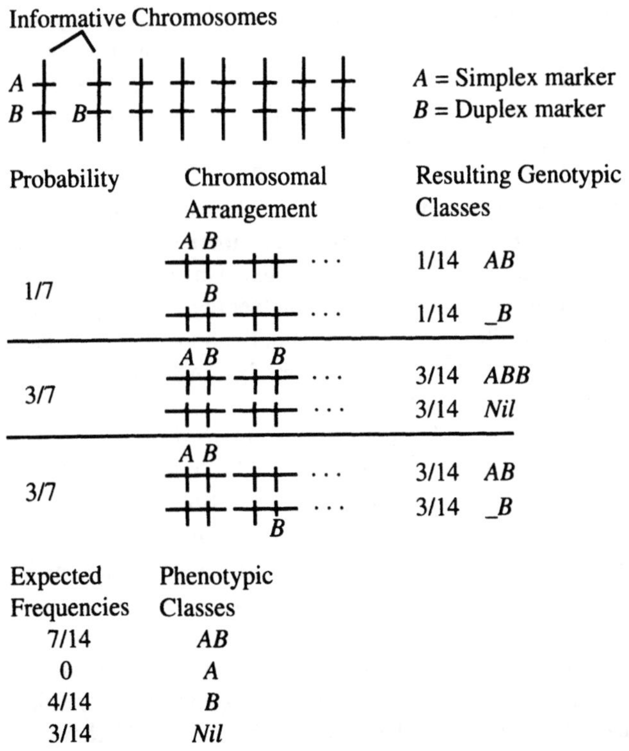

Figure 2. Expected frequencies of phenotypic classes resulting from linkage at 0 cM in 1&2-Asymmetrical Coupling configuration.

informative chromosomes and the other half none of them, resulting in 3/7 × 1/2 = 3/14 probability for each of the classes *ABB* and *Nil*.

(3) The two informative chromosomes will not pair and go to different poles at Anaphase I. In this case the probabilities are the same as in number (2), but the resulting genotypic classes are *AB* and *B*.

The phenotypic class *A* will result only from crossing over between the two markers, but the phenotypic class *B* will result even without recombination, due to difference in the dosage of the two markers. The consequence of this is that only recombinant class *A* is informative for estimating *r*, leading to increased error in estimating the recombination distance.

Pairing Relationships

Chromosome pairing relationships may be established from MD markers. Three linkage groups containing the individual doses of a TD marker, for instance, are likely to be homologous. However, nonhomologous linkage groups may bear homologous segments, resulting from DNA duplications involving nonhomologous chromosomes. Segregation data of MD markers can be used to distinguish these two situations. TD markers are more useful to identify true homologous chromosomes, since it is easier to distinguish disomy from polysomy. In a polysomic octoploid, a TD marker located on nonhomologous chromosomes will show a disomic segregation ratio, that is 7:1, as opposed to the polysomic ratio of 13:1 that is expected when the doses are located on homologous chromosomes. Assuming that homologous segments in nonhomologous chromosomes will not pair, a mapping population of size 180 will give 80% distinction power (at 0.05 level) between disomic and polysomic ratios for triple dose markers (Da Silva, 1993).

In SES 208, DD markers were four times more efficient than the SD markers in identifying pairing partners (homologs). However, with DD markers the distinction between polysomic and disomic segregation requires a much larger population (Da Silva, 1993). For this reason, the homologous groups established may include nonhomologous chromosomes bearing homologous segments resulting from duplications and translocations.

Mapping studies involving populations obtained from crosses between heterozygous parents (Ritter et al, 1990), like used to map the *S. officinarum* and *S. robustum* (Al-Janabi et al, 1994a) genomes, will benefit most from the use of MD markers given the greater proportion of MD markers expected in such populations. In addition, mapping of MD markers allows important linkage information to be obtained from data which otherwise is not useful, without any additional cost associated with laboratory or field work.

The limitations of the methodology are: (1) it is restricted to species with only bivalent formation (absence of double reduction); and (2) estimates of *r* are based on two-point analysis of the expected chromosomal behavior, which tends to provide biased estimates as *r* approaches 0.5 (Welch, 1962). Refinements in the methodology are required, such as improvements on the estimate of *r* and its standard error, as well as the use of three-point linkage tests. However, for detecting pairing partners the ability to discern the true linkages is more important than accurately estimating r. With a minimal coverage of the genome, a small number of MD markers will provide the information necessary to collapse the $2n$ number of linkage groups into the x number (Da Silva, 1993; Da Silva et al, 1995).

The mapping of MD markers complements the SD-marker method in the sense that it allows: (1) the identification of pairing partners; (2) the indirect detection of SD markers linked in repulsion; and (3) the use of data otherwise not available for mapping (30% of the polymorphic loci in a polysomic octoploid).

Type of Inheritance and Level of Euploidy

Two criteria may be used to investigate the type of inheritance (di- or polysomic): (1) marker linkage relationships to compare the expected and observed number of linkages in repulsion-phase; and (2) segregation data to compare expected and observed proportion of SD to MD markers.

Assessing Type of Inheritance Using Linkage Relationships

The mapping population used to construct the linkage map for SES 208 was a backcross, in which the female parent (ADP85-0068, Moore et al, 1989) was a doubled haploid derived from SES 208 (Al-Janabi et al, 1993; Da Silva et al, 1993, 1995). In this situation, if the polyploid is disomic, the expected number of linkages in repulsion phase detectable with SD dominant markers, is a function of r. Linkages in repulsion at $r = 0$ can not be detected because the ADP85-0068 represents one gamete of the male parent (SES 208). As a consequence, only one of the markers involved in the linkage would be polymorphic, whereas the other would be present in both parents, i.e., not available for mapping. For repulsion-phase linkages at $r > 0$, the probability of detecting the linkage is the same as that of having both markers involved in the linkage be polymorphic, which results from recombination during the formation of the gamete that gave origin to ADP85-0068. Because only one of the two resulting recombination classes satisfies the condition for linkage detection (i.e., both markers polymorphic), which is the absence of both markers in ADP85-0068, that probability is $r/2$.

To arrive at the expected proportion of linkages in coupling:repulsion, we must calculate the probability of detecting linkages in coupling-phase. In this situation the event rendering either marker involved in the linkage unavailable for mapping would be recombination between the markers during the formation of the gamete that gave origin to ADP85-0068. Because in this case both recombinant classes would preclude the detection of coupling-phase linkages (one marker would be present in both parents), the probability of linkage detection is $1 - r$.

The probability function for detection of each kind of linkage for $0 < r < 0.25$ (the maximum recombination value used by Al-Janabi et al, 1993; Da Silva et al, 1993, 1995) is given by:

$$R = \int_0^{0.25} \frac{r^2}{4} \qquad [1]$$

and

$$C = \int_0^{0.25} r - \frac{r^2}{2}, \qquad [2]$$

where R and C are probabilities of detection for linkage in repulsion and coupling, respectively. Applying the $R{:}C$ ratio to the total number of two-point linkages observed, one gets the expected proportion of the two kinds of linkage under the disomic inheritance hypothesis, which is then used in a χ^2 test of this hypothesis.

Under polysomic inheritance, the expected number of linkages in repulsion-phase that can be detected is a function of the sample size (i.e., size of mapping population). This value is used in a χ^2 test of the polysomic inheritance hypothesis.

Preliminary data from a double-pseudo-testcross population, derived from a cross between euploid *S. officinarum* 'LA Purple' × *S. robustum* 'Mol 5829', showed that, unlike SES 208, repulsion-phase linkages were detected in both genomes using only 44 progeny (Al-Janabi et al, 1994a). However, repulsion-phase linkages were rare in this preliminary analysis, suggesting a mixture of random and preferential pairing in these species, which is typical of segmental allopolyploids. Because the number of markers and progeny used in the preliminary study were small, further characterization of the distribution of repulsion phase linkages in the genomes of *S. officinarum* and *S. robustum* will need to await the conclusion of ongoing studies (Guimarães et al, 1995). In any event, it suggests that the genetic behavior revealed by analysis of the SES 208 map may not be general to *Saccharum*.

Assessing Type of Inheritance Using Segregation Data

The use of a doubled-haploid parent backcrossed to its progenitor as a mapping population allows the type of inheritance to be assessed using segregation analysis to compare the proportion of polymorphic loci in two marker classes: SD and MD. The number of polymorphic loci required for this task is dictated by the magnitude of the difference between the two expected ratios for those classes. For instance, to differentiate between di- and polysomic octoploids, the two ratios to be distinguished are 2.33:1 and 1.27:1, respectively, and 178 polymorphic loci are required for a 2% probability level (Mather, 1936). Such a large number is not usually available from morphological or isozyme markers. With DNA markers, however, this number of markers has been achieved (Al-Janabi et al, 1993; Da Silva et al, 1993, 1995).

The proportion of polymorphic probes expected to be in single-dose is a function of the segregation at each locus, according to the dosage of the allele detected. In the SES 208 population, this has been demonstrated by calculating the expected proportion of MD markers, recalling that the doubled-haploid is derived solely (i.e., represented one gamete) from the male parent of the mapping population. All polymorphisms with five or more doses in the male are necessarily present in the female since they are present in 100% of the male gametes. Polymorphisms absent in the female must then be in one, two, three, or four doses in the male. By summing the probability of absence for polymorphisms in two, three and four doses (the MD polymorphisms) which are 3/14, 1/14, and 1/70, respectively (in a polysomic octoploid), one obtains the value 0.3 as the expected proportion of MD markers. The subtraction of this value from one, gives the expected proportion of polymorphic fragments that are in SD, that is 0.70.

To arrive at the expected values, under disomic inheritance, we took the expected ratios of two-, three- and four-dose fragments (1/4, 1/8, and 1/16, respectively), which are the three possible cases for a polymorphic band. Adding these values and subtracting from one gives the expected frequency of 0.56 for SD markers. The observed ratio can be compared against the expected ratio for di- and polysomic polyploids. This approach has been used to investigate inheritance in SES 208 both with RFLP- (Da Silva et al, 1993) and PCR-based (Al-Janabi et al, 1993) DNA markers. Segregation analysis of arbitrarily primed PCR (Welsh and McClelland, 1990; Williams et al, 1990) polymorphisms also has been used to assess the type of inheritance of *S. officinarum* (Kehrer, 1994; Al-Janabi et al, 1994a) and of *S. robustum* (Al-Janabi et al, 1994a) using a pseudo-double-backcross, and to assess the type of inheritance in cassava in three intraspecific crosses (Angel et al, submitted).

Estimation of Genome Size and Coverage

To estimate genome size, a method proposed by Hulbert et al (1988) may be used. Assuming a continuous genome, the number of markers expected to fall within a given interval, say 10 cM, can be calculated as:

$$y_x = \frac{n(n-1)}{2} \cdot \frac{2x}{G}, \qquad [3]$$

where y_x is the number of two-point linkages at distance equal to or less than x, n is the number of markers mapped (at a given L.O.D. score); x is the interval size, and G is the estimated genome size. Counting the number of markers linked (two-point linkages) at interval sizes less than or equal to a given value, say 10 cM, discarding the pairs that are not independent events (as the result of being located at homologous chromosomes), and solving for G, one can get an estimate of the genome size. Typically, an estimate is generated with markers ordered at L.O.D. ≥ 3.0, then another with markers ordered at L.O.D. ≥ 2.0, and an average of the estimates is used for a final approximation.

According to Bishop et al (1983), the expected proportion of a genome covered by a linkage map with n randomly placed markers, $E(C_n)$, is given by:

$$E(C_n) = 1 - p_{1,n}. \qquad [4]$$

where $p_{1,n}$ is the probability that a randomly placed point is not covered by n randomly placed markers, and given by:

$$p_{1,n} = \frac{2r}{n+1}\left[\left(1 - \frac{x}{2G}\right)^{n+1} - \left(1 - \frac{x}{G}\right)^{n+1}\right] + \left(1 - \frac{rx}{G}\right)\left(1 - \frac{x}{G}\right)^{n} \qquad [5]$$

where r is the number of chromosomes, x is the length of intervals in cM, and G is the total length of genome in cM.

Quantitative Genetics, Heterozygosity and Genetic Distance

Quantitative Genetics

Many traits of economic interest in plants are thought to be under polygenic control. Such traits are called quantitative traits in contrast to traits controlled by single genes, which are called qualitative. Just as DNA markers did not create linkage mapping, neither did they create studies on quantitative trait loci (QTL). However, the advent of abundant DNA markers has allowed the reduction of quantitative traits into Mendelian loci to be achieved in many plants (Beavis et al, 1991; deVicente and Tanksley, 1993; Doebley and Stec, 1991; Edwards et al, 1987, 1992; Fatokun et al, 1992; Nodari et al, 1993; Paterson et al, 1988, 1991; Stuber at al., 1992), thereby reducing to practice and extending the ideas put forth by Sax (1923), Smith (1937), and Thoday (1961) (Thompson and Thoday, 1979).

The theory of mapping QTLs has also progressed as DNA markers have become increasingly abundant. Futhermore, PCR-based marker methodologies can be fully automated allowing large numbers of individuals to be typed (Sobral and Honeycutt, 1993). Initial QTL detection approaches used one marker at a time (Soller et al 1976; Edwards et al, 1987; Tanksley et al, 1982; Weller et al, 1988). The single-marker approach was plagued by the effects of recombination between the marker and the QTL, which could be partially resolved by the potentially densely populated marker maps that DNA markers allow. Densely populated maps, however, were also the starting point for the second major approach to QTL detection, namely interval mapping (Lander and Botstein, 1986a, 1986b, 1989). Interval mapping has been the recent standard for QTL studies, but it too has problems, particularly in its incapacity to distinguish multiple linked QTL effects, resulting in mislocalization of the QTL (Knott and Haley, 1992), and in the requirement for large population sizes (Van Ooijen, 1992). A third generation of methodologies is now emerging which combines features of single-marker and interval mapping approaches (Zeng, 1993, 1994; Jansen, 1993, 1994; Jansen and Stam, 1994) but involves multiple regression (Cowen, 1989).

QTL detection methodologies have been developed for populations between inbred lines, which are difficult or impossible to make in some outbred species. Beckman and Soller (1988) developed a method to study QTLs in outbred crosses that required the study of three generations. Recently, Haley et al (1994) have generated a least squares method to study QTLs in outbred populations which circumvents some of the limitations of the previous approach. Furthermore, others have been interested in solving the confounding problem of how to declare a QTL without too many false-positives and, at the same time, without missing real QTLs (Rebaï et al, 1994; Churchill and Doerge, 1994; see also chapter by Bridges and Sobral). These advances suggest that QTL mapping in non-inbred populations, such as those available for many complex polyploids, may soon be implemented, just as DNA-marker-assisted linkage mapping has grown to accommodate mapping of polyploids.

In a preliminary evaluation of marker-trait association in sugarcane, Sills et al (1995) studied a sample of 44 F_1 progeny from a population derived by crossing *S. officinarum* with *S. robustum*. The hypothesis is that if Papuan *S. robustum* is the progenitor of Papuan *S. officinarum*, and if the major phenotypic differences between these two sympatric species are stalk fiber and sucrose content, then it follows that analysis of such a population will allow inference of the genetic basis of these differences, which are also of major agronomic interest. QTL analysis has revealed markers significantly associated with the expression of each trait analyzed. Of 18 markers associated with QTL after multiple regression, 10 (56%) descend from *S. officinarum* and 8 (44%) are unique to *S. robustum*. The various multi-locus models explain between 23% and 58% of the total phenotypic variation and 31% to 73% of the R^2_g% for the traits analyzed (Table 7).

The traits analyzed by Sills et al (1995) are quantitative in nature and presumed to be under the influence of multiple loci, each with small effects compared to variation due to other sources. It is, therefore, somewhat surprising that for most traits the genotypes of four or fewer markers are able to account for greater than 50% of the genetic variation, as indicated by R^2_g. The ability to detect QTL depends in part on the magnitude of their effects, the size and genetic structure of the population, and the density of markers across the genome. The high R^2_g values observed suggest that for these traits a small number of loci with large effects control a large proportion of the genetic variation. This would be consistent with few mutations being responsible for the phenotypic differences between *S. robustum* and *S. officinarum*. However, mapping single-dose markers in the F_1 progeny of a cross of heterozygous parents maximizes linkage disequilibrium such that markers represent large genomic regions. Thus, more than one QTL may be represented by a single marker, thereby causing an underestimation of the number of effective factors, and the concomitant incorrect attribution of effects to single loci (Edwards et al, 1987). Furthermore, some marker-QTL associations detected by Sills et al (1995) may be spurious due to the sampling of only 44 progeny which leads to an increased likelihood of type I er-

Table 7. Heritability of Seven Traits in the LA Purple × Mol 5829 Population, and the Effects, Linkage Relationships and Variation Explained by Arbitrarily Primed PCR Markers Significantly Associated with These Traits[a]

Trait	H^2BS	R^2_p	R^2_g	Marker	Parent	Effect[b] ± SE*
SN[c]	0.67	0.34	0.51	*R15.644*	LAP	2.99** ± 1.29
				V17.678	LAP	4.52** ± 1.29
SD	0.85	0.49	0.58	*P10.516*	LAP	−0.21* ± 0.07
				V15.397	LAP	−0.26** ± 0.06
				P20.565	Mol	−0.14** ± 0.05
				P10.516 × *V15.397*		0.20* ± 0.09
WT	0.74	0.23	0.31	*P10.516*	LAP	−5.02* ± 2.47
				V17.678	LAP	6.79** ± 2.47
%T	0.88	0.50	0.57	*R19.603*	LAP	13.81* ± 5.82
				V15.397	LAP	−16.79* ± 5.87
				R14.1741	Mol	12.83* ± 5.76
$(\%S)^{1/2}$	0.72	0.43	0.60	*T12.353*	LAP	1.01* ± 0.41
				U16.1232	LAP	−0.91* ± 0.41
				Q9.492	Mol	1.08* ± 0.40
				V13.749	Mol	−0.99* ± 0.41
%F	0.80	0.58	0.73	*T12.353*	LAP	1.07* ± 0.41
				U1.1066	LAP	−1.37** ± 0.41
				P20.565	Mol	1.28** ± 0.42
				T4.1127	Mol	1.21* ± 0.44
Pol%	0.80	0.43	0.54	*P8.425*	LAP	0.77* ± 0.31
				Q2.1208	LAP	0.63* ± 0.31
				P8.309	Mol	−0.74* ± 0.31
				V12.462	Mol	−1.07* ± 0.31

[a] Analyses for %S based on data transformed to improve normality.
[b] Regression coefficient for given marker determined through multiple regression of trait values on marker genotypes; *,** significant at P = 0.05 and 0.01, respectively.
[c] SN, umber of stalks; SD, stalk diameter in cm; WT, plot weight in Kg; %T, percent tasseled stalks; %S, percent smutted stalks; %F, percent cane fiber; Pol%, pol percent (a measure of stalk sugar).

rors, especially at $\alpha = 0.05$. However, this study should be seen as a primer for QTL detection in complex polyploids.

Heterozygosity and Genetic Distance

DNA markers have been utilized to assess heterozygosity and heterosis in diploids (Melchinger et al, 1990; Stuber et al, 1992) and polyploids (Bonierbale, et al, 1993; Yu and Pauls, 1993; Moser and Lee, 1994; Lu et al, 1994). In the case of sugarcane, DNA markers have been used to assess variability within and among wild species and among commercial varieties and elite genotypes (Burnquist, 1991; Sobral et al, 1994; Al-Janabi et al, 1994b; Honeycutt et al, 1995).

Applying RFLPs to investigate the effect of genetic diversity in maize hybrid vigor, Melchinger et al (1990) concluded that the prediction power is very limited. According to those authors, the low association between genetic distance and heterosis is a consequence of choosing markers arbitrarily placed throughout the genome, resulting in little information on the heterozygosity at QTLs affecting vigor. Moser and Lee (1994) found similar results and have concluded that even though arbitrarily placed markers could provide accurate estimates of genetic divergence among elite lines, it is unlikely that they could predict heterosis or population genetic variance. Stuber et al (1992) provided further support for this conclusion by mapping maize QTLs with RFLPs. They determined the genotypes at the QTLs affecting grain yield of the plants with highest vigor. Interestingly, for all detected QTLs but one, the heterozygotes show superior phenotypes when compared to the homozygotes, suggesting overdominance and control of heterosis by those QTLs. The lesson emerging from studies with diploids seems to be that heterozygosity is important, as long as it is maximized within adapted germplasm or along genealogies (i.e., related plants).

In polyploid plants an individual can have more than two alleles at one locus. An autooctoploid plant, for example, could have as many as eight different alleles at one locus. It has been stated repeatedly that the success of polyploids in nature would be dependent upon the maintenance of hybridity (Barber, 1970; Tal, 1980, Stebbins, 1980). In baker's yeast it has been shown that heterozygosity allied with polyploidy results in increased rates of fermentation and that this increase can be attributed to heterosis or increased heterozygosity, rather than the physical increment in cellular volume (Takagi et al, 1983).

Polysomic inheritance would result in larger heterozygosity in autopolyploids, relative to their diploid counterparts (Muller, 1914; Haldane, 1930; Barber, 1970; Stebbins, 1950, 1980). This has been documented in several autopolyploids (Soltis and Soltis, 1993). Furthermore, polysomic inheritance, coupled with bivalent chromosome pairing, gives a clear route for autopolyploid speciation (Soltis and Rieseberg, 1986) and challenges the notion (Levin, 1983) that autopolyploids are poorly adapted to succeed in nature.

Maximum heterozygosity has been proposed as an important factor for vigor and productivity in autopolyploid species (Bingham, 1980; Hermsen, 1984). According to this hypothesis, vigor in autopolyploids is highly dependent upon intralocus diversity. Heterosis increases progressively with the level of heterozygosity in populations of autopolyploids for biomass production, fertility, and tuber yield (Chase, 1963; Mendiburu and Peloquin, 1977). One can speculate that the presence of more alleles in the same genotype may also minimize the effects of genotype-by-environment interaction, thereby contributing to stability of yield.

The maximum heterozygosity hypothesis has influenced breeding strategies of autopolyploids. Bonierbale et al (1993) have used an RFLP linkage map for potato to assess the relative heterozygosity among plants from different crosses. Their conclusion is that maximum heterozygosity affects the tuber yield components in a favorable manner, when sources of adapted germplasm are combined.

The combination of statistical tools (stepwise multiple regression) and molecular markers may allow the identification of regions in the genome where heterozygosity is crucial for productivity and stability across environments. The strategy would involve: (1) use of a linkage map for the species to sample the genome; (2) genotyping of progenies from a cross between unrelated parents;

(3) regression of the phenotypes on genotypes of each marker; and (4) validation of the results in different crosses. Once heterotic loci are identified, markers can be used to devise crosses aiming at maximizing positive transgressive segregation and to preselect genotypes to be used in the final stages of the breeding cycle, reducing the costs associated with variety development.

Hundreds of crosses are made each year in sugarcane breeding programs. The choice of parents is based mainly on parental phenotypic performance. Because sugarcane F_1 genotypes are perpetuated by vegetative propagation, the dominant component of genetic variance and the specific combining ability of parents are very important genetic characteristics for breeding. In our experience with sugarcane breeding, a very small number of biparental crosses produces outstanding progeny, leading to a relatively large number of elite clones. A method for predicting the specific combining ability of parents would be invaluable in allowing breeders to make fewer crosses (only the potentially outstanding combinations) and raise a larger number of progeny from each cross to exploit transgressive variation.

Our results with *S. spontaneum* suggest that the number of different RFLP bands appearing in a given genotype is in direct relation to the number of different alleles on each locus hybridized (Figure 1). One interesting question that arises from this observation is whether this measurement can serve as a predictor of the specific combining ability of parents in biparental crosses. If loci where heterozygosity is important for vigor and other agronomic traits can be identified via QTL analysis, then DNA markers tightly linked to these loci will allow a genotypic survey of the breeding collection, as an aid for planning crosses. Genetic distances calculated from DNA marker data of loci known to affect the traits of interest (based on QTL analysis) may be used to establish a core germplasm collection and to identify clusters of unrelated varieties. Between-clusters pairwise combinations involving breeding stocks with discrepant phenotypes should maximize polymorphism. Core collections for vegetatively propagated plants are particularly important because seed propagation destroys the genotypes through recombination, and the costs associated with maintaining a vegetative crop in the field can be quite high.

A potential application of DNA markers to polyploids is to use marker-assisted planning of crosses to maximize exploitation of transgressive variation. To address this issue, we remind the reader that in polysomic polyploids, the proportion of segregating loci is more important than the phenotypic difference between two parents: distantly related parents are more likely to have different alleles at loci controlling traits of interest.

Another potential application of DNA markers to sugarcane breeding is for chromosome tagging. Modern sugarcane varieties are interspecific hybrids with varying chromosome numbers (usually $2n > 100$). These hybrids are the result of the process known as nobilization, where *S. officinarum* × *S. spontaneum* F_1 hybrids are back crossed to the "noble cane" (*S. officinarum*), to introgress traits (such as disease resistances, fiber content, etc.) from the other species. D'Hont et al (1993), have been able to construct seven *S. spontaneum*-specific linkage groups in a progeny of an elite clone and have found very low recombination rates, in comparison to maize. These findings, perhaps due to the small size of sugarcane chromosomes and the possible lack of recombination between the *S. spontaneum* and *S. officinarum* genomes, suggest that marking of *S. spontaneum* (and other wild sugarcane relatives) chromosomes, with net positive and negative effects in commercial varieties, may be practical. The approach involves comparative analysis between molecular and agronomic diversity within a simple progeny. Given the few meioses that have occurred since the first hybridizations to date, and the inheritance of large blocks of *S. spontaneum* chromosomes, a few markers may be sufficient to tag the whole chromosomes.

Comparative Mapping

Comparative mapping, which offers a common genetic framework for interpreting genetic information from divergent species is enabled by orthologous DNA markers (see chapter by Bennetzen). From comparative mapping studies the information about chromosome location of loci controlling important traits mapped in one species can be translated and used in another, related species.

Recently, there has been great interest in comparative mapping of plant genomes, especially within the grasses. Although pioneering work in plants was conducted on Solanaceae (Tanksley et al, 1992) and Brassicaceae (Song et al, 1988a and 1988b), perhaps the most exciting results have come from more recent work within the grasses (Doebley et al, 1990, Doebley and Stec, 1991, 1993; Whitkus et al, 1992; Binelli et al, 1992; Berhan et al, 1993; Hulbert et al, 1990; Moore et al, 1993; Kurata et al, 1994; Ahn and Tanksley, 1993; Ahn et al, 1994). The justifiable excitement has been due to the somewhat surprising amount of conservation, at the overall gene order level, that has been observed among distantly related plants, such as rice and maize and Triticaceae (Shields, 1993; Helentjaris, 1993; Bennetzen and Freeling, 1993). Many have noted that this level of conservation suggests that resources used in one grass species can be applied to others, causing considerable synergism among groups and increasing our overall knowledge, both basic and applied, about the many fundamentally important crop species represented within the grasses.

Of particular interest to sugarcane, detailed and multiple comparisons between maize and sorghum, which belong to the same tribe as sugarcane, have been conducted. These have shown that there is a large amount of conservation between these two genomes (Hulbert et al, 1990; Berhan et al, 1993; Binelli et al, 1992; Whitkus et al, 1992), which likely diverged before sugarcane diverged from sorghum (Sobral et al, 1994; Al-Janabi et al, 1994b; Hamby and Zimmer, 1988; Hulbert et al, 1990). The study comparing sorghum and maize, generally considered paleopolyploids, has shown that not only do they share common linkage groups but also that the recombination rate is equivalent in these conserved groups (Whitkus et al, 1992). These findings open the perspective to expedite the mapping of genes controlling important traits in the complex genomes of polyploids by extrapolating information obtained on genetically simpler diploid species.

Recent phylogenetic studies suggest that maternal lineages of sorghum and *Saccharum* may have diverged less than 5 million years ago, if one assumes a molecular clock and a maize-barley divergence of 50 million years ago (Al-Janabi et al, 1994b). These data also suggest that a high level of locus colinearity may exist between sugarcane and sorghum genomes in particular, and that transgressive alleles might be moved from one species to another (Al-Janabi et al, 1994b). Small segments of RFLP locus colinearity between maize and sugarcane has been confirmed by D'Hont et al (1993) and Da Silva et al. (1993) who have mapped a small number of genomic DNA probes of maize onto sugarcane. Efforts are now being made in various laboratories to link the genomes of these and other grass species. Comparative mapping should deliver to sugarcane geneticists an information base of more than 3,000 markers and major agronomic traits derived from six grass species. This is just one powerful example of how a complex polyploid can go from being largely uncharacterized at the genetic level in 1988 to a level of molecular genetic characterization that is in line with many other crops by 1994. In addition, further comparisons within the Andropogoneae and between the Andropogoneae and other grasses are expected to generate significant basic and applied knowledge, especially if results are shared (Bennetzen and Freeling, 1993; Helentjaris et al, 1993). Such knowledge will likely have a significant impact on plant breeding strategies, germplasm characterization, conservation, and use, with significant implications for world agriculture.

Mammalian gene maps have been constructed and studied with two main goals: as a resource for genetic analysis and manipulation, much like plant genome maps, but also to extend the database for evolutionary analysis of the mammalian (human) genome (O'Brien et al, 1993). Within the evolutionary context, exciting studies have also been conducted in plants, resulting in the comparative maps that we have noted. But, additionally, comparisons of domesticated plants and their ancestors or close relatives have yielded exciting insights to processes of morphological evolution and human selection. Perhaps the best example has been the analysis of a maize × teosinte cross in which the major morphological changes between maize and teosinte have been pinpointed to approximately five major genomic regions and genes of which the major effects have been characterized (Doebley et al, 1990; Doebley and Stec, 1991, 1993).

Experience from mammalian systems has shown that two classes of marker loci are required to achieve goals of map saturation and comparative mapping. They are: (1) coding loci that are conserved across that taxa being compared, and (2) loci that are highly polymorphic in the species studied (O'Brien et al, 1993). The first class essentially is composed of cDNA sequences that can be analyzed via specific PCR, DNA sequencing, or Southern hybridization. The second class are minisatellites, microsatellites, single-stranded conformation polymorphisms, arbitrarily primed PCR polymorphisms, etc., and can also be analyzed using

Southern hybridization or PCR, with a variety of gel resolution systems. Each type of locus has specific advantages for constructing maps in new species and extending available maps. The first class of loci is also the basis for reference or anchor markers (O'Brien et al, 1993), which have been invaluable in maximizing the research benefits in comparative mapping in animal systems but have yet to be developed for plant systems (Shields, 1993).

Molecular Cytogenetics

It is likely that multidisciplinary approaches to genetic dissection of polyploids will be the most fruitful, as has been noted by various authors (Stebbins, 1980; Grant, 1981; Soltis and Rieseberg, 1986; Jackson, 1982). For example, the addition of molecular tools to cytogenetic research (Fukui et al, 1994) has provided independent confirmation of the inference, based solely on genetic linkage data, that the basic chromosome complement for SES 208 is $x = 8$ (Al-Janabi et al, 1993; Da Silva et al, 1993, 1995). Independent verification using molecular cytogenetics has been achieved by studying the haploid karyotype of one of the haploids (AP 85-361, $2n = 32$) derived from SES208, which shows bivalent pairing at meiosis (with 16 bivalents). Despite a long history of cytological analysis in sugarcane (Sreenivasan et al, 1987), karyotypes for chromosomes have not existed because of the difficulty in identifying individual chromosomes; even chromosome counts have been uncertain (Fukui et al, 1994). However, through the application of condensation pattern (CP) analysis the molecular karyotype of AP 85-361 has been studied, and it has been found that the 32 chromosomes could be divided into eight groups, each consisting of four chromosomes with similar CPs, suggesting that $2n = 4x = 32$ in AP 85-361 and, consequently, $2n = 8x = 64$ in SES 208 (Fukui et al, 1994). Furthermore, fluorescent in situ hybridization (FISH) with 17*S* and 5*S* rRNA gene probes reveals hybridization to four chromosomes in AP 85-361 (Moore, 1995). FISH and CP analyses are being used to infer the molecular karyotype of *S. officinarum* (Moore, 1995). These results not only demonstrate that integration of molecular genetics and molecular cytogenetics will ultimately allow even the most complex polyploids to be studied, but they also suggest that the taxonomy of the *Saccharum* complex needs revising, and they may provide the tools for aiding in the revision.

Future

Some future challenges for the *Saccharum* complex, our model for complex polyploids, are included in the following research questions:

- What is level of marker colinearity between *S. spontaneum* to *S. robustum* and *S. officinarum*, and are all *Saccharum* genomes based on $x = 8$ as we demonstrated in *S. spontaneum* SES 208?

- What is the genetic basis for quantitative traits (QTL) of agronomic importance, including components of yield for biomass and sucrose?
- Which genomic regions differentiate *S. officinarum* from *S. robustum*? Are these the same as some of the QTL regions?
- What are the linkage relationships of orthologous QTL from sugarcane to sorghum, maize, and more distant grasses?
- Can a set of orthologous PCR-based DNA markers be developed for use by sugarcane breeders to evaluate large segregating populations and wild germplasm?
- Can gene sequences be used to develop a set of orthologous, specific PCR-based markers to serve as anchor loci for within-Andropogoneae (and possibly within grasses) comparisons?

DNA marker technology, although rapidly improving, needs to advance to the point at which technical feasibility, speed, automation, and cost are no longer a hindrance to plant breeding stations. The major bottlenecks seem to be speed, cost, and automation (see chapters by Bult and Bates). These three factors are not independent. However, advances in automation of DNA marker technology are largely driven by sales of products to the medical/pharmaceutical business community. This exclusively market-driven development situation does not incorporate the necessities and limitations of plant breeders or even of most research scientists. Furthermore, it is unlikely that market forces will be sufficient to drive development of technologies that are best suited to plant breeding, particularly in resource-poor but biodiversity-rich tropical countries. Clearly, development of robust, user-friendly, portable, and automated instrumentation is an area where coordinated international investments might allow a large return on investment dollars. Major instrumentation companies are interested in the technologies that might some day deliver the equivalent of a genetic indexing tool similar to a "Star Trek tri-corder" which would allow us "to boldly go" forward. Many, if not most, of the enabling technologies are already in place. However, instrumentation companies are not likely to absorb development costs for what is to them a small, poor market. If we do not act now, only the very richest of breeding efforts will be able to utilize the power of markers to enhance their crops. As usual, the decision is ours, and the consequences will be inherited by our children.

REFERENCES

Ahn S, Tanksley SD (1993): Comparative linkage maps of the rice and maize genomes. *Proc Natl Acad Sci USA* 90:7980–7984

Ahn S, Anderson JA, Sorrells ME, Tanksley SD (1994): Homoeologous relationships of rice, wheat and maize chromosomes. *Mol Gen Genet* (in press)

Al-Janabi SM, Honeycutt RJ, McClelland M, Sobral BWS (1993): A genetic linkage map of Saccharum spontaneum (L.) 'SES 208'. *Genetics* 134:1249–1260

Al-Janabi SM, Honeycutt RJ, Sobral BWS (1994a): Chromosome assortment in *Saccharum. Theor Appl Genet* 89:959–963

Al-Janabi SM, McClelland M, Petersen C, Sobral BWS (1994b): Phylogenetic analysis of organellar DNA sequences in the Andropogoneae:Saccharinae. *Theor Appl Genet* 88:933–944

Allard RW (1966): *Principles of Plant Breeding*. New York: John Wiley & Sons

Angel F, Gomez R, Bonierbale MW, Rodriguez F, Tohme J, Roca WM (1995): Selection of heterozygous parents and single-dose markers for genetic mapping in cassava (submitted)

Avivi L (1976a): The effect of genes controlling different degrees of homoeologous pairing on quadrivalent frequency in induced autotetraploid lines of *Triticum longissimum. Can J Genet Cytol* 18:357–364

Avivi L (1976b): Colchicine induced bivalent pairing of tetraploid microsporocytes in *Triticum longissimum. Can J Genet Cytol* 18:731–738

Barber NH (1970): Hybridization and the evolution of plants. *Taxon* 19:154–160

Beavis WD, Grant D, Albertsen M, Fincher R (1991): Quantitative trait loci for plant height in four maize populations and their associations with qualitative genetic loci. *Theor Appl Genet* 83:141–145

Beckman JS, Soller M (1988): Detection of linkage between marker loci and loci affecting quantitative traits in crosses between segregating populations. *Theor Appl Genet* 76:228–236

Bennetzen JL, Freeling M (1993): Grasses as a single genetic system: genome composition, colinearity and compatibility. *Trends in Genet* 9:259–261

Berhan AM, Hulbert SH, Butler LG, Bennetzen JL (1993): Structure and evolution of the genomes of *Sorghum bicolor* and *Zea mays. Theor Appl Genet* 86: 598–604

Bingham ET (1980): Maximizing heterozygosity in autopolyploids. In: *Polyploidy: Biological Relevance*, Lewis WH, ed. New York: Plenum Press

Binelli G, Gianfranceschi L, Pè ME, Taramino G, Busso C, Stenhouse J, Ottaviano E (1992): Similarity of maize and sorhum genomes as revealed by maize RFLP probes. *Theor Appl Genet* 84:10–16

Bishop DT, Cannings C, Skolnick M (1983): The number of polymorphic DNA clones required to map the human genome. In: *Statistical Analysis of DNA Sequence Data*, Weir BS, ed. New York: Marcel Dekker

Bonierbale MW, Plaisted RL, Tanksley SD (1988): RFLP maps based on a common set of clones reveal modes of chromosomal evolution in potato and tomato. *Genetics* 120:1095–1103

Bonierbale MW, Plaisted RL, Tanksley SD (1993): A test of maximum heterozygosity hypothesis using molecular markers in tetraploid potatoes. *Theor Appl Genet* 86:481–491

Brandes EW (1929): Into primeval Papua by seaplane. *Natl Geo* 56:253–332

Bridges CB (1916): Non-disjunction as proof of the chromosome theory of heredity. *Genetics* 1:1–52

Burnquist WL (1991): Development and application of RFLP technology in sugarcane (*Saccharum* spp.) breeding, (Dissertation). Ithaca N.Y: Cornell University

Chase SS (1963): Analytical Breeding in Solanum tuberosum L.—Ascheme using parthenotes and other diploid stocks. *Can Genet Cytol* 5(4):359–363

Churchill GA, Doerge RW (1994): Empirical threshold values for quantitative trait mapping. *Genetics* 138:963–971

Clausen J, Keck DD, Hiesey WM (1945): Experimental studies on the nature of species. II. Plant evolution through amphidiploidy and autoploidy with examples from the *Madiinae.* Washington: Carnegie Institute, Publ. 564

Cowen NM (1989): Multiple linear regression analysis of RFLP data sets used in mapping QTLs. In: *Development and Application of Molecular Markers to Problems in Plant Genetics*, Helentjaris T, Burr B eds. New York: Cold Spring Harbor Laboratory Press

Crawford DJ, Smith EB (1984): Allozyme divergence and intraspecific variation in *Coreopsis grandiflora* (Compositae). *Syst Bot* 9:219–225.

Da Silva JAG (1993): A Methodology For Genome Mapping of Autopolyploids and Its Application to Sugarcane (*Saccharum* spp) (Dissertation). Ithaca, NY: Cornell University

Da Silva JAG, Sorrells ME, Burnquist WL, Tanksley SD (1993): RFLP linkage map of *Saccharum spontaneum. Genome* 36:782–791

Da Silva JAG, Honeycutt RJ, Burnquist W, Al-Janabi SM, Sorrells ME, Tanksley SD, Sobral BWS (1995): *Saccharum spontaneum* L. 'SES 208' genetic linkage map combining RFLP- and PCR-based markers. *Mol Breed* 1:165–179

DeVincente MC, Tanksley SD (1993): QTL analysis of transgressive segregation in an interspecific tomato cross. *Genetics* 134:585–596

Doebley J, Stec A (1991): Genetic analysis of the morphological differences between maize and teosinte. *Genetics* 129:285–295

Doebley J, Stec A (1993): Inheritance of the morphological differences between maize and teosinte: comparison of results for two F_2 populations. *Genetics* 134:559–570

Doebley J, Stec A, Wendel J, Edwards M (1990): Genetic and morphological analysis of a maize-teosinte F_2 population: implications for the origin of maize. *Proc Natl Acad Sci USA* 87:9888–9892

D'Hont A, Lu Y, de Leon DG, Grivet L, Feldmann P, Lanaud C, Glaszmann JC (1993): A molecular approach to unraveling the genetics of sugarcane, a complex polyploid of the Andropogoneae tribe. *Genome* 37:222–230

Edwards MD, Stuber CW, Wendel JF (1987): Molecular-marker-facilitated investigations of quantitative trait loci in maize. I. Numbers, genomic distribution and types of gene action. *Genetics* 116:113–125

Edwards MD, Helentjaris T, Wright S, Stuber CW (1992):): Molecular-marker-facilitated investigations of quantitative trait loci in maize. *Theor Appl Genet* 83:765–774

Fatokun CA, Menancio-Hautea DI, Danesh D, Yound ND (1992): Evidence for orthologous seed weight genes in cowpea and mung bean based on RFLP mapping. *Genetics* 132:841–846

Fisher RA (1949): The linkage problem in a tetrasomic wild plant, *Lythrum Salicaria. Proc. 8th Intern. Congr. Genet* 225–233

Fukui K, Mhmido N, Ha S, Moore PH (1994): Analysis and utility of chromosome information. 67. Complete identification of wild sugarcane chromosomes. *Japan J Breed* (Suppl 2):29

Gebhardt CE, Ritter E, Debener T, Schachtschabel U, Walkemeier B, Urig H, Salamini F (1989): RFLP-analysis and linkage mapping in *Solanum tuberosum. Theor Appl Genet* 78:65–75

Grant V (1981): *Plant Speciation.* New York: Columbia University Press

Grattapaglia D, Sederoff R (1994): Genetic linkage maps of *Eucalyptus grandis* and *Eucalyptus urophylla* using a pseudo-testcross: mapping strategy and RAPD markers. *Genetics* 137:1121–1137

Grattapaglia D, Wilcox P, Chaparro JX, O'Malley D, McCord S, et al (1991): A RAPD map of loblolly pine in 60 days, (abstract 2224). In: *Third International Congress of the International Society for Plant Molecular Biology*, Tucson Az

Guimarães C, Sills GR, Honeycutt RJ, Sobral BWS (1995): unpublished data

Haldane JBS (1930): The theoretical genetics of autopolyploids. *J Genet* 22: 359–372

Haley CS, Knott SA, Elsen J-M (1994): Mapping quantitative trait loci in crosses between outbred lines using least squares. *Genetics* 136:1195–1207

Hamby, RK, Zimmer, EA (1988): Ribosomal RNA sequences for inferring phylogeny within the grass family (*Poaceae*). *Pl Syst Evol* 160:29–37

Helentjaris T (1993): Implications for conserved genomic structure among plant species. *Proc Natl Acad Sci USA* 90:8308–8309

Hermsen JGT (1984): Nature, Evolution and breeding of polyploids. *Iowa State J Res* 58:411–412

Honeycutt RJ, Jannoo N, Burnquist WB, Sobral BWS (1995): unpublished data

Hulbert SH, Richter TE, Axtell JD, Bennetzen JL (1990): Genetic mapping and characterization of sorghum and related crops by means of maize DNA probes. *Proc Natl Acad Sci USA*. 87:4251–4255

Jackson RC (1982): Polyploidy and diploidy: New perspectives on chromosome pairing and its evolutionary implications. *Amer J Bot* 69:1512–1523

Jackson RC, Casey J (1980): Cytogenetics of polyploids. In: *Polyploidy: Biological Relevance*, Lewis WH, ed. New York: Plenum Press

Janaki-Ammal EK (1936): Cytogenetic analysis of Saccharum spontaneum L. 1. Chromosome studies in Indian formas. *Indian J Agric Sci* 6:1–8

Jansen RC (1993): Interval mapping of multiple quantitative trait loci. *Genetics* 135:205–211

Jansen RC (1994): Controlling type I and type II errors in mapping quantitative trait loci. *Genetics* 138:871–881

Jansen RC, Stam P (1994): High resolution of quantitative traits into multiple loci via interval mapping. *Genetics* 136:1447–1455

Kam-Morgan LNW, Gill BS (1989): DNA restriction fragment length polymorphisms: a strategy for genetic mapping of D genome of wheat. *Genome* 32:724–732

Kehrer RL (1994): A RAPD Analysis of the Segregation Patterns in the cross of *Saccharum officinarum* (La Purple) with *Saccharum robustum* (Molokai 5829) (MSc Thesis). Salt Lake City, Utah: Brigham Young University

Kihara H, Ono T (1926): Chromosomenzahlen und systematische Gruppierung der *Rumex*-Arten. *Zeitschr Zellforsch* 4:475–481

Knott SA, Haley CS (1992): Aspects of maximum likelihood methods for mapping quantitative trait loci in line crosses. *Genet Res* 60:139-151

Kurata N, Moore G, Nagamura Y, Foote T, Yano M, Minobe Y, Gale M (1994): Conservation of genome structure between rice and wheat. *Bio/Technology* 12:276–278

Lander ES, Botstein D (1986a): Mapping complex genetic traits in humans: new methods using a complete RFLP linkage map. *Cold Spring Harbor Symp on Quantitative Biol* 51:49–62

Lander ES, Botstein D (1986b): Strategies for studying heterogeneous genetic traits in humans by using a linkage map of restirction fragment length polymorphisms. *Proc Natl Acad Sci USA* 83:7353–7357

Lander ES, Botstein D (1989): Mapping Mendelian factors underlying quantitative traits using RFLP linkage maps. *Genetics* 121:185–199

Levin DA (1983): Polyploidy and novelty in flowering plants. *Amer Nat* 122:1–25

Lewis WH (1980): *Polyploidy: Biological Relevance*. New York: Plenum Press

Little TM (1945): Gene segregation in autotetraploids. *Bot Rev* 11:60–85

Little TM (1958): Gene segregation in autotetraploids. II. *Bot Rev* 24:318–339

Mac Key J (1987): Implications of polyploidy breeding. *Biol Zent bl* 106:257–266

Mather K (1935): Reductional and equational separation of the chromosomes in bivalents and multivalents. *J Genet* 30:53–78

Mather K (1936): Segregation in autopolyploids. *J Genet* 32:287–314

Melchinger AE, Lee M, Lamkey KR, Woodman WL (1990): Diversity of restriction fragment length polymorphism: relation to estimated genetic effects in maize inbreds. *Crop Sci* 30:1033–1040

Mendiburu AO, Peloquin SJ (1977): The significance of 2*n* gametes in potato breeding. *Theor Appl Genet* 49:53–61

Moore G, Gale MD, Kurata N, Flavell RB (1993): Molecular analysis of small grain cereal genomes: Current status and prospects. *Bio/Technology* 11: 584–589

Moore PH (1995): personal communication.

Moore PH, Fitch MMM (1990): Sugarcane (*Saccharum* spp.) anther culture studies. In: *Biotechnology in Agriculture and Forestry*. Volume 12, Haploids in Crop Improvement, Bajaj YPS ed. Heidelberg: Springer-Verlag

Moore PH, Nagai C, Fitch MMM (1989): Production and evaluation of sugarcane haploids. *Proc Intl Soc Sugar Cane Technol* 20:599–607

Moser H, Lee M (1994): RFLP variation and genealogical distance, multivariate distance, heterosis, and genetic variance in oats. *Theor Appl Genet* 87: 947–956

Mukherjee, SK (1957): Origin and distribution of *Saccharum*. *Bot Gaz* 119:55–61

Muller HJ (1914): A new mode of segregation in Gregory's tetraploid Primulas. *Amer Nat* 48:508–512

Nodari RO, Tsai SM, Guzman P, Gilbertson RL, Gepts P (1993): Toward an integrated linkage map of common bean. III. Mapping genetic factors controlling host-bacteria interactions. *Genetics* 134:341–350

O'Brien SJ, Womack JE, Lyons LA, Moore KJ, Jenkins NA, Copeland NG (1993): Anchored reference loci for comparative mapping in mammals. *Nature Genetics* 3:35103–112

Paterson AH, Lander ES, Hewitt JD, Peterson S, Lincoln SE, Tanksley SD (1988): Resolution of quantitative traits into mendelian factors using a complete linkage map of restriction fragment length polymorphisms. *Nature* 335:721–726

Paterson AH, Damon S, Hewitt S, Zamir JD, Rabinowitch HD, Lincoln SE, Lander ES, Tanksley SD (1991): Mendelian factors underlying quantitative traits in tomato: comparison acros species, generations, and environments. *Genetics* 127:181–197

Price S (1963): Cytogenetics of modern sugar canes. *Econ Bot* 17:97–106

Rebaï A, Goffinet B, Mangin B (1994): Approximate thresholds of interval mapping tests for QTL detection. *Genetics* 138:235–240

Reinisch AJ, Dong J-m, Brubaker CL, Stelly DM, Wendel JF, Paterson AH (1994): A detailed RFLP map of cotton, *Gossypium hirsutum* × *Gossypium barbadense*: chromosome organization and evolution in a disomic polyploid genome. *Genetics* 138:829–847

Reiseberg LH, Van Fossen C, Desrochers AM (1995): Genomic reorganization accompanies hybrid speciation in wild sunflowers. *Nature* (in press)

Riley R, Chapman V (1958): Genetic control of the cytologically diploid behavior of hexaploid wheat. *Nature* 182:713–715

Ritter E, Gebhardt C, Salamini F (1990): Estimation of recombination frequencies and construction of RFLP linkage maps in plants from crosses between heterozygous parents. *Genetics* 125:645–654

Sax K (1923): The association of size differences with seed-coat pattern and pigmentation in *Phaseolus vulgaris*. *Genetics* 8:552–560

Sears ER, Okamoto M (1958): Intergenomic chromosome relationships in hexaploid wheat. *Proceedings of the International Congress on Genetics.*

Sharp LW (1934): *Introduction to Cytology*. New York: McGraw-Hill

Shields R (1993): Pastoral synteny. *Nature* 365:297–297

Sills GR, Bridges W, Al-Janabi SM, Sobral BWS (1995): Genetic analysis of agronomic traits in a cross between sugarcane (*Saccharum officinarum* L.) and its presumed progenitor (*S. robustum* Brandes & Jesw. ex Grassl). *Mol Breed* (in press)

Smith HH (1937): The relation between genes affecting size and color in certain species of *Nicotiana*. *Genetics* 22:361

Sobral BWS, Honeycutt RJ (1993): High output genetic mapping in polyploids using PCR-generated markers. *Theor Appl Genet* 86:105–112

Sobral BWS, Honeycutt RJ (1994): Genetics, plants, and the polymerase chain reaction. In: *The Polymerase Chain Reaction*, Mullis KB, Ferré F, Gibbs A eds. Boston: Birkhauser

Sobral BWS, Braga DPV, LaHood ES, Keim P (1994): Phylogenetic analysis of chloroplast restriction enzyme site mutations in the Saccharinae Griseb. subtribe of the Androponeae Dumort. tribe. *Theor Appl Genet* 87:843–853

Soller M, Brody T, Genizi A (1976): On the power of experimental design for detection of linkage between marker loci and quantitative loci in crosses between inbred lines. *Theor Appl Genet* 47:35–39

Soltis DE, Rieseberg LH (1986): Autopolyploidy in *Tolmeia menziesii* (Saxifragaceae): Genetic insights from enzyme electrophoresis. *Amer J Bot* 73:310–318

Soltis DE, Soltis PS (1993): Molecular data and the dynamic nature of polyploidy. *Crit Rev Pl Sci* 12:243–273

Song KM, Osborn TC, Williams PH (1988a): *Brassica* taxonomy based on nuclear restriction fragment length polymorphisms (RFLPs). I. Genome evolution of diploid and amphidiploid species. *Theor Appl Genet* 75:784–794

Song KM, Osborn TC, Williams PH (1988b): *Brassica* taxonomy based on nuclear restriction fragment length polymorphisms (RFLPs). II. Preliminary analysis of subspecies within *B. rapa* (syn. *campestris*) and *B. oleracea. Theor Appl Genet* 76:593–600

Song KM, Osborn TC, Williams PH (1990): *Brassica* taxonomy based on nuclear restriction fragment length polymorphisms (RFLPs). III. Genome relationships in *Brassica* and related genera and the origin of *B. oleracea* and *B. rapa* (syn. *campestris*). *Theor Appl Genet* 79:497–506

Sorrells ME (1992): Development and application of RFLPs in polyploids. *Crop Sci* 32:1086–1091

Sreenivasan TV, Ahloowalia BS, Heinz DJ (1987): Cytogenetics. In: *Sugarcane Improvement through Breeding*. New York: Elsevier

Stebbins GL (1947): Types of polyploids: Their classification and significance. *Adv Genet* 1:403–429

Stebbins, GL (1950): *Variation and evolution in plants*. New York: Columbia University Press

Stebbins GL (1980): Polyploidy in plants: unresolved problems and prospects. In: *Polyploidy: Biological Relevance*, Lewis WH, ed. New York: Plenum Press

Stuber CW, Lincoln SE, Wolff DW, Helentjaris T, Lander ES (1992): Identification of genetic factors contributing to heterosis in a hybrid from two elite inbred lines using molecular markers. *Genetics* 132:823–839

Takagi A, Harashima A, Oshima Y (1983): Construction and characterization of isdogenic series of *Saccharomyces cerevisiae* polyploid strains. *Appl Env Microbiol* 45:1034–1040

Tal M (1980): Physiology of polyploids. In: *Polyploidy: Biological Relevance*, Lewis WH, ed. New York: Plenum Press

Tanksley SD, Medina-Filho H, Rick CM (1982): Use of naturally occurring enzyme variation to detect and map genes controlling quantitative traits in an interspecific backcross of tomato. *Heredity* 49:11–25

Tanksley SD, Ganal MW, Prince JP, deVicente MC, Bonierbale MW, Broun P, Fulton TM, Giovannoni JJ, Grandillo S, Martin GB, Messeguer R, Miller JC,

Miller L, Paterson AH, Pineda O, Röder MS, Wing RA, Wu W, Young ND (1992): High density molecular linkage maps of the tomato and potato genomes. *Genetics* 132:1141–1160
Thoday JM (1961): Location of polygenes. *Nature* 191:368–370
Thompson JN, Thoday JM (1979): *Quantitative Genetic Variation*. New York: Academic Press
Timmis JN, Rees H (1971): A pairing restriction at pachytene upon multivalent formation in autotetraploids. *J Hered* 26:269–275
Uhl CH (1992): Polyploidy, dysploidy, and chromosome pairing in *Echeveria* (Crassulaceae) and its hybrids. *Am J Bot* 79:556–566
Van Ooijen JW (1992): Accuracy of mapping quantitative trait loci in autogamous species. *Theor Appl Genet* 84:803–811
Wall AM, Riley R, Gale MD (1971): The position of a locus on chromosome 5B of *Triticum aestivum* affecting homoeologous meiotic pairing. *Genet Res* 18:329–333
Welch JE (1962): Linkage in autotetraploid maize. *Genetics* 47:367–396
Weller JI, Soller M, brody T (1988): Linkage analysis of quantitative traits in an interspecific cross of tomato (*Lycopersicon esculentum* × *Lycopersicon pimpinellifolium*) by means of genetic markers. *Genetics* 118:329–339
Welsh J, McClelland M (1990): Fingerprinting genomes using PCR with arbitrary primers. *Nucl Acids Res* 18:7213–7218
Whitkus R, Doebley J, Lee M (1992): Comparative mapping of sorghum and maize. *Genetics* 132:1119–1130
Williams JGK, Kubelik AR, Livak KJ, Rafalski JA, Tingey SV (1990): DNA polymorphisms amplified by arbitrary primers are useful as genetic markers. *Nucl Acids Res* 18:6531–6535
Wu KK, Burnquist W, Sorrells ME, Tew TL, Moore PH, Tanksley SD (1992): The detection and estimation of linkage in polyploids using single-dose restriction fragments. *Theor Appl Genet* 83:294–300
Yu K, Pauls KP (1993): Rapid estimation of genetic relatedness among heterogeneous populations of alfalfa by random amplification of bulked genomic DNA samples. *Theor Appl Genet* 86:788–794
Zeng Z-B (1993): Theoretical basis of separation of multiple linked gene effects on mapping quantitative trait loci. *Proc Natl Acad Sci USA* 90:10972–10976
Zeng Z-B (1994): Precision mapping of quantitative trait loci. *Genetics* 136:1457–1468
Zivy M, Devaux P, Blaisonneau J, Jean R, Thiellement H (1992): Segregation distortion and linkage studies in microspore-derived double haploid lines of *Hordeum vulgare* L. *Theor Appl Genet* 83:919–924

ACKNOWLEDGMENTS

We graciously acknowledge financial support for various aspects of our research by the following: International Consortium for Sugarcane Biotechnology and Pioneer Hi-Bred International (Des Moines, IA). We thank sugarcane breeders throughout the world for assistance in collecting and shipping germplasm samples for our studies. We thank Michael McClelland (California Institute of Biological Research, La Jolla—CIBR), Rhonda Honeycutt (CIBR), and Gavin Sills (CIBR) for critical reading of this manuscript and many fruitful discussions. We also thank Paul Moore (USDA-ARS, Aiea, Hawaii) for sharing results prior to publication. Finally, we thank Rhonda Honeycutt (CIBR), Michael McClelland (CIBR), Roger Beachy (Scripps, La Jolla), Ron Sederoff (North Carolina State University Forestry Biotechnology Group, Raleigh), Daniella Braga (Copersucar Technology Center, Brazil) and Richard Jefferson (Center for Application of Molecular Biology to International Agriculture, Canberra, Australia—CAMBIA) for friendship and encouragement. One of us (BWSS), would also like to leave a short poem to transmit a feeling to go with concepts we have tried to transmit herein:

THE CROSSING
crash course in living
sweating through the bathroom
walking down the desert beachfront
sandy fingertips bearing down on the outer fringes of imagination
peeling the scars off reality
plunging into the divine depths
that echo through hearts left standing
in the drizzling cold of loneliness
staring into space
no expression found on face
faceless trials
in featureless paths
waterhole discovery in the calm of night
searing pain of dry
aching mouth filling passionate pasts
crawling through doors blown open by the passage of time
eroded curvatures
roads flung open in search
lost in rear view imaginations
imagine!
when we were children
we filled the prophecies of earth and tree
expanded horizons
went out to see
eyes cast
dice thrown
pillows still warm from tender loving sleep
but that was before we were taught the impossible
that we did not have life ahead of us
that the dreams of our imagination could not be taught

found, felt
run wild
so run
wild
beauty
saying all
wordless charm
phonetic aberration
whatever the poetry peels off
is worth loosing
for the love of enchantment!

2

Validation Strategies for Analysis of Quantitative Trait Loci Using Markers

WILLIAM C. BRIDGES, JR. AND BWS SOBRAL

Introduction

The basic theory of using genetic markers to manipulate the loci controlling a trait of interest to plant geneticists was introduced by Sax (1923) over 70 years ago. The application of this theory since that time has been limited by the lack of available segregating genetic markers. Recent advances in methods for assaying DNA polymorphisms have produced hundreds of segregating genetic markers in many species. These advances have allowed the application and further development of the theory of Sax. The genetic markers have been used: (1) as X variables in linear and nonlinear models to determine which markers are near (with reference to genetic recombination) loci controlling a trait of interest; (2) as indirect selection criteria; and (3) in traditional linkage analysis to arrange them into dense genetic linkage maps. Information from the map can be used further to map genetically the trait loci and to determine starting points for finding the trait loci using physical mapping.

The primary focus of this chapter will be on the issues involved in determining markers that are near trait loci and using these markers as selection criteria. We will review briefly the standard techniques for finding markers near trait loci in plant species and point out an inherent problem (multicollinearity). We will demonstrate a method (cross validation) for researchers to identify this problem in their experiments. Finally, we will speculate on how the information we learn in cross validation can be used when markers are to be used as indirect selection criteria, and we will consider the use of cross validation in another technique for finding markers near trait loci in plant species, namely the use of pedigree data.

The Impact of Plant Molecular Genetics
BWS Sobral, Editor

Determining Which Markers Are Near Trait Loci

The technique for finding which of a given set of markers are near trait loci involves the following steps. A cross of two parents (preferably inbreds) is used to produce an F_2 or backcross (BC) set of progeny. The use of non-inbred parents introduces the complications of greater than two alleles, unknown phase, and less than maximum disequilibrium. These complications cause difficulty in the steps outlined below for finding important markers. The magnitude of these difficulties, and exactly how they impact the detection of important markers in plant species that are not amenable to inbred line development, is a topic of tremendous interest to researchers. It is an interesting area for additional research.

Progeny are scored for the trait of interest (denoted by Y) and the markers available (denoted X_i with $i = 1$ to m for a set of **m** markers). A linear model is defined to establish a relationship between the trait value (Y) and the each of the markers (X_is). The form of the model is

$$Y_{jk} = \beta_0 + \beta_1 X_{ij} + \beta_2 X_{ij}^2 + \epsilon_{jk}$$

where $j = 0,1,2$ denotes the three possible X_i values for each marker (we will assume the parents are inbred and the F_2 is used), $k = 1$ to n_{ij} denotes the number of progeny in marker form j, β_1 and β_2 are the additive and dominance effects, respectively, of the locus near the marker defining X_i, and ϵ is the residual.

This model is analyzed separately (sometimes described as one-at-a-time) using least squares or maximum likelihood for each of the **m** markers to produce **m** individual analyses. The choice between least squares and maximum likelihood is based on the nature of the residuals. If the residuals are normally distributed and have equal variance in all levels of X, then the least squares and maximum likelihood analyses are equivalent. If the residuals have heterogeneous variance, then the maximum likelihood analysis can be advantageous. Also, if the model becomes more complicated (i.e., the nonlinear models discussed below), then maximum likelihood can be used. The effect of nonnormal data on marker analysis, and how to correct for it, is an area of important research potential.

If upon analysis, a significant relationship is found between Y and X_i for an individual model, then we usually conclude that a trait locus is near the marker that defines the X_i in that model. This method has been used by many researchers; however, one of the first reports of this method can be found in Soller and Beckman (1983). The significant X_is are sometimes used in a prediction equation for the trait value. The predicted values from this equation can be used as an indirect selection criterion (i.e., marker assisted selection) (Lande and Thompson, 1990; Knapp, 1994).

The methods discussed above are called the single marker method. The single marker method has been since modified to correct for bias due to recombination between the marker and the trait loci (r), using flanking markers and nonlinear models (Lander and Botstein, 1989; Knapp et al, 1990), and this modification is called the flanking marker method.

The one-at-a-time analyses described above (both the single and flanking marker methods) can be used effectively if the trait is controlled by only a few loci (i.e., a qualitative trait), and there are not too many markers to evaluate. However, neither of the one-at-a-time analyses work if the trait is controlled by many loci (i.e., a quantitative trait). Quantitative trait loci (hereafter denoted QTL) **must** be evaluated simultaneously (Bridges, 1992; Knapp et al, 1992) to determine the relative main effects of the QTL and the effects of interactions among the QTL on the quantitative trait value. Unfortunately, using markers to truly evaluate the QTL simultaneously is almost intractable, so an approximation of the simultaneous evaluation using some form of multiple regression on the markers has been proposed. One of the first uses of multiple regression on markers has been reported by Bridges et al (1988) and Nienhuis et al (1987, 1988) and recent modifications for the use of multiple regression have been reported by Bridges (1992), Zeng (1993, 1994), Jansen (1993), and Doerge et al (1994). The methods proposed to perform this approximation all basically involve putting the single markers into a multiple regression model. The model is of the form

$$Y = \beta_0 + X_1 + X_2 + \ldots + X_m + \epsilon$$

where X_i represents the terms $\beta_1 X_i$ and $\beta_2 X_i^2$ for marker i.

This model is analyzed using least squares or maximum likelihood, and if X_is are found to have a significant relationship with Y, then the markers that define those X_is are said to be near QTL. As mentioned above, we can use these X_is in a prediction equation for marker assisted selection.

There are some slight differences in the multiple regression methods used by different researchers. Some researchers use pairs of flanking markers in the multiple regression model instead of the single markers because they are concerned with the bias due to r. It should be noted that the bias in the estimate of the QTL effect caused by using one-at-a-time analyses is greater than any bias due to r. Some researchers use different methods to determine how individual markers are chosen for inclusion in the final model (i.e., variable selection methods and type of sum of squares used). Also, researchers differ in whether or not second order terms (quadratics or dominance effects and interaction or epistasis effects) are included in the model. However, to restate, the methods differ little.

Multiple regression approximations are certainly more useful than one-at-a-time analyses for determining the markers near QTL (i.e., QTL mapping), but the results from the multiple regression are still suspect due to multicollinearity caused by missing marker (and QTL) genotypes. This can be illustrated with the following example from Bridges (1992). Suppose that the set of progeny under study contains only observations for the genotypes indicated with an asterisk:

X_1 values

X_2 values		0	1	2
	0	*		
	1		*	*
	2			*

Markers that define X_1 and X_2 appear to be correlated due to the sample of genotypes in our set of progeny, even if the markers are not actually linked. In this situation, it is difficult to determine the relative effects of a possible QTL near X_1 and a possible QTL near X_2 on Y (the quantitative trait value) even using multiple regression. Another difficulty with data such as these is that if there is only a QTL (or a qualitative trait locus) near X_1, both X_1 and X_2 would appear to be significant, and we would not be able to determine from the data which marker the QTL is actually near.

These issues of missing genotypes and multicollinearity are not new problems in multiple regression, but the magnitude of the multicollinearity is unusual. Suppose that we are evaluating 50 markers, and each has three possible genotypes in an F_2 sample. We have 3^{50} genotypes to evaluate to determine the nature of the relationship between all possible markers and the quantitative trait value. Since most experiments involve 100 to 200 plants, we have a *severe* missing genotype problem, which introduces the multicollinearity. There are some possible solutions to this problem. One is to determine the fraction of the full factorial that the genotypes for which we have observations represent and estimate some main effects ignoring interactions (or epistasis). Another solution is to use contrasts among the genotypes to determine simple effects of X_1 within X_2 and vice versa. Still another solution is to use a method of multiple regression that minimizes the effects of multicollinearity (i.e., ridge regression). Each of these solutions is difficult to implement and has certain drawbacks. Before any of these are attempted, the researcher should try to determine how much of a problem the missing genotypes and multicollinearity has introduced in their QTL analysis.

Cross Validation

The best method to determine the effect of any multicollinearity on the QTL analysis is to conduct a new study. The researcher can then compare the two studies to determine if the same markers are found to be related to the quantitative trait value in both studies. If the same markers are in fact found, the researcher has evidence that there is no problem with multicollinearity in the analysis, the results generalize, and the model can be reliably applied to other samples. If different markers are found, the researcher has evidence of multicollinearity problems in the analyses, and the results are specific to that sample. However, this approach is expensive and sometimes not even possible. The method of cross validation attempts to assess the effect of multicollinearity with a single study.

The steps in a cross validation analysis are relatively simple. A useful reference for this technique is Kleinbaum et al (1988). Randomly split the data into two halves (in fact, cross-validation is sometimes called split-sample validation). Call them subsets A and B. Use multiple regression to determine the markers near the QTL in subset A and determine the markers near the QTL in subset B.

Compare the results of the two models. If the analyses include the same set of markers, then you have some evidence that the markers you have chosen are in fact the ones near the QTL, and multicollinearity is not an issue in the analysis. Unfortunately, there are almost always differences in the markers deemed to be near the QTL in the two different sets of data. This result indicates that it is difficult to determine the markers that are actually near the QTL and therefore map the QTL from a single study. Results from several studies must be used to find markers that are consistently shown to be associated with the QTL, and these are the markers that should be used to map the QTL. The actual number of studies required and how confident you need to be in the association of the marker and QTL before you begin to map is an interesting area of research that is just beginning to be explored.

The fact that a single study cannot be used for mapping QTL doesn't exclude the use of a single study to build a valid prediction equation and to use the markers as indirect selection criteria. If this is the objective, all the researcher needs to determine is if the equation will consistently predict the quantitative trait value (even if the researcher doesn't have all the correct markers near QTL). To accomplish this, use the markers found to be associated with QTL in one subset (e.g. A) in a prediction equation for Y and determine how well the equation performs in the other subset (e.g. B). The formal steps are as follows. Calculate the R^2 value for subset A. Next predict the Y values for subset B using the prediction equation from subset A. Calculate the Pearson product moment correlation between the actual Y values and the predicted Y (based on the model from A) for subset B. The square of the correlation is called cross validation R^2 and is denoted $R^2{}_*$. The difference between the two R^2 values ($R^2 - R^2{}_*$) is called the shrinkage in R^2 (S_R) on cross validation. $R^2{}_*$ is typically a less biased estimator of the population squared multiple correlation than the positively biased R^2, and therefore the shrinkage value is almost always positive. The process is repeated with the subsets switched to produce a second value of shrinkage.

Interpretation of the shrinkage values is basically descriptive in nature. Small values of the shrinkage indicate the QTL model we have can consistently predict the quantitative trait value, and therefore the model is useful in marker assisted selection. The larger the shrinkage values become, the more unreliable the QTL model. The decision as to what constitutes a large or small value is a judgement call for the researcher since there are no tabulated quantiles to determine the significance of the shrinkage values.

Using only half the data as a subset may introduce small sample size issues when determining terms to include and exclude from the model. A strategy to overcome this issue is to choose n-1 of progeny to build a model and predict the Y value for the one held out progeny. Each progeny is used as the hold out so that n different models are used to produce the n predicted values. This method is often called jackknifing. This strategy does not have to be so extreme. Two-thirds of the sample can be used to build a model and predict the remaining one-third.

Example

A hypothetical data set has been generated to demonstrate the comparison of the models for two subsets and the calculation of the shrinkage statistics. The characteristics of the data set are as follows. The trait under consideration is controlled by 16 QTL (denoted $L_1, L_2, \ldots, L_{16}$) with two alleles each. A set of 16 markers is considered (denoted $X_1, X_2, \ldots, X_{16}$) that happen to be exactly on the corresponding QTL ($L_1, L_2, \ldots, L_{16}$) so there is no problem with bias due to **r**. A random sample of 200 F_2 progeny are randomly generated from the 3^{16} (43,046,721) possible genotypes, and so the opportunity for missing genotype and multicollinearity problems is present in the data set. Note that we only generate the data set to demonstrate and discuss cross validation, not to simulate many data sets in an attempt to study the overall properties of cross validation. The quantitative trait value for each progeny is a simple linear function of the individual additive effects of the 16 loci (no dominance or epistasis) plus an error deviate. Loci L_1, L_2, L_3, and L_4 have additive effects of $\pm$ 8, loci L_5, L_6, L_7, and L_8 have additive effects of $\pm$ 4, loci L_9, L_{10}, L_{11}, and L_{12} have additive effects of $\pm$ 2, and loci L_{13}, L_{14}, L_{15}, and L_{16} have additive effects of $\pm$ 1. The error deviates are selected from a normal distribution with a variance such that $\sigma_g^2/\sigma_p^2 = 0.5$.

A multiple regression analysis of all 200 F_2 progeny is performed. The initial model involves all 16 markers, and backwards elimination (at $\alpha = 0.05$) is used to arrive at a final model of markers associated with the quantitative trait value. The final model is given below and is denoted as the overall model since it involves all the data.

Overall Model

$$\begin{aligned} Y = &-107.34 + 8.41X_1 + 7.24X_2 + 10.53X_3 + 7.97X_4 \\ &+ 5.24X_5 + 2.95X_6 + \_\_\_\_X_7 + 4.07X_8 \\ &+ 3.94X_9 + \_\_\_\_X_{10} + \_\_\_\_X_{11} + \_\_\_\_X_{12} \\ &+ \_\_\_\_X_{13} + \_\_\_\_X_{14} + \_\_\_\_X_{15} + 2.93X_{16} \end{aligned}$$

where underscores indicate the marker is not included in the final model. Realize (for discussion later) that the expected value for the coefficient for X_1 to X_4 is 8 based on the additive effect of the QTL (L_1 to L_4) near these markers; the expected value for X_5 to X_8 is 4; the expected value for X_9 to X_{12} is 2; and the expected value for X_{13} to X_{16} is 1. The R^2 for this model is 0.49.

Splitting the data into two sets (A and B) of 100 progeny and performing multiple regression as above with cross validation resulted in the following:

Set A Model

$$\begin{aligned} Y = &-115.24 + 9.45X_1 + \_\_\_\_X_2 + 11.12X_3 + 7.65X_4 \\ &+ 6.98X_5 + 5.44X_6 + \_\_\_\_X_7 + \_\_\_\_X_8 \end{aligned}$$

$$+ 4.55X_9 + \_\_\_X_{10} + 6.25X_{11} + \_\_\_X_{12}$$

$$+ \_\_\_X_{13} + \_\_\_X_{14} + \_\_\_X_{15} + 6.12X_{16}$$

$$R^2 = 0.45 \qquad R^2_* = 0.28 \qquad S_R = 0.17$$

Set B Model

$$Y = -85.40 + 9.23X_1 + 9.58X_2 + 10.93X_3 + 5.44X_4$$

$$+ 6.89X_5 + \_\_\_X_6 + \_\_\_X_7 + \_\_\_X_8$$

$$+ \_\_\_X_9 + \_\_\_X_{10} + \_\_\_X_{11} + \_\_\_X_{12}$$

$$+ \_\_\_X_{13} + \_\_\_X_{14} + \_\_\_X_{15} + \_\_\_X_{16}$$

$$R^2 = 0.60 \qquad R^2_* = 0.22 \qquad S_R = 0.38$$

Discussion

The primary result of the cross validation analysis is that two different sets of markers are found to be associated with the quantitative trait value in the two halves of the overall data set. In an actual data set, this result would force the researcher to question the validity of the set of markers in the overall QTL model (i.e., would another set of 200 progeny produce a different QTL model, and if so, which would be correct?). Determining if a QTL is actually near markers where the results are inconsistent, for example X_{11}, would be extremely difficult in an actual data set. Recall that the model inconsistencies are caused by the missing genotypes and the introduced multicollinearity. The cross validation simply brings the problem to the notice of the researcher; it does not solve the problem.

The magnitude of the shrinkage estimates also calls the use of marker assisted selection into question. The overall QTL model has an R^2 of approximately 50%. However, the shrinkage estimates suggest that using the model to predict the quantitative values in another data set does not work all that well (i.e., the prediction only holds for the data set we have). Therefore, the use of the predicted values from the QTL model as indirect selection criteria is problematic. An important implication is that markers used in a QTL model should be reevaluated each generation for use in marker assisted selection.

In this example, the true genetic model is known, and it is clear where important QTL are undetected in the models and where the estimated coefficients associated with the markers in the QTL model are inconsistent with the value of the actual QTL. The magnitude of the problems introduced by missing genotypes and multicollinearity can be seen. It should also be noted that these problems occur in this data set involving the most simple genetic case (no dominance and epistasis). This problem of inconsistent sets of markers could be expected to be even worse if there were a more complex genetic control of the trait.

An important comment about cross validation is that it works in methods of QTL detection besides those described above. A specific method of tremendous interest (and research potential) to plant breeders is the analysis of nuclear pedigrees used in humans to detect QTL near markers (Lalouel, 1992). The basic approach is to gather information on a trait and marker scores from several offspring, parents, and grandparents in a pedigree. The Elston-Stewart algorithm (Elston and Stewart, 1971) is used to calculate a measure (usually a LOD score) (Botstein et al, 1980; Ott, 1991) of the association between each of the **m** markers being considered and the trait value of interest. A large LOD score (usually greater than 3) is interpreted as evidence that a QTL is near the marker being considered.

The pedigrees used in line development in plants are similar to the nuclear pedigrees in humans. The analysis methods used in human pedigrees are beginning to be applied by some researchers to plant pedigrees for QTL detection, and the results appear to be promising (Sobral and Bridges, 1993). There are some problems to be overcome before this approach is completely applicable to plants.

The first difficulty is that each marker is tested individually (one-at-a-time) for association with the trait value. This is not a tremendous obstacle in human research where most of the interest is in disease traits that are simply inherited (i.e., qualitative traits). In plants, many traits being considered are quantitative in nature, and the fact that the method considers markers one-at-a-time is a major disadvantage as discussed earlier. The methods used in humans must be improved to actually use the markers simultaneously (analogous to the multiple regression approach, rather than simply adjusting the significance level of the LOD score), before the methods will have applicability for most plant traits.

A second problem with the application of the methods used in human pedigrees is that the plant pedigrees often involve selection. Only certain individuals are advanced in the pedigree. This means that many siblings are missing; only the best ones are kept and available for scoring. This missing data due to selection rather than randomness is difficult to correct for in a pedigree analysis and can be a source of bias in the detection of association between the markers and trait values.

Assuming these problems can be resolved, then cross validation is a useful method in these analyses. Simply split the members of the pedigree into two groups and perform a QTL detection analysis in each group. There are some interesting research questions at this point. How exactly should the pedigree be split and how do we calculate a measure of cross validation shrinkage? While there are several research questions to be addressed, the use of pedigrees for QTL detection appears to offer great potential for plant genetics researchers.

In summary, the general conclusion of this paper is that cross validation should be part of our QTL mapping efforts. It provides an indicator of the impact that missing genotypes and multicollinearity have on our QTL analysis and marker assisted selection. It also demonstrates to the researcher that using the results of one study to find important markers and map QTL is a questionable practice. The particular cross validation analysis for the example in this paper shows that the set of markers used in a QTL model is incorrect, and the predictability of the model is overstated (i.e., we can see the actual difficulties introduced by the multicollinearity since we know the actual genetic model).

Several topics for future research have been mentioned, and these are each important components of a complete and usable theory for the use of markers to study QTL. To review, these future research topics are: (1) the impact of non-inbred parents on QTL analysis; (2) the effect of non-normal data on QTL analysis; (3) further development of the methods to correct for multicollinearity in this setting (i.e., fractional factorials, ridge regression, . . .); and (4) development of pedigree analysis in plant species and appropriate cross validation strategies. A final topic for future research is the extension of cross validation to provide some measure of the confidence we have in the predictions of marker assisted selection. Specifically, the possible use of jackknifing or bootstrapping (Knapp et al, 1989) to calculate a measure of the standard error of the marker assisted selection predictions should be explored.

REFERENCES

Botstein D, White R, Skolnick M, Davis R (1980): Construction of a genetic linkage map in man using restriction fragment length polymorphisms. *Am J Hum Genet* 32:314–331

Bridges WC (1992): New challenges for statistical consultants—analysis of linkage map data. In: *Proceedings of the Statistical Education Section.* American Statistical Association

Bridges WC, Knapp SJ, Stuber CW, Edwards MD (1988): Molecular marker facilitated investigations of quantitative trait loci: Epistasis and alternative model strategies. *Agron Abst* 80:75

Doerge RW, Zeng Z-B, Weir BS (1994): Statistical issues in the search for genes affecting quantitative traits in populations. In: *Analysis of Molecular Marker Data, Proceedings of the Joint Plant Breeding Symposium Series,* August 5–6, 1994, Corvallis, OR. Crop Science Society of America and American Society of Horticultural Science

Elston RC, Stewart J (1971): A general model for the genetic analysis of pedigree data. *Hum Hered* 21:523–542

Jansen RC (1993): Interval mapping of multiple quantitative trait loci. *Genetics* 135:205–211

Kleinbaum DG, Kupper LL, Muller KE (1988): *Applied Regression Analysis and Other Multivariate Methods.* Boston: PWS-Kent Publishing

Knapp, SJ (1994): Selection using molecular marker indexes. In: *Analysis of Molecular Marker Data, Proceedings of the Joint Plant Breeding Symposium Series* (supplement), August 5–6, 1994, Corvallis, OR. Crop Science Society of America and American Society of Horticultural Science

Knapp SJ, Bridges WC, Yang M-H (1989): Nonparametric confidence interval estimators for heritability and expected selection response. *Genetics* 121: 891–898

Knapp SJ, Bridges WC, Birkes D (1990): Mapping quantitative trait loci using molecular marker linkage maps. *Theor Appl Genet* 79:583–592

Knapp SJ, Bridges WC, Liu B-H (1992): Mapping quantitative trait loci using nonsimultaneous and simultaneous estimators and hypothesis tests. In: *Plant Genomes: Methods for Genetic and Physical Mapping*, Beckman JS, Osborn TC, eds. Dordrecht, the Netherlands: Kluwer Academic Publishers

Lande R, Thompson R (1990): Efficiency of marker-assisted selection in the improvement of quantitative traits. *Genetics* 124:743–756

Lander ES, Botstein D (1989): Mapping mendelian factors underlying quantitative traits using RFLP linkage maps. *Genetics* 121:185–199

Lalouel J-M (1992): Linkage analysis in human genetics. In: *Plant Genomes: Methods for Genetic and Physical Mapping*, Beckman JS, Osborn TC, eds. Dordrecht, the Netherlands: Kluwer Academic Publishers

Nienhuis J, Bridges WC, Ruggero B, Schaefer A (1988): Comparison of statistical techniques for relating quantitative trait variation to molecular marker. *Agron Abst* 80:90

Nienhuis JT, Helentjaris T, Slocum M, Rugger B, Schaefer A (1987): Restriction fragment length polymorphism analysis of loci associated with insect resistance in tomato. *Crop Sci* 27:797–803

Ott J (1991): *Analysis of human genetic linkage*. Baltimore MD: Johns Hopkins University Press

Sax K (1923): The association of size differences with seed-coat pattern and pigmentation in *Phaseeolus vulgaris*. *Genetics* 8:552–560

Sobral BWS, Bridges WC (1993): unpublished data.

Soller M, Beckmann JS (1983): Genetic polymorphism in varietal identification and genetic improvement. *Theor Appl Genet* 47:179–190

Zeng Z-B (1993): Theoretical basis of precision mapping of quantitative trait loci. *Proc Natl Acad Sci* USA 90:10972–10976

Zeng Z-B (1994): Precision mapping of quantitative trait loci. *Genetics* 136: 1457–1468

3

Complex Trait Dissection in Forest Trees Using Molecular Markers

DAVID M. O'MALLEY

Trees are the dominant plant life covering 4 billion hectares of the earth, and forests are vital plant communities that sustain a great diversity of life, as well as supply fuel, fiber and building materials for human needs (Laarman and Sedjo, 1992). Too often people's current needs have overwhelmed nature's ability to renew the forest. Currently, the loss of tropical forests and forest soils could account for approximately 20% of global annual CO_2 emissions, while temperate forests are a net sink for CO_2 (Wisniewski et al, 1993, Dixon et al, 1994). There is need for reforestation and sustainable forest management practices throughout the world, but especially in the low latitudes where 0.75 billion hectares have been cleared. All but a small portion (<10%) of this land has been degraded and abandoned, or put into marginal agriculture or under inefficient forest management. Conservation of forest resources requires a deeper understanding of the ecological processes that affect forests as well as improved systems for the intensive production of forest products on a shrinking land base.

Forest trees include some of the oldest, largest, and most diverse living organisms. Trees have a long life span, a large mass at maturity, and high levels of genetic variability. However, tree species could evolve slowly compared with other organisms with shorter generation times, and long generation times make genetic studies and breeding a slow process. Plantations of improved trees remain in the field for many years, exposed to extremes of climate and the challenges of insects and disease. Thus tree breeding must have a long-term time horizon and an ecological perspective. Large tracts of forest in North America and Europe have symptoms of forest decline that could be related to air pollution from industrial activities and transportation (Wellburn, 1994). Forests could be susceptible to rapid changes in global climate resulting from the current increasing trends for some atmospheric gases. There is a need for more efficient and faster methods to increase forest productivity through tree breeding.

Tree breeding programs have become an important part of intensive forestry.

The Impact of Plant Molecular Genetics
BWS Sobral, Editor

The major economic traits of forest trees, growth and volume, wood properties, and stem form have been treated as quantitative traits and analyzed in common garden studies (Zobel and Talbert, 1984, Namkoong and Kang, 1990). In these studies, phenotypic variation is partitioned into genetic and environmental components. Genetic control is summarized as narrow-sense heritability, h^2, the ratio of the additive genetic variance to total phenotypic variance. The heritability of many quantitative traits in forest trees is low, but substantial gains have been achieved through selection and breeding. Hybrid breeding has yielded large gains in some species that can be vegetatively propagated (e.g., eucalypts). However, Mendelian analysis has not been useful for forest tree improvement. Forest trees have few simply-inherited polymorphisms, multi-generation pedigrees are uncommon, and inbreeding depression has prevented the development of genetically uniform lines (Franklin, 1969). Compared with crop plants such as maize or tomato, the power of conventional genetic analysis has not been brought to bear on any forest trees.

Complex trait dissection using molecular markers (Lander and Schork, 1994) is a new paradigm for genetic analysis in humans that makes it possible to study existing trait variation in populations and families without long-term controlled breeding experiments. Complex trait dissection could also transform forest genetics, where the opportunities for conventional genetic analysis are limited. Outbred organisms such as humans and forest trees have a plentiful supply of simply inherited genetic markers based on naturally occurring DNA sequence variation. Genetic analysis in many organisms has now shown that much of the apparently continuous variation of quantitative traits can be attributed to a small number of genes with large effects on the phenotype (quantitative trait loci, QTL). Thus, phenotypic variation can generally be associated with genetic markers that have simple patterns of inheritance in 2-generation or 3-generation pedigrees. These new approaches can complement existing tree breeding efforts, but are not a replacement for conventional approaches.

This review considers what impact genomic mapping and complex trait analysis could have on forest genetics. Earlier progress in forest genetics occurred with little knowledge of specific gene effects for economically important traits, but in the future, these traits could be subject to manipulation through genetic engineering. However, the first application of molecular genetics in forestry could be the use of genomic mapping in tree breeding (Kirk, 1994). The central issue addressed here is how molecular markers could be used to circumvent the problems that longevity and an outcrossed breeding system pose for genetic analysis. While forest trees have the cellular machinery common to all plants, forest geneticists often have different goals and different approaches than common for crop plant species. There is a parallel comparison with the contrasting approaches in human and animal genetics. For example, understanding the genetic basis for complex traits such as human behavior variation could make a significant contribution to human health, and this knowledge is indispensable (Plomin et al, 1994). However, the identification and cloning of the genes responsible for variation in behavior can be accomplished more efficiently using the mouse (Takahashi et al, 1994). Similarly, complex trait dissection in forest trees could lead to a better understanding of the role of genetics in tree growth, maturation, and response to the environment. This new knowledge could lead to more efficient methods for tree breeding and help to guide efforts for forest conservation.

In this review, current work in forest genetics is discussed, and research problems with special importance to forest trees are identified. Genomic mapping in plants to dissect complex traits has been reviewed by Tanksley (1993). Haines (1994) provided a more comprehensive review of forest biotechnology, including genomic mapping. After summarizing progress in genetic markers and methods for complex trait dissection, the potential to use these methods to study the genetic basis for growth and volume traits, maturation, and adaptation in forest trees is explored. Genetic dissection of disease and pest resistance is also discussed, and the results of some studies on forest trees are summarized. The objective is to try to frame some of the issues in forest genetics where progress could be made using molecular markers.

Genetic Markers and Maps

Genetic markers are DNA sequence polymorphisms that can be readily assayed without obtaining the explicit DNA sequence and that show Mendelian inheritance. For genomic mapping, the ideal genetic marker is codominant, multiallelic, and hyper variable (i.e., segregates in almost every family). In most crops, simply inherited morphological polymorphisms provided the first linkage maps, but in forest trees, the first maps were constructed using allozymes (Conkle, 1981). Allozyme markers are codominant and multiallelic, but the number of allozyme markers is small, the heterozygosity per locus is limiting, and genetic variation is detected at the level of gene products.

Restriction fragment length polymorphisms (RFLPs) are genetic markers that are obtained by using restriction endonucleases to precisely cleave a genomic DNA fragment containing a particular gene sequence. RFLPs provide useful genetic markers for forest trees (Bradshaw et al, 1994; Devey et al, 1994; Jermstad et al, 1994). Many RFLP probes yield simple patterns of variation that are readily interpreted as a single locus containing a specific gene corresponding with the probe sequence. RFLPs in pine are as variable as allozymes and are often inherited as codominant markers. RFLPs require large amounts of genomic DNA and are laborious to carry out, compared with polymerase chain reaction methods.

The Polymerase Chain Reaction (PCR) has provided a new way to obtain genetic markers, based on amplification of specific DNA fragments from small quantities of genomic DNA template. A large amount of detailed DNA sequence information is required for this approach. PCR-amplified markers can be based on anonymous genomic DNA fragments that vary in size (codominant inheritance), or can be amplified from some individuals but not others (dominant), or can be cleaved differentially by restriction enzymes (Bradshaw et al, 1994). Bradshaw et al (1994) and Voo et al (1995) each reported a marker developed from the DNA sequences of a known gene in *Populus* and *Pinus*, respectively.

PCR can also readily provide many genetic markers based on amplification from genomic DNA template using a single short primer (Randomly Amplified Polymorphic DNA (RAPD), Williams et al, 1990; Arbitrarily Primed Polymerase Chain Reaction (AP-PCR), Welsh and McClelland, 1990). These markers are anonymous DNA sequences flanked by the primer sequence in opposite orientation. The mode of inheritance for RAPDs is usually dominant; the sequence either amplifies or not, and one copy cannot be distinguished readily from two copies. With care, large numbers of RAPD markers are easily resolved and high quality genomic maps can be generated (Grattapaglia and Sederoff, 1994; Plomion et al, 1995a, 1995b). Another PCR-based anonymous marker system, amplified fragment length polymorpisms (AFLPs) is based on restriction digestion, ligation of linkers, and PCR-amplification (Zabeau, 1993).

Microsatellite markers are based on the PCR-amplification of a genomic region containing a simple repeated sequences (Morgante and Olivieri, 1993). The length of these repeated sequences often varies, and the markers are codominant. Smith and Devey (1994) reported several microsatellite sequences for *Pinus radiata*. They documented inheritance for two microsatellite markers and estimated the heterozygosity for these two markers in the *Pinus radiata* population to be 0.60–0.65. Thus, these markers should be segregating in the majority of trees studied and could be used to establish synteny for linkage groups established using anonymous markers and megagametophytes. However, microsatellite markers require considerable effort for development.

Families

Segregating of markers in 3-generation outbred pedigrees can be used to construct linkage maps and associate markers with phenotypes to identify genomic regions associated with quantitative effects or qualitative differences (Haley et al, 1994; Groover et al, 1994). Inbred lines of homozygous individuals are unavailable for forest trees, so the standard analyses of backcross and inbred F_2 families used for genetic studies of experimental organisms are not feasible. The F_1 individual of an inbred pedigree is either homozygous or heterozygous for any gene, thus segregating QTL effects are readily traced with markers. Outbred pedigrees are more complex because the family that is analyzed has two parents, and each of these parents could have a heterozygous or two types of homozygous genotypes for one gene with two alleles. For two genes, each with 2 alleles, there are 81 possible genotypic configurations among the grandparents. Not all of these configurations are informative for linkage analysis, and combined analysis of markers and traits following so many different possible genetic models of inheritance is a difficult statistical problem. However, 3-generation outbred pedigrees have several advantages, including known linkage phase for codominant markers and the opportunity to test various genotypic combinations for departures from additive gene action.

F_2 families can be generated by selfing a tree, but selfing often produces low seed set and distorted segregation ratios (Franklin, 1969). Inbred F_2 families have been used especially to study interspecific hybridization (Bradshaw et al, 1995). F_2 families in trees are more complex than inbred F_2 families in crop plants because there are two variable grandparents rather than two inbred lines at the top of the pedigree, and these individuals usually cannot be assumed to be homozygotes selected in opposite directions.

An alternative two generation approach is the analysis of markers that segregate 1:1 in a full sib family following a pseudotestcross model (Grattapaglia and Sederoff, 1994). Anonymous marker methods such as RAPD generate can generate large numbers of 1:1 segregations where a marker locus is in the heterozygous condition in one parent, but homozygous null in the other parent. This circumvents one of the problems with dominance, but determining which linkage groups from the two parents are homologous requires some codominant markers. The mode of gene action cannot be readily determined from pseudo-testcross model involving outbred parents unless homology of linkage groups is established. Determining homologous linkage groups is also a problem for genetic analysis of F_2 families obtained by selfing heterozygous individuals.

Gymnosperms present a special situation for genetic analysis (Grattapaglia et al, 1992). The seeds of conifers contain megagametophyte tissue, which is haploid and derived from a single megaspore. Allozyme markers segregate 1:1 in megagametophytes from a heterozygous tree. Megagametophytes contain enough DNA template for hundreds of PCR reactions, but only enough DNA for a few Southern blots. Conifer megagametophytes have been used for linkage analysis using RAPDs (Plomion et al, 1995a, 1995b). One advantage of genetic analysis using megagametophytes is that segregations can be obtained from open pollinated seeds from any tree (the paternal contribution in the embryo does not matter). Segregation analysis of megagametophytes follows a phase unknown model analogous to a backcross, but parental and recombinant types can be readily inferred if there are several linked markers. Marker segregation in megagametophytes can be unambiguously associated with the phenotype of the corresponding seedling, allowing trait dissection in half-sib families.

To overcome the time limitation involved with new experiments, genetic analysis in trees should take advantage of any cross in which breeders have identified useful germplasm. To accomplish this, a set of codominant multiallelic markers that are well-defined and segregating in almost every cross would be ideal but laborious to develop. Anonymous marker systems such as RAPDs are useful now for genetic dissection, but establishing syntenic groups for anonymous markers in different individuals has been a limitation. However, when a common set of RAPD primers are assayed in unrelated individuals within a species, common markers with similar linkage arrangements can be identified that allow syntenic groups to be established (Wilcox, 1995). Integrating anonymous marker maps with databases will be easier when more cDNA-based expressed sequence tagged sites and microsatellites are available.

Complex Trait Dissection

In classical genetic analysis, morphological differences are attributed to a single gene if there is no significant departure from Mendelian expectations for the segregating polymorphism and if the polymorphism is inherited over generations. Nondeparture from Mendelian expectations in two generations (parents to progeny) is often not compelling evidence that a single gene is responsible for the phenotypic difference. Mendelian ratios are not obtained for some polymorphisms because some genotypes are lethal, have low viability, or reduced penetrance (Franklin, 1969). However, these polymorphisms could still be attributed to a single gene if the inheritance of the trait difference over several generations could be explained by a more complex genetic model that incorporated additional parameters (Robinson, 1971). Sometimes a polymorphism is controlled by interacting loci (i.e., epistasis) resulting in more complex patterns of inheritance. Thus, even in the absence of Mendelian ratios, simple inheritance can be inferred over several generations, but almost any ratio of phenotypes in a 2-generation pedigree could be explained away by constructing a complicated genetic model.

Complex trait analysis gives geneticists working with undomesticated, outbred, or long-lived organisms new opportunities for genetic analysis that previously were not feasible (Lander and Schork, 1994). The association of segregating genetic markers with phenotypic differences can be compelling evidence for major gene control of the trait even in a two generation pedigree. Following the marker:trait association over three generations provides evidence, equivalent to classical genetic analysis, that a major gene underlies the phenotypic differences segregating in the pedigree.

The association of a genetic marker with a segregating morphological polymorphism can be especially useful for woody plants. Complex trait analysis can be carried out using the segregation of markers spaced at regular intervals on a genomic map to systematically test for associations with the phenotype. However, for species where genomic maps and well-characterized markers are unavailable, bulked segregant analysis using RAPD markers (Michelmore et al, 1991) provides a useful alternative to genomic maps. For example, dioecy is often controlled by more than locus in plants, sometimes with multiple alleles, and usually without a well-differentiated sex chromosome (Durrand and Durand, 1990; Irish and Nelson, 1989). Mulcahy et al (1992) associated RAPD markers with gender in the herbaceous *Silene latifolia* using bulked segregation analysis. Hormaza et al (1994) used this approach to find a RAPD marker associated with gender in *Pistacia vera*, a dioecious tree cultivated for nut production. Pistachio trees remain juvenile for five to eight years, and the marker allows the gender of individuals to be determined at the seedling stage. This marker will be useful for breeding. A marker for gender could be also be commercially useful for the production of only pollen parents or only seed parents in some woody species.

Selection for Growth and Volume Traits

Quantitative genetics is based on the resemblance among relatives that is due to the genes inherited from common parents or grandparents (Falconer, 1989). The phenotype is the product of both genetic and environmental influences. The extent of genetic control is assessed in common garden studies from the phenotypic correlation of individuals within families. The genetic variation in traits that is transmitted from one generation to the next is called additive genetic variation. Nonadditive genetic variation is caused by specific genotypes or genetic interactions (i.e., dominance and epistasis). Narrow-sense heritability (h^2) is the ratio of the additive genetic variance ($\sigma^2{}_A$) to total phenotypic variance ($\sigma^2{}_P$). Additive genetic variation determines the response to selection in breeding and in models of evolution. Therefore, a critical issue is the extent to which $\sigma^2{}_A$ is oligogenic and can be attributed to major gene effects.

Growth and volume in forest trees are quantitative traits that have a low heritability, usually less than 1/2 (Zobel and Talbert, 1984). The heights and diameters of forest trees are strongly correlated, and together these measurements provide a prediction of the volume of wood produced by a tree. Breeding for growth and volume in forest trees has usually followed a population approach based on additive genetic variation because most tree species show strong inbreeding depression. Family selection in early generations of tree breeding has resulted in substantial gains for growth and volume, but inbreeding increases rapidly with intense family selection. In advanced generations, a combined selection approach is needed in which within family and among family selection are balanced (Falconer, 1989). However, within family phenotypic selection is less efficient for low heritability traits than among family selection based on progeny tests. A combined selection strategy could make use of a complementary mating scheme where half-sib families would be used to estimate breeding values of parents for family selection, and a half-diallele of full-sib families would be generated for within family selection (McKeand and Bridgwater, 1992). Advanced generation tree breeding plans could also incorporate a replicate population approach in which the population is subdivided into small units to ensure that unrelated matings can be made for production of seeds.

The efficiency of within family selection could be increased using molecular markers. Lande and Thompson (1990) evaluated the efficiency of within family marker assisted selection as a ratio of selection response based on markers plus phenotype to phenotypic selection alone. They showed that efficiency was highest when h^2 was low, and that efficiency strongly depended upon the portion of additive genetic variation explained by markers. In crop plants, genetic dissection is often carried out with a narrow perspective of phenotypic variation within an inbred pedigree. Moreno-Gonzalez (1993) showed how additive and dominance effects of QTLs could be estimated from various mating designs.

In forest trees, additive genetic variation is usually measured for a more diverse and largely unrelated population of individuals, and selection is generally based on individual breeding values. The breeding value of an individual is twice the average effect transmitted by one gamete from that individual to offspring when the individual is mated to a random sample of gametes from the population (Falconer, 1989). Breeding value is expressed as two times the mean deviation of these half-sib offspring from the population mean. The variance of individual breeding values is σ^2_A. QTL effects can be related to population level σ^2_A through the concept of breeding value. Average effect QTLs defined in half-sib families are single locus components of breeding value, and they were called "chromosome substitution effects" by Dentine and Cowan (1990) and Dekkers and Dentine (1991).

Based on the infinitesimal model for quantitative trait inheritance, the additive genetic variation (segregation variance) transmitted by the common parent to a half-sib family could be small (e.g., ~14% of the total within half-sib family phenotypic variance for $h^2 = 1/2$, O'Malley and McKeand, 1995). This small expected effect suggests that markers segregating in the common parent of a half-sib family could explain only a small amount of phenotypic variance of a half-sib family, and QTL detection will be challenging. However, half-sib families contain a large amount of genetic variation. Genes with large average effect are likely to be at low frequency, have a dominant mode of inheritance, and have a large effect on the phenotype (Falconer, 1989). The average effect of individual QTLs detected in a large population of dairy cattle ranged from 40%–179% of the expected segregation variance transmitted by the common parent (Georges et al, 1995). From the perspective of σ^2_A measured at the population level, QTL effects detected in a specific family (e.g., a full-sib cross of two parents) are not a useful predictor of breeding value. For example, a QTL with large effect in a full-sib family could already be at high frequency in the breeding population, and hence have little breeding value. While population-level average effects could be detected in half-sib families, markers associated with these effects would be selected within full-sib families. Thus, integration of MAS with conventional tree breeding approaches could require a different perspective than conventional crop plant approaches.

Genetic information from half-sib families of forest trees is important for at least two reasons. First, QTLs defined in half-sib families can be directly related to breeding value and additive genetic variation. Second, time is a critical factor in tree breeding, and ways to circumvent the generation length problem are valuable. Half-sib families can provide a retrospective analysis of QTL effects based on already mature trees. Earlier genetic tests generally contain full-sib families that are too small for marker:trait association studies, but large half-sib families can sometimes be pieced together from half-diallele testing designs. For high heritability traits such as wood specific gravity, trees from several locations could be used. Furthermore, large half-sib families are sometimes available in genetic tests or in operational family block plantings. These planting could offer an opportunity to discover marker:trait associations in older trees. Markers for mature performance could be identified using methods similar to Dentine and Cowan (1990), and then these markers could be selected in juveniles.

Juvenile:Mature Correlation

Mature trees differ from juvenile trees in several important traits, and the lack of understanding of the maturation process is a major impediment to tree breeding (Greenwood and Volkaert, 1992). Mature trees are capable of producing pollen and seeds, but they are usually more difficult to propagate vegetatively than are seedlings. Mature trees also have different wood properties compared to juveniles (Zobel and van Buijtenen, 1989). In loblolly pine, the transition from juvenile wood to mature wood occurs at approximately ten years of age. Juvenile wood has shorter cells with larger lumens and thinner cell walls. Juvenile wood has larger microfibril angles, more compression wood, lower specific gravity, higher lignin content, and lower cellose content. Selection of individuals for breeding is delayed until the trees reach an age at which trait values are strongly correlated with trait values at harvest age. For pine, selection age can be several years old. The cycle of breeding and selection could be shortened if mature performance could be predicted earlier and if juvenile trees could be stimulated to produce reproductive structures. The low heritability of growth and volume traits in trees could be due, in part, to a shift in genetic control during the maturation process in trees.

The juvenile forms of many organisms must cope with different challenges and sometimes even live in different environments than their mature parents. The transition from juvenile to mature phases of the life cycle can be accompanied by dramatic morphological changes (e.g., the metamorphosis of amphibians and lepidopterans). Williams (1987) noted that juvenile pine seedlings produce primary needles during the free-growth phase that occurs directly after seedling germination, but the needles on successive flushes of growth and on mature trees occur in fascicles. She showed that shoot elongation in loblolly pine seedlings following free growth was more strongly correlated with performance at selection age than total height of seedlings. Several studies in pine have obtained a variety of results for the value of shoot growth components as predictor performance at selection age (Greenwood and Volkaert, 1992).

The allometry of plant growth provides another explanation for the low correlation between juvenile and mature performance. As trees age, there is a progressive decrease in the rate at which height increases relative to diameter (i.e., increments of diameter growth correspond with smaller and smaller increments of height growth). This ontogenetic change has been interpreted in terms of optimal mechanical design based on engineering principles (Niklas, 1994). For example, if diameter is held constant, then height is eventually limited by the weight of the stem that would cause elastic buckling. Height should scale as the 2/3 power of diameter if elastic self-similarity governs plant design. Alternative scaling relationships of heights and diameters are predicted by stress-similarity for self loading and geometric self-similarity for wind loading (exponents of 1/2 and 1 respectively). As trees mature, the scaling exponent decreases from above 1 to 1/2 (Niklas, 1995). The change in mechanical design suggested by these scaling changes could reflect different requirements for height growth and tree architecture at different stages in the life cycle of trees. Height could be attained at different costs of biomass and different stem strengths for the different designs. The allometry of trees could also change in response to variation in environmental conditions such as stand density and shading (Bonser and Aarssen, 1994). Balocchi et al (1993) showed that the contributions of the components of genetic variation in height growth changed with age in loblolly pine, with σ^2_D largest in early years, but σ^2_A eventually exceeding then far surpassing the σ^2_D. Their results and the ontogenetic changes during tree growth and maturation suggest that QTL effects could be specific to different phases of the life cycle of trees.

Adaptation

An important issue in forest genetics is matching genotypes to sites (Zobel and Talbert, 1984; Namkoong and Kang 1990). The ranking of families can change from one test planting to another due to genotype by environment interactions, (GxE). Breeding zones are regions, or environments within regions delimited by elevation or edaphic factors, where family rankings are generally stable. A breeding zone must be large enough to justify breeding effort, so a precise match of genotypes to sites is not feasible. Forestry is generally practiced on marginal lands not suited to agriculture where trees can harvest light energy for many years with minimal input from man. Wood is harvested when enough biomass has been produced to justify the cost. Alternatively, intensive forestry is practiced in plantations on productive sites using genetically improved seedlings. Therefore, a better understanding of GxE and the more fundamental issue of adaptation is important for tree breeding.

Can complex trait analysis identify major gene effects useful for understanding adaptation in forest trees? Adaptation is the process of transgenerational change in which organisms become better suited to their environment through the alteration of features or functions that solve or improve solutions to problems posed by the environment, of integrating metabolism, or of enhancing reproduction (Burian, 1992). Orr and Coyne (1992) reviewed the genetics of adaptation. The critical question they identified is "How often does adaptation involve major genes?" The argument against major genes is based on the conformity of parts, the idea that mutations with large effect (macromutations) are disadvantageous because they are likely to disrupt the physiological and developmental integration of an organism. Adaptation has been assumed to follow the polygenic model of inheritance that has been the basis of quantitative genetics.

Recently, Lai et al (1994) associated DNA sequence variation with phenotypic effects at the *scabrous* locus (*sca*) in a natural population of *Drosophila melanogaster*. *sca* is a gene implicated in nervous system development that contributes to genetic variation in numbers of abdominal and sternopleural bristles. They found several alleles at intermediate frequency that had large effects. These results suggest that the genetic architecture of quantitative trait loci is compatible with a role for major genes in adaptation. Major genes have been implicated in some cases of adaptation involving visual differences among individuals and species (e.g., mimicry, Charlesworth, 1994), but visual polymorphisms could be a special case. The genetic basis for morphological differentiation of closely related species has been examined in several cases and has often been found to involve a small number of major genes (Paterson et al, 1988; Doebley and Stec, 1993). An important perspective has been obtained from studies of insect adaptation to insecticide (McKenzie and Batterham, 1994). Resistance response involving major genes could be favored by high selection intensity, and a polygenic response could be favored by low or moderate selection intensity. Adaptation in response to intense selection is more likely to depend upon the genetic variation already present in a population. Thus, genomic mapping and complex trait dissection could be useful for studying the process of adaptation.

The dominant feature of the life history of trees is longevity. Trees must cope with environmental changes and challenges that occur on a time scale much longer than other plants, but within a single generation for trees. The mechanisms that allow organisms to survive to great age are not understood (Partridge and Barton, 1993). A flexible response to variation in environmental conditions encountered during the long life span of individuals could be advantageous. The interaction of genotype with environment to produce the phenotype is conceptualized as the norm of reaction of an individual (Schlichting and Pigliucci, 1995). An individual genotype's norm of reaction could be assessed if the individual's genotype could be cloned and the ramets assessed in different environments or across an environmental gradient. An important question is: can selection in two or more different environments (either spatially or temporally) result in adaptive phenotypic plasticity (Via et al, 1995)? While phenotypic plasticity has been difficult to define precisely, the concept is well illustrated by an aquatic plant, *Ranunculus flammula*, where an individual can produce either submerged leaves or morphologically different emergent leaves on the same plant, depending on water levels and plant height (Cook and Johnson, 1968).

Adaptive phenotypic plasticity is currently a matter of intense interest in studies of ecology and evolution (Via et al, 1995). Two genetic mechanisms have been proposed for phenotypic plasticity. Allelic sensitivity occurs when the effect of an allele varies in different environments. Gene regulation can causes suites or cascades of genes to be expressed in some environments but not in others. These two mechanisms are not mutually exclusive, but they could intergrade if variants at a regulatory locus cause different levels of gene product to be expressed in different environments. Two types of phenotypic plasticity are recognized, graded response and discrete (or switched) response (Via et al, 1995).

Individual forest trees are exposed to wide extremes of climate during their life span, and mechanisms that adjust physiology and development to these conditions could be important for survival and growth. One example of phenotypic plasticity in woody plant response to predictable environmental change is the annual cycle of leaf loss in the autumn and leaf production in the spring for deciduous trees. Bradshaw et al (1995) found six QTLs for phenological variation in an interspecific F_2 family of *Populus*, suggesting the potential importance of environmental regulation for forest trees. Doebley (1993) speculated that some genes responsible for QTL effects could be environmentally responsive regulatory genes. Genetic variation at these loci could redirect plant growth and development to better suit current environmental circumstances. The *teosinte-branched* gene in maize is responsible for a shrubby appearance when plants are grown in the open, but a single straight stem is produced when the plants are shaded or grown in crowded conditions (Doebley et al, 1995). Differences in stand density or shading also affect the allometry of tree growth (Bonser and Aarssen, 1994). Response to unpredictable change associated with year to year differences in growth conditions (e.g., drought) could be best studied in mapping populations established on contrasting sites (e.g., QTLs that explain growth at one site but not the other). Alternatively, the record of annual growth reflected in tree rings could provide yearly components of growth for analysis.

One of the ways in which forest trees could differ from crop plants is in the relative importance of developmental differences in gene expression and in responsiveness to environmental stimuli. Genotype by environment interaction has been described for QTL effects identified in studies of crop plants (Hayes et al, 1993; Paterson et al, 1991). While some QTL effects varied across environments, the majority were stable. QTL detection experiments are subject to type I and type II statistical error (Jansen, 1994; Churchill and Doerge, 1994), and distinguishing QTL effects that show interactions with the environment from statistical error will require careful experimental design. These statistical, developmental, and environmental issues could be critical for forest trees, making analysis very complex. Alternatively, QTLs for summary growth measurements taken at maturity could be sufficient to explain much of the genetic variation in growth and volume. More studies are needed to test these ideas.

Biotic Interactions

Biotic interactions should be more intense for trees compared to other plants because longevity exposes individuals to large numbers and many generations of pests and pathogens. Feeny (1976) proposed that organisms such as trees are apparent to pests and pathogens. Highly apparent organisms should have a generalized defense against pests and pathogens (Feeny, 1976; Rhoades and Cates, 1976). How could specificity of plant defense response evolve in such long-lived plants confronted by such diverse pests and pathogens? Lerdau et al (1994) reviewed the tradeoff between the cost of defense and plant growth in the context of plant monoterpenes. Monoterpenes are feeding deterrents to many generalist animals, but they sometimes are used by specialized herbivorous insects to recognize their host plants. These chemicals tend to be constitutively produced by many forest tree species, although the quantities vary during the growing season.

Induced monoterpenes differ among geographic regions, suggesting a specificity of response to feeding by specialized insects.

Highly specific interactions among crop plants and pathogens have been described by the gene-for-gene model in which each resistance gene in the host can be overcome by a corresponding virulence gene in the pathogen (Thompson and Burdon, 1992). Resistance in plants can involve any number of genes, from a single major gene to many resistance loci. Simply inherited resistance has been easy to manipulate in breeding programs; however, sensational epidemics have occurred as a consequence of the pathogen population overcoming the resistance by evolving new virulence. Polygenic resistance has been postulated to be more durable (Robinson, 1987). The lack of evidence for simply inherited resistance in forest trees has supported a widely held view that resistance in forest trees is durable and polygenic, unlike crop plants where gene-for-gene systems of host resistance and pathogen virulence have been described that are often unstable (Robinson, 1987; Thompson and Burdon, 1992).

The lack of evidence for simply inherited disease resistance in forest trees could be due to the limited potential for Mendelian analysis in forest trees and to the complex life cycles of many forest pathogens. Mendelian inheritance of resistance to white pine blister rust was described for sugar pine by Kinloch et al (1970), and the frequency of individuals with fusiform rust galls in some full-sib families of loblolly pine suggested major resistance genes (Kinloch and Stonecypher, 1989). Nelson et al (1993) studied phenotypic interactions of clonally propagated slash pine genotypes with isolates of the fusiform rust pathogen, and concluded that the pattern of the interactions was compatible with a small number of resistance genes in the host. Wilcox (1995) associated RAPD markers with the presence/absence or fusiform rust disease in loblolly pine, and followed the inheritance of a major fusiform rust resistance locus in loblolly pine pedigree through three generations. The resistance locus segregated in full-sib families inoculated with one single aeciospore line of the pathogen, *Cronartium quercuum* f. sp. *fusiforme*, but was overcome by another line, demonstrating that virulence in the pathogen varied (Kuhlman et al, 1995).

The genetic analysis of fusiform rust resistance in loblolly pine is an excellent example of complex trait analysis. A RAPD marker closely associated with the resistance locus accounted for all virtually all resistant individuals, but approximately 20% of the individuals lacking the marker failed to develop the disease. Some plants that are inoculated with pathogens can fail to develop disease for environmental reasons, even though these plants are susceptible (escapes). To map the resistance locus, Wilcox (1995) used a maximum likelihood model that included two parameters: the recombination fraction with a genetic marker, and the frequency of escapes. This analysis located the marker approximately 2 cM from the resistance locus and quantified the precision of the inoculation procedure. When the trait was treated as a locus, the resistance appeared to be approximately 14 cM from the RAPD marker most highly correlated with the disease trait. In further studies, the RAPD marker was associated with resistance in half-sib families grown in the field for seven years and inoculated by the local population of the pathogen under natural conditions. This result would have been impossible if the marker and resistance locus were common in loblolly pine and if the corresponding virulence locus in the pathogen were at high frequency. There is evidence for additional simply inherited resistance loci in loblolly pine that are the subject of ongoing research.

New opportunities exist for studying and understanding disease resistance in trees through complex trait analysis using molecular markers, as demonstrated with fusiform rust resistance in loblolly pine (Wilcox, 1995). Shepherd et al (1995) have reported mapping RAPD markers that are closely associated with resistance to beetles in eucalyptus hybrids, and that the chemical basis of the resistance is variation for an aromatic oil, showing that similar opportunities could exist for studying insect resistance using complex trait analysis. Gene-for-gene coevolution (Thompson and Burdon, 1992) has rarely been described for insect-plant pairs of species (the Hessian fly on wheat is a notable exception), but powerful tools to detect these genetic systems have not been previously available, especially for natural populations. Understanding forest pathosystems is essential for plantation forestry, especially in tropical regions, and these studies could provide insight on durable systems of resistance for crop plants. Disease and pests play a role in one of the mechanisms hypothesized to account for biodiversity in tropical forests (Gilbert et al, 1994; Burkey, 1994).

Complex Trait Dissection and Population Studies

Until recently, the major use of genetic markers in forestry involved the application of allozymes to population genetic studies (Adams et al, 1992). These studies were both at the applied level (e.g., pollen movement and genetic diversity in seed orchards), and in natural populations. Current knowledge of the distribution of genetic variation within and among populations that has been obtained using essentially neutral markers shows high levels of genetic variation, but most of the variation resides within populations (Hamrick et al, 1992). Westfall and Conkle (1992) provide an alternative multivariate perspective that suggests a greater degree of differentiation. However, little is known about the distribution of adaptive genetic variation except from common garden studies that do show extensive differentiation over short distances in some locations.

Much effort has been expended studying the adaptive significance of allozyme variation. In their review, Bush and Smouse (1992) concluded that the adaptive distance model, which hypothesizes that homozygote fitnesses are proportional to their frequencies in the population, could account for a large portion (~1/4) of the variation in growth in conifers. Much of this effect can be attributed to rare alleles at only a few loci (different allozymes in different studies). They suggested that the effects of rare alleles are disadvantageous over some portions of the life cycle, but are advantageous at other stages. Genetic dissection studies using molecular markers can now address these issues by directly mapping the effect. Concordance of the genomic map locations of allozymes and trait effects could resolve some of issues concerning the adaptive significance of allozymes.

Complex Trait Analysis in Forest Trees

A remarkable amount of progress recently has been made for genomic analysis in forest trees and other woody species. Lawson et al (1995) have located QTLs for several traits on the genomic map for the 'Rome Beauty' × 'White Angel' family of apple. A QTL for branching habit has been detected that was tightly associated with timing of initial vegetative growth, suggesting pleiotropic effects of a single locus. A QTL for root suckering has been found, as well as for the timing of reproductive budbreak. QTLs for the timing of phenological events have adaptive significance in undomesticated tree species. Bradshaw et al (1995) have analyzed several traits in an inbred F_2 family obtained from an interspecific cross of *Populus*. They have found two QTLs for diameter growth. Very close to the map location of each of these QTLs, a QTL for branch traits has been found, sylleptic branching in one case and branch leaf area in the other case. This genetic result parallels the correlation observed at the phenotypic level between sylleptic branch leaf area and diameter growth and suggests pleiotropic effects of a single underlying gene in each of the two genomic regions. Five QTL have been found for spring leaf phenology. A significant QTL effect for height growth has been detected in year 2, but not in year 1, although markers in a different linkage group have shown a moderately strong association with year 1 height (LOD 2.38).

QTLs for wood quality could be especially valuable for marker assisted selection trees because the trait is economically important and is difficult to assay (Williams and Neale, 1992). Groover et al (1994) detected significant associations ($P < 0.05$) between five genomic regions and wood specific gravity in a three generation outbred pedigree of loblolly pine. One QTL was highly significant ($P < 0.0002$). Their analysis also revealed evidence for nonadditive gene action and genotype by environment interaction for these QTLs. The five QTLs together explained approximately 23% of the phenotypic variation in wood specific gravity in this loblolly pine family. Grattapaglia et al (1994) detected QTLs for wood properties and diameter (centimeters at breast height, CBH) segregating in an interspecific hybrid half-sib family of *Eucalyptus*. Grattapaglia et al (1994) reported five putative QTLs ($P < 0.02$) for wood specific gravity in seven year old mature trees that accounted for 21%–25% of the phenotypic variance. The QTLs effects are average effects transmitted from the maternal parent to the F_1 population sired by approximately 15 pollen parents, thus the extent of genetic control for wood specific gravity could be double this value in a full sib cross, based on additive effects. A putative QTL was detected for % pulp, but small sample size limited detection for this trait. The % pulp QTL could account for 10% of the phenotypic variance. Three QTLs were detected for % bark, explaining approximately 12% of the phenotypic variance. Three QTLs were detected for diameter, explaining 11%–14% of the variation in diameter. This analysis showed that growth and volume traits could be detected in mature trees, in spite of the potential for developmental and environmental effects to obscure the relationship.

Conclusions

Complex trait dissection has provided the tools to address critical issues for understanding the breeding and evolutionary biology of forest trees. Genome analysis in plants can now carry out a powerful program of forward genetics where variation in phenotypes can be related to the genes responsible for those differences through methods of positional cloning (Tanksley, 1993; Tanksley et al, 1995). While gene identification and gene isolation can now be contemplated for forest trees, genetic markers enable studies on the nature and extent of genetic control during tree development, and on tree response to both temporal and spatial environmental heterogeneity. These issues have been difficult to investigate using the methods of quantitative genetics. The results of Grattapaglia et al (1994) show that QTLs for diameter in mature eucalypts can be identified, so some genes must contribute to variation in growth over many years. However, analysis of annual increments of growth could reveal more genetic control than suggested by the low estimates of heritability for most quantitative traits. Early growth results from *Populus* and some preliminary results from pine suggest different QTLs could be important in different years (Bradshaw et al, 1995). Validation of QTL effects is an important issue for understanding these effects.

Major genes that control phenology (flushing date in spring, onset of dormancy in fall) could also have effects on growth and volume traits. Genetic dissection studies of phenological variation in trees from different latitudes or elevations could greatly increase understanding of provenance variation. Allozyme studies have shown that most of the selectively neutral genetic variability is located within populations, with little variation among populations. However, variation in genes controlling variation in phenology are likely to impact tree growth, as well as vary among populations.

The new approach to forest genetics made possible by molecular markers has the potential to dramatically increase our understanding of genetic control of growth and differentiation in trees, and could increase our ability to capture genetic gains and thereby increase forest productivity. Complex trait dissection for traits has already revealed the major genes for disease resistance and dioecy in some trees. Marker assisted breeding strategies can be designed for situations like these, on a case by case basis. The identification of QTLs for yield traits has the potential to increase the efficiency of within family selection in tree breeding. How to systematically exploit this knowledge in a breeding program is not yet understood. For a tree breeding program that is based on breeding value and that maintains a highly diverse, outbred breeding population, average effect QTLs could be defined in half-sib families. These QTLs could be considered components of the additive genetic variance of the whole breeding population. Marker assisted selection following Lande and Thompson (1990) could be used in a prospective way, as a tool to increase the precision of phenotypic selection within families. Alternatively, marker:trait associations could be used retrospectively to predict which seeds or seedlings are likely to have a favorable phenotype when they are adults. Vegetative propagation (Gupta et al, 1993) of selected individuals can capture nonadditive genetic variation. Other strategies to use markers to identify useful genetic effects have been described by de Vicente and Tanksley (1993) and Stuber et al (1992). Strauss et al (1992) has provided a pessimistic view of the prospects for the application of marker assisted selection in forest trees, but has emphasized the value of these new tools for understanding the biology and genetics of forest trees. Now, the tools for complex trait analysis are in hand, and the intricacy of the issues is more clear.

ACKNOWLEDGMENTS

I thank Barbara Crane, Ben Liu, Glen Dale, Ron Sederoff, and Christophe Plomion for helpful discussion and suggestions. This work was partially supported by USDA Plant Genome NRICGP grants 91-37300-6341 and 92-37300-7549, by the NCSU Forest Biotechnology Industrial Associates, and by the NCSU Industry Cooperative Tree Improvement Program.

LITERATURE CITED

Adams WT, Strauss SH, Copes DL, Griffin AR, eds. (1992): Proceedings of the International Symposium on Population Genetics of Forest Trees. 1990 Jul 31–Aug 2; Corvallis, OR. *New Forests* 6:1–420

Balocchi C, Bridgwater FE, Zobel BJ, Jahromi S (1993): Age trends in genetic parameter for tree height in a nonselected population of loblolly pine. *Forest Sci* 33:231–251

Bonser SP, Aarssen LW (1994): Plastic allometry in sugar maple (*Acer saccharum*): adaptive responses to light availability. *Am J Bot* 81:400–406

Bradshaw HD Jr, Villar M, Watson BD, Otto KG, Stewart S, Stettler RF (1994): Molecular genetics of growth and development in *Populus*. III. A genetic linkage map of a hybrid poplar composed of RFLP, STS, and RAPD markers. *Theor Appl Genet* 89:167–178

Bradshaw HD Jr, Stettler RF (1995): Molecular genetics of growth and development in *Populus*. IV. Mapping QTLs with large effects on growth, form and phenology traits in a forest tree. *Genetics* 139:963–973

Burian RM (1992): Adaptation: historical perspectives. *In*: *Keywords in Evolutionary Biology*, Keller EF, Lloyd EA, eds. Cambridge: Harvard University Press

Burkey TV (1994): Tropical tree species diversity: a test of the Janzen-Connell model. *Oecologia* 97:533–540

Bush RM, Smouse PE (1992): Evidence for the adaptive significance of allozymes in forest trees. *New Forests* 6:179–196

Charlesworth B (1994): The genetics of adaptation: lessons from mimicry. *Am Nat* 144:839–847

Churchill GA, Doerge RW (1994): Empirical threshold values for quantitative trait mapping. *Genetics* 138:963–971

Conkle MT (1981): Isozyme variation and linkage in six conifer species. In: Conkle MT (Technical Coordinator) *Proceedings of the Symposium on Isozymes of North American Forest Trees and Forest Insects.* July 27, 1979; Berkeley, CA. Berkeley, California: Forest Service, U.S. Department of Agriculture. Gen Tech Rep PSW-48

Cook SA, Johnson MP (1968): Adaptation to heterogeneous environments. I. Variation in heterophylly in *Ranunculus flammula* L. *Evolution* 22:496–516

Dekkers J, Dentine MR (1991): Quantitative genetic variance associated with chromosomal markers in segregating populations. *Theor Appl Genet* 81:212–220

Dentine MR, Cowan CM (1990): An analytical model for the estimation of chromosome substitution effects in the offspring of individuals heterozygous at a segregating marker locus. *Theor Appl Genet* 79:775–780

Devey ME, Fiddler TA, Liu BH, Knapp SJ, Neale DB (1994): An RFLP linkage map for loblolly pine based on a three-generation outbred pedigree. *Theor Appl Genet* 88:273–278

deVicente MC, Tanksley SD (1993): QTL analysis of transgressive segregation in an interspecific tomato cross. *Genetics* 134:585–96

Dixon RK, Brown S, Houghton RA, Solomon AM, Trexler MC, Wisniewski J (1994): Carbon pools and flux of global forest ecosystems. *Science* 263: 185–190

Doebley J (1993): Genetics, development and plant evolution. *Curr Opin Genet Dev* 3:865–72

Doebley J, Stec A (1993): Inheritance of the morphological differences between maize and teosinte: comparison of results for two F_2 populations. *Genetics* 134:559–70

Doebley J, Stec A, Gustus C (1995): *teosinte branched 1* and the origin of maize: evidence for epistasis and the evolution of dominance. *Genetics* 141:333–346

Durrand R, Durrand B (1990): Sexual determination and sexual differentiation. *Crit Rev Plant Sci* 9:295–316

Falconer DS (1989): *Introduction to Quantitative Genetics*. 3rd ed. Essex, England: Longman

Feeney P (1976): Plant apparency and chemical defense. *Rec Adv Phytochem* 10: 1–40

Franklin EC (1969): Mutant forms found by self-pollination in loblolly pine. *J Hered* 60:315–320

Georges M, Nielsen D, Mackinnon M, Mishra A, Okimoto R, Pasquino AT, Sargeant LS, Sorensen A, Steele MR, Zhao X, Womack JE, Hoeschele I (1995): Mapping quantitative trait loci controlling milk production in dairy cattle by exploiting progeny testing. *Genetics* 139:907–920

Gilbert GS, Hubbell SP, Foster RB (1994): Density and distance-to-adult effects of a canker disease of trees in a moist tropical forest. *Oecologia* 98:100–108

Grattapaglia D (1994): Genetic mapping of quantitatively inherited economically important traits in *Eucalyptus* (Dissertation). Raleigh, NC: North Carolina State University

Grattapaglia D, Sederoff R (1994): Genetic linkage maps of *Eucalyptus grandis* and *Eucalyptus urophylla* using a pseudo-testcross: mapping strategy and RAPD markers. *Genetics* 137:1121–1137

Grattapaglia D, Chaparro J, Wilcox P, McCord S, Werner D, Amerson H, McKeand S, Bridgwater F, Whetten R, O'Malley D, Sederoff R (1992): Mapping in woody plants with RAPD markers: application to breeding in forestry and horticulture. In: *Proceedings of the Symposium on Applications of RAPD Technology to Plant Breeding*. 1992 Nov 1; Minneapolis, MN. Alexandria, VA: Am Soc Hort Sci

Grattapaglia D, Bertolucci FL, Penchel R, Sederoff R (1994): Molecular genetic mapping of economically important traits in Eucalyptus grandis. In: *TAPPI Proceedings 1994 Biological Sciences Symposium*. Oct 3–6; Minneapolis, MN. Atlanta, GA: TAPPI Press

Greenwood MS, Volkaert HA (1992): Morphophysiological traits as markers for the early selection of conifer genetic families. *Can J For Res* 22: 1001–1008

Groover A, Devy M, Fiddler T, Lee J, Megraw T, Mitchell-Olds T, Sherman B, Vujcic C, Williams C, Neale D (1994): Identification of quantitative trait loci influencing wood specific gravity in an outbred pedigree of loblolly pine. *Genetics* 138:1293–1300

Gupta PK, Pullman G, Timmis R, Kreitinger M, Carlson WC, Grob J, Welty E (1993): Forestry in the 21st century: the biotechnology of somatic embryogenesis. *Bio/Technology* 11:454–459

Haines R (1994): *Biotechnology in Forest Tree Improvement*. Rome: FAO

Haley CS, Knott SA, Elsen J (1994): Mapping quantitative trait loci in crosses between outbred lines using least squares. *Genetics* 136:1195–1207

Hamrick JL, Godt MJW, Sherman-Broyles SL (1992): Factors influencing levels of genetic diversity in woody plant species. *New Forests* 6:95–124

Hayes PM, Liu BH, Knapp SJ, Chen F, Jones B, Blake T, Franckowiak J (1993): Quantitative trait locus effects and environmental interaction in a sample of North American barley germ plasm. *Theor Appl Genet* 87:392–401

Hormaza JI, Dollo L, Polito VS (1994): Identification of a RAPD marker linked to sex determination in *Pistacia vera* using bulked segregation analysis. *Theor Appl Genet* 89:9–13

Irish EE, Nelson T (1989): Sex determination in monoecious and dioecious plants. *Plant Cell* 1:737–734

Jansen RC (1994): Controlling type I and type II errors in mapping quantitative trait loci. *Genetics* 138:871–881

Jermstad KD, Reem AM, Henifin JR, Wheeler NC, Neale DB (1994): Inheritance of restriction fragment length polymorphisms and random amplified polymorphic DNAs in coastal Douglas-fir. *Theor Appl Genet* 89:758–766

Kinloch BB, Walkinshaw CH (1991): Resistance to fusiform rust in southern pines: how is it inherited? In: *Proceedings of the IUFRO Rusts of Pine Working Party Conference.* Sept 18–22, 1989; Banff, Alberta, Canada. Inf Rep NOR-X-317, Forestry Canada NW Region

Kinloch BB, Parks GK, Fowler CW (1970): White pine blister rust: simply inherited resistance in sugar pine. *Science* 167:193–195

Kirk TK (1994): Technical overview of forest biotechnology research in the US. In: *TAPPI Proceedings of the 1994 Biological Sciences Symposium.* 1994 October 3–6; Minneapolis, MN. Atlanta, GA: TAPPI Press

Kuhlman EG, Amerson HV, Wilcox PL (1995): Recent research on fusiform rust disease. In: *Proceedings of the 4th IUFRO Rusts of Pines, Working Party Conference.* Kaneko F, Katsuya K, Kakishima M, Ono Y, eds. October 2–7, 1994; Tsukuba, Japan. Ibaraki, Japan: Forestry and Forest Products Research Institute

Laarman JG, Sedjo, RA (1992): *Global Forests.* New York: McGraw-Hill

Lai C, Lyman RF, Long AD, Langley CH, Mackay TFC (1994): Naturally occuring variation in bristle number and DNA polymorphisms at the *scabrous* locus of *Drosophila melanogaster. Science* 266:1697–1702

Lande R, Thompson R (1990): Efficiency of marker-assisted selection in the improvement of quantitative traits. *Genetics* 124:743–756

Lander ES, Schork NJ (1994): Genetic dissection of complex traits. *Science* 265:2037–2048

Lawson DM, Hemmat M, Weeden NF (1995): The use of molecular markers to analyze the inheritance of morphological and developmental traits in apple. *J Amer Soc for Hort Sci* 120:532–537

Lerdau M, Litvak M, Monson R (1994): Plant chemical defense: monoterpenes and the growth-differentiation balance hypothesis. *TREE* 9:58–61

McKeand SE, Bridgwater FE (1992): Third-generation breeding strategy for the North Carolina State University—Industry Cooperative Tree Improvement Program. In: *Proceedings of the IUFRO Conference S2.02-08, Breeding Tropical Trees,* Lambeth CC and Dvorak W, eds. 1992 Oct. 8–18; Cartagena and Cali, Colombia. Raleigh, NC: CAMCORE, North Carolina State University

McKenzie JA, Batterham P (1994): The genetic, molecular and phenotypic consequences of selection for insecticide resistance. *Tr Ecol Evol.* 9:166–169

Michelmore RW, Paran I, Kesseli RV (1991): Identification of markers linked to disease resistance genes by bulked segregation analysis: a rapid method to detect markers in specific genomic regions by using segregating populations. *Proc Nat Acad Sci USA* 88:9828–9832

Moreno-Gonzalez J (1993): Efficiency of generations for estimating marker-associated QTL effects by multiple regression. *Genetics* 135:223–31

Morgante M, Olivieri AM (1993): PCR-amplified microsatellites as markers in plant genetics. *Plant J* 3:175–182

Mulcahy DL, Weeden NF, Kesseli R, Carrol SR (1992): DNA probes for the Y-chromosome of Silene latifolia, a dioecious angisperm. *Sex Plant Reprod* 5: 86–88

Namkoong G, Kang H (1990): Quantitative genetics of forest trees. *Plant Br Rev* 8:139–188

Nelson CD, Doudrick RL, Nance WL, Hamaker JM, Capo B (1993): Specificity of host:pathogen interaction for fusiform rust disease on slash pine. In: *Proceedings of the 22nd Southern Forest Tree Improvement Conference.* 1993 June 14–17; Atlanta, Georgia. Publication No. 44 of the Southern Forest Tree Improvement Committee, National Technical Information Services, Springfield, VA

Nelson CD, Nance WL, Doudrick RL (1993): A partial genetic linkage map of slash pine (*Pinus elliottii* Engelm. var. *elliottii*) based on random amplified polymorphic DNAs. *Theor Appl Genet* 87:145–151

Niklas KJ (1994): *Plant Allometry.* Chicago: University of Chicago Press

Niklas KJ (1995): Size-dependent allometry of tree height, diameter and trunk-taper. *Ann Bot* 75:217–227

O'Malley DM, McKeand SE (1994): Marker assisted selection for breeding value in forest trees. *Forest Genet* 1:231–242

Orr HA, Coyne JA (1992): The genetics of adaptation: a reassessment. *Am Nat* 140:725–742

Partridge L, Barton NH (1993): Optimality, mutation and the evolution of ageing. *Nature* 362:305–311

Paterson AH, Lander ES, Hewitt JD, Peterson S, Lincoln SE, Tanksley SD (1988): Resolution of quantitative traits into Mendelian factors by using a complete linkage map of restriction fragment length polymorphisms. *Nature Uk* 335:721–726

Paterson AH, Damon S, Hewitt JD, Zamir D, Rabinowitch HD, Lincoln SE, Lander ES (1991): Mendelian factors underlying quantitative traits in tomato: comparison across species, generations, and environments. *Genetics* 127:181–197

Plomin R, Owen MJ, McGuffin P (1994): The genetic basis of complex human behaviors. *Science* 264:1733–1739

Plomion C, O'Malley DM, Durel CE (1995a): Genomic analysis in maritime pine (*Pinus pinaster*) Comparison of two RAPD maps using selfed seeds and open-pollinated seeds of the same individual. *Theor Appl Genet* 90:1028–1034

Plomion C, Bahrman N, Durel CE, O'Malley DM (1995b): Genomic mapping in maritime pine (*Pinus pinaster*) using RAPD and protein markers. *Heredity* 74:661–668

Rhoades DF, Cates RG (1976): A general theory of plant anti-herbivore chemistry. *Rec Adv Phytochem* 10:168–213

Robinson R (1971): *Gene Mapping in Laboratory Mammals*, Part A. London: Plenum

Robinson RA (1987): *Host Management in Crop Pathosystems.* New York: MacMillan

Schlichting CD, Pigliucci M (1995): Gene regulation, quantitative genetics and the evaluation of reaction norms. *Evol Ecol* 9:154–168

Shepherd M, Chaparro J, Dale G, Jefferson L, Duong H, Bogel H, Walsh J, Gibbings M, Teasdale R (1995): Mapping insect resistance and essential oil traits in a tropical eucalyptus hybrid. In: *Eucalypt Plantations: Improving Fiber*

Yield and Quality, Potts et al, eds. Proceedings of the CRC-IUFRO Conference, 1995 Feb 19–24; Hobart, Australia. Hobart, Australia: CRC for Temperate Hardwood Forestry

Smith DN, Devey ME (1994): Occurrence and inheritance of microsatellites in *Pinus radiata*. *Genome* 37:977–983

Strauss SH, Lande R, Namkoong G (1992): Limitations of molecular-marker-aided selection in forest tree breeding. *Can J For Res* 22:1050–1061

Stuber CW, Lincoln SE, Wolff DW, Helentjaris T, Lander ES (1992): Identification of genetic factors contributing to heterosis in a hybrid from two elite maize inbred lines using molecular markers. *Genetics* 132:823–39

Takahashi JS, Pinto LH, Vitaterna MH (1994): Forward and reverse genetic approaches to behavior in the mouse. *Science* 264:1724–1733

Tanksley SD (1993): Mapping polygenes. *Ann Rev Genet* 27:205–233

Tanksley SD, Ganal MW, Martin GB (1995): Chromosome landing: a paradigm for map-based cloning in plants with large genomes. *TIG* 11:63–68

Thompson JN, Burdon JJ (1992): Gene-for-gene coevolution between plants and parasites. *Nature* 360:121–125

Via S, Gomulkiewicz R, De Jong G, Scheiner SM, Schlichting CD, Van Tienderen PH (1995): Adaptive phenotypic plasticity: consensus and controversy. *TREE* 10:212–217

Voo KS, Whetten R, O'Malley DM, Sederoff RR (1995): 4-Coumarate coA ligase from loblolly pine xylem. *Plant Phys* 107: (in press)

Wellburn A (1994): *Air Pollution and Climate Change*. Essex, England: Longman

Welsh J, McClelland M (1990): Fingerprinting genomes using using PCR with arbitrary primers. *Nucl Acids Res* 19:303–306

Westfall RD, Conkle MT (1992): Allozyme markers in breeding zone designation. *New Forests* 6:279–309

Wilcox PL (1995): Genetic dissection of fusiform rust disease resistance in loblolly pine (Dissertation). Raleigh, NC: North Carolina State University

Williams CG (1987): The influence of shoot ontogeny on juvenile-mature correlations in loblolly pine. *Forest Sci* 33:411–422

Williams CG, Neale DB (1992): Conifer wood quality and marker-aided selection: a case study. Marker aided selection: a tool for the improvement of forest tree species. *Can J For Res* 22:1009–1017

Williams JGK, Kublelik AR, Livak KJ, Rafalski JA (1990): DNA polymorphisms amplified as arbitrary primers are useful genetic markers. *Nucl Acids Res* 18: 6531–6535

Wisniewski J, Dixon RK, Kinsman JD, Sampson RN, Lugo AE (1993): Carbon dioxide sequestration in terrestrial ecosystems. *Clim Res* 3:1–5

Zabeau M (1993): Selective restriction fragment amplification: a general method for DNA fingerprinting. European Patent Application. Publication No. 0534858 A1

Zobel BZ, Talbert JT (1984): *Applied Tree Improvement*. New York: Wiley and Sons

Zobel BZ, van Buijtenen JP (1989): *Wood Variation: Its Causes and Control*. Berlin: Springer-Verlag

4

The Use of Comparative Genome Mapping in the Identification, Cloning and Manipulation of Important Plant Genes

JEFFREY L. BENNETZEN

Our understanding of plant genome organization originated with cytogenetic and recombinational mapping studies. These investigations demonstrated that most genes are located in fixed positions in the euchromatic regions of chromosomes, and that this euchromatin is interspersed with heterochromatic (highly condensed) regions that are relatively deficient in both gene activity and recombinational exchanges. Subsequent studies utilizing genomic DNA renaturation have shown that much of the DNA in higher plant genomes is repetitive (Flavell et al, 1974), and that these repetitive sequences are interspersed in some manner with low copy number (gene-containing) sequences. Beyond this, our knowledge of plant genome organization has been largely limited to examination of single genes, tandem gene families, or particular classes of repetitive DNA.

The development of new technologies for the physical mapping (pulsed field gel electrophoresis) and molecular cloning (artificial chromosome libraries) of large regions of contiguous DNA now have allowed investigators to look at the arrangement of sequences at a level between that of single genes and cytogenetic structures. Similarly, the advent of DNA marker technology has expanded greatly the sensitivity of recombinational mapping. With these tools in hand, and continuing to increase in power, we now are beginning to understand the organization of plant genomes at a detailed level. Many of the observations have been surprising, and they also have opened up new avenues for the study and manipulation of plants. However, several very basic questions still need to be answered, and potential problems loom on the horizon. This chapter will discuss the key observations that have been made, the major questions that persist, and the potential value of using this information to synergize plant research and crop improvement.

The Impact of Plant Molecular Genetics
BWS Sobral, Editor

Parallel Genomes

The initial studies of Tanksley and coworkers indicated that closely related plant species (e.g., tomato, potato, and pepper) have been highly conserved in gene content (Bonierbale et al, 1988; Tanksley et al, 1988; Zamir and Tanksley, 1988). DNA probes representing portions of genes have been conserved at a hybridizational level, while other sequences (low copy number or repetitive) have not been conserved even between these closely related species. Moreover, the chromosomal positions of genes also have been maintained between tomato and potato (Bonierbale et al, 1988; Gebhardt et al, 1991), although many chromosomal rearrangements have differentiated these two species from pepper (Tanksley et al, 1988).

Hulbert et al (1990) have determined that sorghum and maize, among the grasses, also have nearly identical gene content by hybridization criteria. These low copy number sequences have been highly conserved between maize and sorghum, while the dispersed repetitive sequences in maize have been found to be either missing or highly diverged in sorghum (Bennetzen et al, 1994; Hulbert et al, 1990). Most DNA markers that have been linked in maize have been also linked in sorghum, and in a similar order (Figure 1). This has been confirmed and the similarities extended between maize and sorghum (Binelli et al, 1992; Hulbert et al, 1990; Melake-Berhan et al, 1993; Pereira et al, 1994; Ragab et al, 1994; Whitkus et al, 1992) and other members of the tribe Andropogoneae (D'Hont et al, 1994).

Further investigations of similarity in gene content and gene order have extended this high level of conservation across a broad spectrum of grasses, as exemplified by comparisons of rice, wheat, and maize (Ahn and Tanksley, 1993; Ahn et al, 1993; Devos et al, 1993; 1994; Kurata et al, 1994). Gene orders are not identical across the grasses; inversions and translocations involving large portions of chromosome arms often differentiate the genetic maps of the cereals. However, within any given arm, gene orders are found to be highly conserved despite the more than 60 million years of independent descent that separate these grass species (Doebley et al, 1990). Given that plants are continuously exposed to numerous exogenous (e.g., radiation) and endogenous (e.g., transposable elements, ectopic recombination) factors that will rearrange genomes, it is surprising that gene order is so well conserved. This suggests either that induced chromosomal rearrangements are rarer than experimental evidence suggests or that there is a stronger selection for gene order than has been demonstrated in the laboratory. Regardless of the reason(s) for this conservation of gene content and map position, the observation indicates that individual grass genomes can be viewed as manifestations of a single grass genetic system (Bennetzen and Freeling, 1993).

Implications of the Similarity of Gene Content in the Grasses

The similarity in gene content of the various grasses indicates that truly new grass genes are rarely created within evolutionary timeframes of a 10^7 year magnitude.

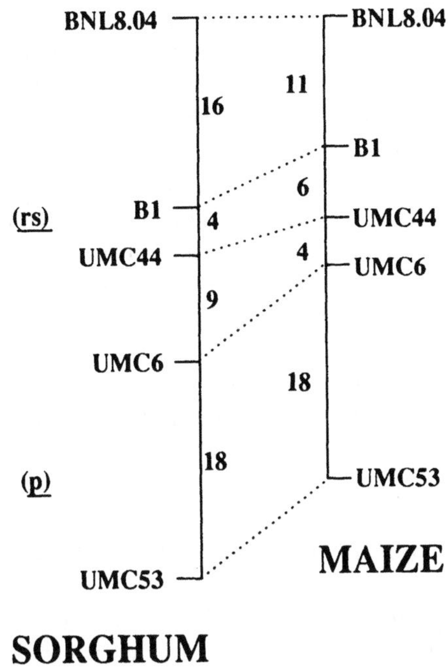

Figure 1. Comparison of the maps of a region of the maize and sorghum genomes constructed with DNA markers from maize. Designations (*rs*) and (*p*) indicate morphological markers for seedling color and plant color, respectively, that we have mapped to the approximate positions shown (Salimath, Subramanian and Bennetzen, 1995). We have also mapped quantitative traits for seed and leaf polyphenols in sorghum to the *B1*/UMC53 interval (Weerasuriya, 1995). Vertical lines indicate linkage groups. Numbers between the lines indicate approximate map units between adjacent markers.

As shown both by conservation of hybridizational homology and variation in the copy number of gene-specific probes, most new genes probably arise from gene duplication and/or minor sequence modification of currently existing genes. However, a relatively small number of gene alterations can lead to major modifications in plant morphology and growth habit (Doebley and Stec, 1991).

The commonality of grass gene content indicates that most, or all, genes required for any grass biochemical pathway are present in all grass species. These genes and pathways may differ in their timing, level, or tissue specificity, especially since such regulatory changes may be both easy to induce and likely to provide a selectively-significant level of physiological change. Hence, single genes from one grass species are likely to find genes in other grass species that support their activity. For instance, transfer of regulatory or signal-transduction genes (e.g., genes involved in pigment production or disease resistance) from one grass species into another by transformation usually leads to productive activation and interaction with endogenous genes in the transformed plant. A pathway believed to be unique to a particular species, for example, may have only one genetic difference (e.g., an altered substrate specificity or tissue-specificity) that differentiates it from other species not believed to have this physiological process.

The combination of similar gene content and similar physiological pathways within the grasses predicts that transfer of genes from one grass species to another often will lead to the productive function of transgenes, and this prediction has been supported by transgenic studies. There is every reason to believe that this productive expression of transgenes will also be extended to more complex, quantitative traits such as yield and abiotic stress resistance. Previous predictions have held that multigenic traits like drought resistance, for instance, would be too difficult to transfer to other species because a whole array of genes must be identified and transferred. However, if the entire set of required genes is present in each species, then all that needs to be identified and transferred are the key (or rate limiting) genes (or gene) that account for a significant percentage of one plant species' superior drought tolerance. Hence, similar genes in different species can be viewed as cross-species alleles that may be directly utilized for crop improvement or study. From this perspective, the geneticist and genetic engineer can search the entire set of grass species for the "best" allele needed for the improvement of a particular pathway or process. As described below, the limiting step in this process is to determine the relative "allelism" of genes in different species, and the collinearity of grass genomes provides a central tool for these determinations.

Implications and Uses of Grass Genome Collinearity

Identification of novel and allelic genes in different grass species. The current and ever-expanding power of genetic mapping with DNA markers has allowed identification of genetic regions that make even minor contributions to the within-species variability in a particular physiological/biochemical process. These quantitative trait loci (QTL), like major or qualitative genes, are both identified and positionally localized when large populations are mapped with a reasonable number of well-distributed markers. In sorghum, for instance, we have mapped quantitative traits for leaf and seed polyphenol production and qualitative traits for seedling and plant color to particular genomic regions (Figure 1).

Not surprisingly, since anthocyanidin pigments are themselves polyphenolic compounds, one major QTL for polyphenolics maps in a region of the sorghum genome to which we have also mapped seedling plant color (*rs*) and adult plant color (*p*) loci (Figure 1) (Weerasuriya, 1995). One of these single genes, *rs*, maps at approximately the same position as a DNA marker, the maize *B1* gene. *B1*, in maize, is a positive regulator of plant (and sometimes seed) pigment production; the similarity in map position and phenotypic contribution of the maize and sorghum genes suggest that *B1* and *rs* may be allelic. The second sorghum gene, *p*, maps about 15 cM from *rs* in a position where a homoeologous maize gene involved in pigment production has not been identified. Although sorghum *p* may thus be viewed as a novel locus, it is likely that maize has this same gene. In any single species, natural or induced mutations are likely to uncover only a subset of the loci required for a particular biochemical pathway. Some genes will be missed by chance, especially in species that have not been extensively investigated, and many will not be identified by mutation due to their being duplicated in that species. For instance, we have found that we were unable to isolate sorghum mutations that do not produce alcohol dehydrogenase in the pollen because the pollen-specific *Adh* locus of sorghum is duplicated, while the orthologous gene of maize is present in only a single copy (Liu and Bennetzen, 1995).

In other studies on this theme, Fatokun et al (1992) have shown that QTL affecting seed weight map to homoeologous positions in the mung bean and cowpea genomes, while Pereira and Lee (1995) have determined that some of the QTL for plant height in sorghum map to the same approximate positions as dwarfing mutations of maize.

Additional mapping of qualitative and quantitative traits on comparable grass maps will uncover similar map positions for many related traits in the different species. Often, this will be an indication of cross-species alleles. Hence, if an investigator plans to improve drought tolerance in maize using sorghum genes, for instance, one approach would be to identify sorghum drought tolerance QTL that map in the same place as maize drought tolerance QTL. Because sorghum has greatly superior tolerance to water and heat stress, it is probable that the sorghum QTL allele provides better drought resistance than would the same maize QTL. Of course, some sorghum genes would not provide improved tolerance, partly because the high incidence of transgressive segregation tells us that species or individuals that do not exhibit a particular trait often carry superior/hidden alleles that condition that trait. In the long run, the potential presence of and access to superior alleles for any trait in the entire range of grass species is an enormous asset. At least initially, utilization of these genes in crop improvement would require a liberal dose of trial and error in transgenic systems.

The identification of novel genes would be a particularly powerful and exciting outcome of parallel grass genome mapping. These genes would be identified as novel by their unique map position relative to the position of similar genes in other species that condition a similar process. As mentioned above, it is likely that these novel genes usually would be present in other grass species but missed due to limitations in genetic analysis. By using the full array of grass species, it thus may be possible to identify all of the genes involved in a given physiological process because the genes that are duplicated or otherwise phenotypically hidden in one species may not be hidden in another species. Hence, this result of comparative mapping would prove a boon to physiologists who desire genetic tools for every step in a pathway they study. Although no single species may provide this resource, use of the full range of grasses may provide access to mutant alleles for every structural and regulatory gene.

The comparative mapping approach should be a central tool for the identification of truly new genes, ones that have arisen in the ancestor of a subset of grasses. If such genes exist, their study would enlighten our understanding of a basic evolutionary process, e.g. are new genes created by assembly from pieces of other genes, do they arise from pseudogenes, or are they truly created *de novo* from relatively unconstrained sequences like introns or gene-flanking sequences? When such a gene conditions an agronomically useful phenotype, like disease resistance, then its engineering into new species may be especially valuable, perhaps providing a wholly new mechanism to accomplish a particular process.

Genome Mapping

The high level of conservation of gene sequences allows probes from one grass species to be used in the DNA-marker-based mapping of other grass species (Ahn and Tanksley, 1993; Ahn et al, 1993; Binelli et al, 1992; Devos et al, 1993, 1994; D'Hont et al, 1994; Gill et al, 1991; Hulbert et al, 1990; Jena et al, 1994; Kurata et al, 1994; Melake-Berhan et al, 1993; Pereira et al, 1994; Ragab et al, 1994; Whitkus et al, 1992). This obviously saves effort in the preparation and screening of libraries of potential probes. More importantly, an anchor set of restriction fragment length polymorphism (RFLP) probes that have been mapped in a large number of grass species can be used. This guarantees that the broadest coverage of the genome in a yet unmapped species will be accomplished with the minimum number of probes. This should save at least a factor of two in the time and cost required to generate a first map. For regions that require more detailed mapping, additional probes can be identified by homoeology with other maps, saving tremendous amounts of time and expense in a fine-mapping project.

Once mapped with these anchor probes, the newly mapped genome can now be directly compared to that of all other similarly mapped grass genomes. In this way, we should be able to establish a phylogeny of the chromosomal rearrangements that have punctuated the evolution of the grasses. Our understanding of plant evolution would be greatly enhanced by identification of the nature and frequency of such changes and may also provide some foundation on which to investigate a possible selective advantage for conservation of gene order.

Given that the use of common anchor probes in grass genome mapping would be quicker, more cost efficient and more informative, it is difficult to justify the generic preparation and use of any additional RFLP probes from any plant species in any study from this day forward. As with several other aspects of an integrated grass genome approach, those species that have so far received the least investigation would benefit the most.

Chromosome Walking

The genomes of higher plants vary by more than two orders of magnitude in DNA content (Arumuganathan and Earle, 1991; Michaelson et al, 1991). Part of this variation is due to different levels of ploidy, particularly in the grasses where most species are polyploid. Even some apparently diploid species, like maize and sorghum, apparently are derived from tetraploid ancestors (Helentjaris et al, 1988; Hulbert et al, 1990). Variation in repetitive DNA content is also a major factor. Repetitive DNA makes up less than 15% of the nuclear DNA in some plants with small genomes, while plant species with large genomes can have more than 60% of their genomic DNA composed of diverse repetitive elements (Flavell et al, 1974; Leutwiler et al, 1984).

Both hybridization and clone analyses indicate that higher plant repetitive DNAs are mostly interspersed with single copy DNAs and that mixed repetitive blocks apparently average more than 50 kb in size in maize and sorghum (Bennetzen et al, 1994; Springer et al, 1994) (Figure 2). The large size and high repetitive DNA content of many higher plants could make gene cloning by chromosome walking a difficult and often fruitless undertaking. However, if smaller genomes are collinear, then a gene targeted in one species may be most easily cloned by chromosome walking in a parallel genome from a small genome grass. For instance, the *A1* and *Sh2* loci map about 0.1 to 0.2 cM apart, and they have been found to be separated by about 140 kb (Civardi et al, 1994). We have now cloned homologues of these genes from two smaller-genome grasses, rice and sorghum. Our preliminary results indicate that *A1* and *Sh2* homologues are separated by less than 30 kb in both rice and sorghum (SanMiguel et al, 1995). Hence, if this microcollinearity proves the rule, then chromosome walking to large-genome genes in small-genome plants would convert this technology from barely feasible to routine.

Beyond the technology of chromosome walking itself, one limitation in using this approach to map-based gene isolation is identifying DNA probes that are tightly linked to the targeted gene. Flanking markers are necessary, and the more closely linked these are then the fewer steps will be needed in a walk (Martin et

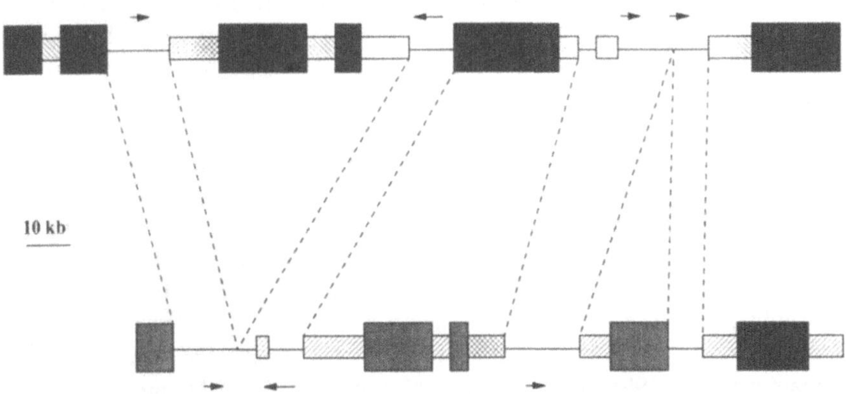

Figure 2. Model for comparison of the genomes of two grass species. The upper figure represents a portion of the chromosomal DNA of a grass species with a large genome, while the lower figure represents the homoeologous region of a grass species with a somewhat smaller genome. The thick boxes are highly repetitive DNA clusters, the medium boxes are middle repetitive DNA clusters, and the thin line indicates low copy number DNA. The arrows above and below the lines indicate transcripts, showing that most of the transcripts in this region are from low copy number sequences and are mostly conserved between the two species. The different fills of the repetitive blocks for the two species are meant to show that they will rarely cross-hybridize between species, although there will be rare exceptions (black boxes, cross-hatched boxes). The dotted lines connect the boundaries of highly-conserved gene-containing sequences. Note that the repetitive DNA blocks do not necessarily vary in size between large-genome and small-genome grasses, as we have seen between maize and sorghum. Moreover, some genes that are separated by long stretches of repetitive DNA in one species may not be separated by long stretches of repetitive DNA in a related species. Middle repetitive DNAs are found mixed in with both highly repetitive blocks and in low-copy number regions, sometimes within the transcribed or regulatory regions of genes.

al, 1992). Once again, parallel maps can provide access to these markers with a limited amount of effort. By comparing maps, even at the crude current level of grass map integration, we have been able to identify more than 40 DNA markers for almost any 10 cM region of any mapped grass genome. These probes, from different grass species, can be obtained through the mail. Although many will be duplicates of essentially the same probe from different mapping groups and/or different mapping species, it is likely that one or more DNA markers within 0.5 cM of your targeted gene will be acquired by this process. If these probes are linked tightly enough in the specific mapping population that targets the sought-after gene (Hulbert and Bennetzen, 1991; Martin et al, 1992), then these probes could allow the first step (and perhaps the only required step) in a chromosome walk, thereby saving the investigators weeks, months, or even years in searching for such probes by other approaches.

The last, and often most difficult, step in cloning by chromosome walking would be precise identification of the targeted gene within the cloned interval. One definitive approach to this process is through expression in a transgenic plant. Comparative genome analysis can also contribute greatly to bringing this step to a successful conclusion. If, for instance, a contiguous region of 100 kb has been cloned and is known to contain the targeted gene, then individual fragments of this region could be cloned into a transformation vector and transferred to a transformable plant species. The similarity in the gene content of grass genomes suggests that the transferred gene should function relatively well in any heterologous grass species, so the investigator can choose as host that grass species most amenable to transformation with the resources at hand. Because the use of genomic subclones would require many overlapping fragments to be transformed, investigators commonly first identify cDNAs homologous to genes in the cloned interval. This approach may not succeed, however, because some genes would be expressed at levels too low to allow routine identification of a cDNA clone. Moreover, we have recently observed that the majority of the identified transcribed sequences in some grass genomic regions come from interspersed highly repetitive DNAs that are not likely to contribute mapped genetic functions (Avramova et al, 1995). One way to circumvent these possible problems would be to hybridize the cloned interval with the homoeologous interval from another grass species. In theory, the stringent conservation of genes and the poor conservation of other low copy number and repetitive elements between plant species (Hulbert et al, 1990; Zamir and Tanksley, 1988) suggest that only the gene-containing fragments would cross-hybridize in this interval-probe experiment.

Possible Problems, Necessary Investigations

The current enthusiasm for the use of parallel genome analysis stems from a number of potential opportunities, ranging from studies of evolution to the engineering of improved crops (Bennetzen and Freeling, 1993). At the moment, the possible value of this approach to map-based gene isolation has drawn the most attention and activity. Various projects are currently ongoing to use chromosome walking in sorghum or rice to clone genes in maize, barley or wheat. One possible flaw in this approach derives from our lack of understanding of the limits of collinearity in grass genomes.

Even at a crude map level, apparent exceptions to collinearity are often observed between species or between mapping populations within a single species. In comparing one of our maps of sorghum (Melake-Berhan et al, 1993) to the maize genetic map and to another sorghum genetic map (Whitkus et al, 1992), we found apparent rearrangements of markers U22, U88, and U139 (Figure 3). Most disturbingly, these apparent discontinuities in the map could not be explained by any obvious single rearrangement. However, when a more advanced maize map was produced (Maize Genetics Cooperative Newsletter, 1993), U139 was found to map to two different positions on the same maize chromosome arm. Comparisons to the sorghum maps suggest that U88 and U22 might also be included in this distant-tandem duplication in maize and in sorghum. Making the

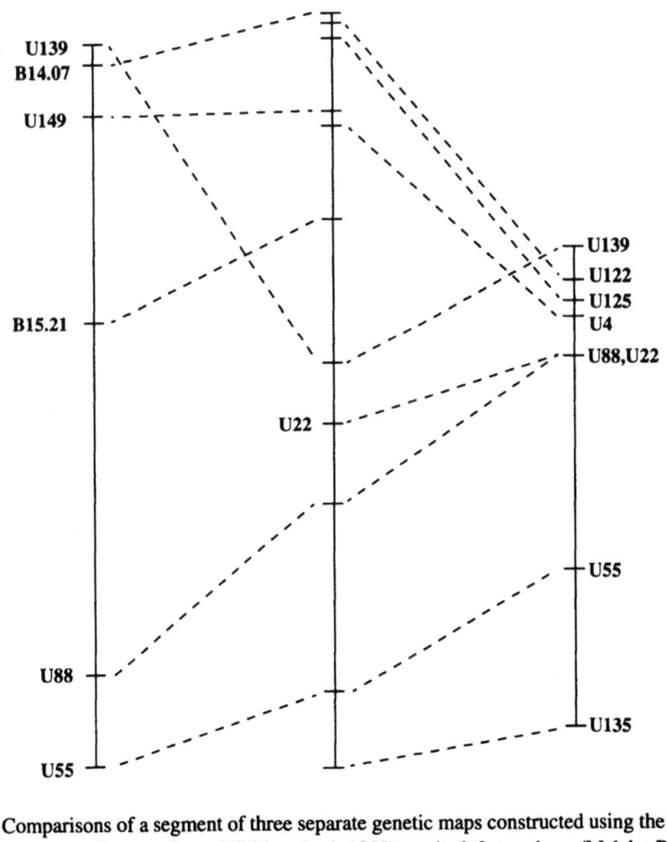

Figure 3. Comparisons of a segment of three separate genetic maps constructed using the same DNA markers. The maps are of sorghum (Whitkus et al, 1992) on the left, sorghum (Melake-Berhan et al, 1993) in the center, and maize on the right. Dashed lines connect positions of the same markers, and the maps are drawn to approximate scale.

assumption that the duplicated versions of U88 and U22 have not yet been mapped in maize, and knowing that only one of the two U139 hybridization bands has been mapped in each sorghum population, then a fully collinear set of maps can be drawn from the same data (Figure 4). We often have seen these distant-tandem duplications, in which the same DNA probe detects fragments separated by a few cM on the same chromosome arm in maize (Sanz-Alferez et al, 1995), and there is evidence that they may be common in other genomes. Because of these factors and others that will confuse mapping analyses, some exceptions to collinearity of maps would be due to limitations in the quality or comprehensiveness of the mapping data and not to actual rearrangements.

The presence of linked and unlinked gene duplications would complicate any map-based cloning program, particularly if the duplications cover long intervals containing many genes. Further experiments, both recombinational mapping and model chromosome walks, need to be conducted to address this issue.

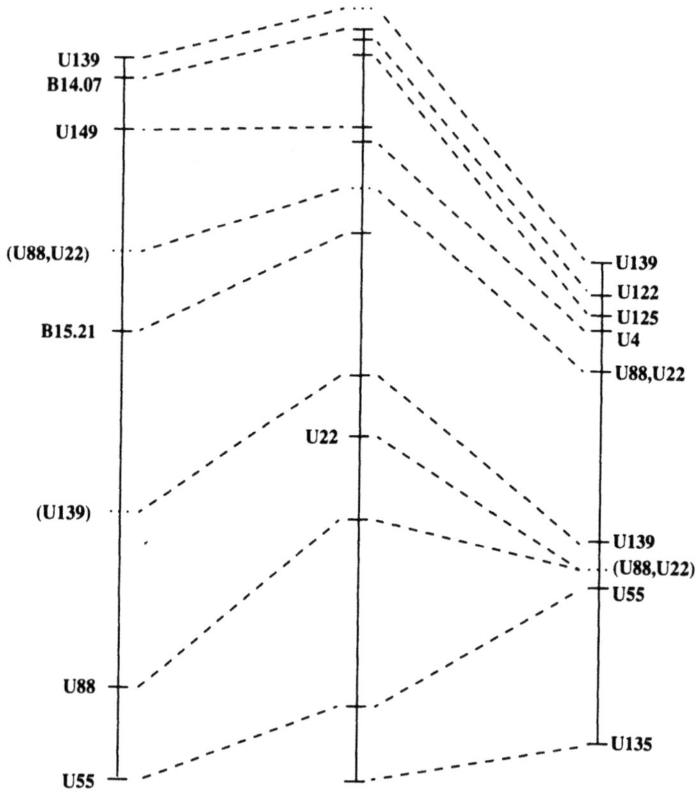

Figure 4. Comparisons of a segment of three separate genetic maps constructed using the same DNA markers after identification of a distant-tandem duplication in this region. The sorghum maps (left and center) are as in Figure 3, but the maize map now contains an additional map position for U139 and proposed new map positions for U22 and U88 (see text).

Specifically, we need to find out how often this abundance of gene duplication leads to a chromosome walk that leaps from chromosome to chromosome or across segments on a chromosome arm in a manner not expected (and, perhaps, initially undetected) by the investigator.

Both targeted and model chromosome walks need to be undertaken to address the question of microcollinearity. Rearrangements, particularly inversions, duplications, and translocations, are observed at the level of the recombinational map. It is possible, even likely, that small versions of these rearrangements would be even more abundant. If micro-rearrangements are frequent, then parallel chromosome walks in the grasses would seldom be useful. In addition, it would be useful to determine how often large blocks of repetitive DNA would halt a chromosome walk in a particular species. These blocks appear to average over 50 kb, ranging up to as much as 200 kb, in maize and sorghum (Bennetzen et al, 1994; Springer et al, 1994). The similar size of these repeat blocks in sorghum, despite its smaller genome size, suggests that the size of such clusters of dispersed repeats may have a selected component relative to genome organization (Avramova et al, 1995; Bennetzen et al, 1994). The possibility exists that these repetitive DNA blocks might be present in different locations in different species, and that parallel chromosome walks might be best accomplished by a genomic cross-referencing procedure in which the investigator takes steps in one species until halted, then takes the next steps in a species in which this step is not blocked, and so on (Figure 2).

Tests are needed of the potential of using multiple grass species to allow a complete genetic dissection of specific biochemical pathways, and to identify new genes and cross-species alleles. This means broad mapping of similar morphological and physiological traits (both qualitative and quantitative) across several species with markers that allow comparisons of map positions. Of course, once anchor probes have been mapped in a species, any additional mapped traits would immediately be comparable.

Finally, a minimal organizational structure for the integration of grass genome studies is needed. The possible value of an International Grass Genome Initiative (IGGI) has been discussed at workshops at the Plant Genome II and Plant Genome III conferences and at a special Banbury conference held in November of 1994 for this purpose. Beyond high general enthusiasm for the concept, the summarized conclusion of all these gatherings and discussions has been that an international effort should be mounted to bring an IGGI program into existence. A first effort has been made in this regard with the funding by the USDA of an IGGI component. The major contributions of this USDA organizational grant will be in the choice and dissemination of anchor probes for grass genome mapping (Susan McCouch, Cornell), the development and distribution of software for integration of grass genome maps and of databases with grass genome information (Sam Cartinhour, USDA/NAL), publication and distribution of a comparative grass genome newsletter (Jeff Bennetzen, Purdue), and the organization of additional IGGI-related conferences (Ron Phillips, Minnesota). Individuals interested in helping to develop these resources, or in making use of them, are encouraged to contact the responsible parties. In addition, international efforts to develop IGGI are also ongoing, some in coordination with our USDA-funded program.

Prospects

The potential of comparative grass genome analyses to enhance basic and applied plant science is enormous. This goes beyond the obvious improvements in our ability to clone genes by map-based techniques, to identify the significant steps in genome evolution, and to identify/isolate genes that can be used in trans-species crop improvement. With a parallel genome viewpoint, studies on individual grass species would not need to be isolated investigations. Observations in one species would be directly pertinent to our understanding of the same gene's role in another species, once we have confirmed the orthology of the observations by homoeology of map positions. Hence, the grasses can be utilized for their combined strengths as basic systems of biological study. The excellent chromosomal mechanics of wheat and maize, the outstanding mutant collections in barley, and the small genomes and relative ease of transformation of rice and sorghum are all assets that can be used appropriately in investigations of grass biology. In addition, parallel investigations of these phylogenetically-characterized species would place these studies in an evolutionary perspective that would allow investigators to gain insights into not only how and what processes plants carry out, but why they have evolved these particular processes (Bennetzen and Freeling, 1993; Kellogg and Birchler, 1993). Taken together, the grasses will prove as strong a model system as any higher eukaryote.

Although initial studies in dicotyledonous species (Tanksley et al, 1988) have suggested that the genomes of broad leaf plants are not as highly conserved as are those of the grasses, there will be cases where parallel analyses can be utilized between closely related species or along short stretches of chromosomes (Bonierbale et al, 1988; Fatokun et al, 1994; Gebhardt et al, 1991; Kowalski et al, 1994; Teutonico and Osborn, 1994). It will be particularly exciting to see whether some regions of both monocotyledonous and dicotyledonous chromosomes will exhibit collinearity, as one might expect. In cases where this holds true, then much of what applies to the above discussion of grass genomes also could apply across the broader plant kingdom.

ACKNOWLEDGMENTS

The writing of this manuscript, and the unpublished results reported herein, were supported by grants from the McKnight Foundation and the USDA/NRIGP (award #94-37300-0299). I would like to thank Dr. Michael Lee for helpful discussions and permission to cite a manuscript in press.

REFERENCES

Ahn S, Tanksley SD (1993): Comparative linkage maps of the rice and maize genomes. *Proc Natl Acad Sci USA* 90:7980–7984

Ahn S, Anderson JA, Sorrells ME, Tanksley SD (1993): Homoeologous relationships of rice, wheat and maize chromosomes. *Mol Gen Genet* 241:483–490

Arumuganathan K, Earle ED (1991): Nuclear DNA content of some important plant species. *Plant Mol Biol Rep* 9:208–218

Avramova Z, SanMiguel P, Georgieva E, Bennetzen JL (1995): Matrix attachment regions and transcribed sequences within a long chromosomal continuum containing maize *Adh1*. *Plant Cell* (in press)

Bennetzen JL, Freeling M (1993): Grasses as a single genetic system: genome composition, collinearity and compatibility. *Trends Genet* 9:259–261

Bennetzen JL, Schrick K, Springer PS, Brown WE, SanMiguel P (1994): Active maize genes are unmodified and flanked by diverse classes of modified, highly repetitive DNA. *Genome* 37:565–576

Binelli G, Gianfranceschi L, Pe ME, Taramino G, Busso C, Stenhouse J, Ottaviano E (1992): Similarity of maize and sorghum genomes as revealed by maize RFLP probes. *Theor Appl Genet* 84:10–16

Bonierbale MW, Plaisted RL, Tanksley SD (1988): RFLP maps based on a common set of clones reveals modes of chromosomal evolution in potato and tomato. *Genetics* 120:1095–1103

Civardi L, Xia Y, Edwards KJ, Schnable PS, Nikolau BJ (1994): The relationship between genetic and physical distances in the cloned *a1-sh2* interval of the *Zea mays* L. genome. *Proc Natl Acad Sci USA* 91:8268–8272

Devos KM, Chao S, Li QY, Simonetti MC, Gale MD (1994): Relationship between chromosome *9* of maize and wheat homoeologous group *7* chromosomes. *Genetics* 138:1287–1292

Devos KM, Millan T, Gale MD (1993): Comparative RFLP maps of homeologous group 2 chromosomes of wheat, rye and barley. *Theor Appl Genet* 85:784–792

D'Hont A, Lu Y-H, de Leon DG, Grivet L, Feldmann P, Lanaud C, Glaszmann JC (1994): A molecular approach to unraveling the genetics of sugarcane, a complex polyploid of the Andropogoneae tribe. *Genome* 37:222–230

Doebley J, Stec A (1991): Genetic analysis of the morphological differences between maize and teosinte. *Genetics* 129:285–295

Doebley J, Durbin M, Golenberg EM, Clegg MT, Ma DP (1990): Evolutionary analysis of the large subunit of carboxylase (*rbc*L) nucleotide sequence among the grasses (Gramineae). *Evolution* 44:1097–1108

Fatokun CA, Menancio-Hautea DI, Danesh D, Young ND (1992): Evidence for orthologous seed weight genes in cowpea and mung bean based on RFLP mapping. *Genetics* 132:841–846

Flavell RB, Bennett MD, Smith JB, Smith DB (1974): Genome size and proportion of repeated nucleotide sequence DNA in plants. *Biochem Genet* 12:257–269

Gebhardt C, Ritter E, Barone A, Debener T, Walkemeier B, Schachtschabel U, Kaufmann H, Thompson RD, Bonierbale MW, Ganal MW, Tanksley SD, Salamini F (1991): RFLP maps of potato and their alignment with the homoeologous tomato genome. *Theor Appl Genet* 83:49–57

Gill KS, Lubbers EL, Gill BS, Raupp WJ, Cox TS (1991): A genetic linkage map of *Triticum tauschii* (DD) and its relationship to the D genome of bread wheat (AABBDD). *Genome* 34:362–374

Helentjaris T, Weber DL, Wright S (1988): Identification of the genomic locations of duplicate nucleotide sequences in maize by analysis of restriction fragment length polymorphisms. *Genetics* 118:353–363

Hulbert SH, Bennetzen JL (1991): Recombination at the *Rp1* locus of maize. *Mol Gen Genet* 226:377–382

Hulbert SH, Richter TE, Axtell JD, Bennetzen JL (1990): Genetic mapping and characterization of sorghum and related crops by means of maize DNA probes. *Proc Natl Acad Sci USA* 87:4251–4255

Jena KK, Khush GS, Kochert G (1994): Comparative RFLP mapping of a wild rice, *Oryza officinalis*, and cultivated rice, *O. sativa*. *Genome* 37:382–389

Kellogg EA, Birchler JA (1993): Linking phylogeny and genetics: *Zea mays* as a tool for phylogenetic studies. *Syst Biol* 42:415–439

Kowalski SP, Lan T-H, Feldmann KA, Paterson AH (1994): Comparative mapping of *Arabidopsis thaliana* and *Brassica oleracea* chromosomes reveals islands of conserved organization. *Genetics* 138:499–510

Kurata N, Moore G, Nagamura Y, Foote T, Yano M, Minobe Y, Gale M (1994): Conservation of genome structure between rice and wheat. *Bio/Technology* 12:276–278

Leutwiler LS, Hough-Evans BR, Meyerowitz EM (1984): The DNA of *Arabidopsis thaliana*. *Mol Gen Genet* 194:15–23

Liu C-N, Bennetzen JL (1995): unpublished observations

Maize Genetics Cooperation Newsletter (1993), Coe EH (ed.), 67:133–167

Martin GB, Ganal MW, Tanksley SD (1992): Construction of a yeast artificial chromosome library of tomato and identification of cloned segments linked to two disease resistance loci. *Mol Gen Genet* 233:25–32

Melake-Berhan A, Hulbert SH, Butler LG, Bennetzen JL (1993): Structure and evolution of the genomes of *Sorghum bicolor* and *Zea mays*. *Theor Appl Genet* 86:598–604

Michaelson MJ, Price HJ, Ellison JR, Johnston JS (1991): Comparison of plant DNA contents determined by feulgen microspectrophotometry and laser flow cytometry. *Am J Bot* 78:183–188

Pereira MG, Lee M (1995): Identification of genomic regions affecting plant height in sorghum and maize. *Theor Appl Genet* 90:380–388

Pereira MG, Lee M, Bramel-Cox P, Woodman W, Doebley J, Whitkus R (1994): Construction of an RFLP map in sorghum and comparative mapping in maize. *Genome* 37:236–243

Ragab RA, Dronavalli S, Saghai Maroof MA, Yu YG (1994): Construction of a sorghum RFLP map using sorghum and maize DNA probes. *Genome* 37:590–594

Salimath S, Subramanian V, Bennetzen JL (1995): unpublished observations

San Miguel P, Chen M-C, Woo S-S, Zhang H, Wing R, Bennetzen JL (1995): unpublished observations

Sanz-Alferez S, Richter TE, Hulbert SH, Bennetzen JL (1995): The *Rp3* disease resistance gene of maize: mapping and characterization of introgressed alleles. *Theor Appl Genet* 91:25–32

Springer PS, Edwards KJ, Bennetzen JL (1994): DNA class organization on maize *Adh1* yeast artificial chromosomes. *Proc Natl Acad Sci USA* 91:863–867

Tanksley SD, Bernatsky R, Lapitan NL, Prince JP (1988): Conservation of gene repertoire but not gene order in pepper and tomato. *Proc Natl Acad Sci USA* 85:6419–6423

Teutonico RA, Osborn TC (1994): Mapping of RFLP and qualitative trait loci in *Brassica rapa* and comparison to the linkage maps of *B. napus*, *B. oleracea*, and *Arabidopsis thaliana*. *Theor Appl Genet* 89:885–894

Weerasuriya YM (1995): The construction of a molecular linkage map, mapping of quantitative trait loci, characterization of polyphenols, and screening of genotypes for *Striga* resistance in sorghum (Dissertation). Lafayette, IN: Purdue University.

Whitkus R, Doebley J, Lee M (1992): Comparative genome mapping of sorghum and maize. *Genetics* 132:1119–1130

Zamir D, Tanksley SD (1988): Tomato genome is comprised largely of fast-evolving, low copy-number sequences. *Mol Gen Genet* 213:254–261

5

The Potential Impacts of Apomixis: A Molecular Genetics Approach

RICHARD A. JEFFERSON, ROSS BICKNELL

Introduction

Any volume purporting to describe the impacts of plant molecular genetics can currently deal only with anticipated impacts, rather than actual and measurable effects. It will still be years until a substantial change in agricultural practice and economic return ensue from these research efforts. Thus, it is relevant here to speculate on and outline strategies for achieving the best possible outcome from development of a new technology. In this chapter, we will describe some of the avenues that may be productive for development of apomixis as a powerful new technology, and speculate on impacts that could be achieved with proper attention (Bicknell, 1994c; Jefferson, 1992)

What is Apomixis?

Apomixis is the naturally occurring ability of many plants to produce unreduced seeds parthenogenetically—without fertilization of the female gamete. Apomixis exists in hundreds of plant species distributed among many plant taxa, and is accomplished through a wide range of mechanisms (Gustafsson, 1946; Richards, 1986; Asker and Jerling, 1992).

Very few crop species have substantial apomictic characters. Over the last thousands of years today's crop species have been chosen from among the numerous edible or fibrous proto-crops by farmers. The criteria for such selection almost certainly would have involved the ability of the proto-crop to segregate variation—to improve under mass selection—the very property that apomixis in most of its guises prevents. Thus our small collection of modern day crops probably represents a biased population in favor of sexuality.

The Impact of Plant Molecular Genetics
BWS Sobral, Editor

Potential Impacts of Apomixis

When apomixis is generated with a very high degree of flexibility, the impacts on agriculture could be profound in nature and extremely broad in their scope. A few of the changes that could ensue are:

- immediate fixation of heterozygous genotypes, including those made through wide crosses, allowing single plant evaluation and making possibilities for new breeding strategies and methods in sexual and vegetatively propagated crops;
- plant breeding could become readily responsive to microenvironments, cropping conditions, pathogen populations, and markets, stimulating diverse strategies for agroecosystem management and optimization;
- preparation of very large numbers of hybrid cultivars from almost every crop species into which the trait is introduced;
- propagation of hybrid seed directly by the farmer without the need for inbreds or male steriles and without recourse to frequent seed purchases;
- true-seed propagation of traditionally vegetatively-propagated crops, with concommitant elimination or reduction of disease, substantial increases in germplasm flows, and potential expansion of growing regions;
- elimination of anthesis and fertilization-related crop losses (eg. anthesis drought, heat stress, pollinator failures, submergence or pathogenesis), a major cause of reductions in crop yield and reliability;
- substantial increases in yield in some crops due to increased photosynthate availability through elimination of male flowers/flower parts;
- breeding specifically for endosperm performance in locally adapted varieties without compromising agronomic traits in nonautomous apomicts;

This list of possible impacts is so striking, and so comprehensive that apomixis, when developed to achieve the goals on the list, may well be the most important target for concerted international agricultural research.

Molecular Biology as an Approach to *De Novo* Development of Apomixis

The introduction of apomixis into a target species can be achieved by two possible mechanisms, introgression from an apomictic relative, or transgenesis of an engineered genetic construct. Traditional approaches for developing apomictic crops involve the introgression of apomictic traits or tendencies from wild or related material into crop species. Results have, in general, been disappointing and limited in their potential. Existing apomictic tendencies, available in the gene pools of crop relatives, are neither farmer nor breeder-controllable in a useful manner, are often inefficient (with limited penetrance), are frequently associated with pseudogamy, requiring pollination, and of questionable generality. What is the likelihood then, of finding a truly useful mode of apomixis in a wild relative of a crop?

Unlike introgression, molecular biology offers the potential to develop one or more generic mechanisms for apomixis, crafted to allow field-level control of the trait which is then applied to numerous species. In a more fundamental sense, molecular biology may also provide an avenue for elucidating the underlying biology of the phenomenon, and for tailoring it for human requirements, and for the vicissitudes of the environment and circumstances.

Although molecular biology and genetic engineering are often presented as powerful new tools to introduce an existing trait into sexually incompatible species, the real power of these approaches lies elsewhere. The ability to understand the underlying mechanism(s) of the biological processes involved through combined molecular, cellular, and genetic analysis opens up numerous possibilities for experimentation, enhancements and alterations. These need not, and most probably will not, arise through simple introduction of preexisting alleles or genes from other sources. Rather, adjustments and modifications will be made to native processes by molecular intervention.

In just the last few years, for example, conditional nuclear male sterility has been genetically engineered into numerous crops by transforming plants with chimeric genes that direct and limit the expression of toxic gene products into the male flower parts. This gives rise to a dominant nuclear male sterile phenotype with efficiencies approaching 100%. When combined with a genetically engineered restorer line that encodes a specific inactivator of the toxin, a universal and completely reversible two line male sterility system is achieved. There is every reason to believe that this mechanism, with few variations or exceptions, will be applicable to most crop species. Currently it is effective in both monocots and dicots, both autogamous and allogamous species. This has been carried out largely by teams in the private sector with the goal of developing commercial and proprietary hybrid production for many crops (Mariani et al, 1991). However, while elegant, it is only the beginning of what can be done by adjusting the sexual systems of plants through molecular intervention. Hybrids, while alone very important, are greatly increased in value if the hybrid trait can be fixed in a heterozygous form and thus breed true. More important, an ability to perform single-plant level evaluation would eliminate the major bottleneck of hybrid cultivar development.

Achieving such modifications in transgenic plants will probably need the full arsenal of existing molecular methodologies for ectopic and heterochronic gene expression and will doubtless use dominant suppression mechanisms such as antisense and cosuppression.

Integrated Strategy to Develop Apomixis Through a Transgenic Mechanism

Envision a scenario in which a single apomictic gene cassette that produces several linked phenotypes can be generated through molecular biology. This cassette would function as a dominant trait in virtually all crops to which it is introduced in order to:

(1) block entry of the embryo mother cell into meiosis, and thus to produce a fully functional diploid female gamete;

(2) block the development of male sexual structures;

(3) allow fully autonomous development of the embryo and endosperm without pollination;

(4) have this condition be fully penetrant; and

(5) have the condition be dominant but conditional, when the default state is apomictic, but upon application of a nonproprietary, inexpensive compound, the trait is fully suppressed so that crosses can be performed in either direction.

This scenario is the best case situation, and it should be borne in mind for the research and development of apomixis.

In nature, most apomixis coexists with some degree of sexuality. There are, however, very few, if any, cases in which there are defined and manipulable conditions to interchange these two states at will. Thus, while environmental extremes such as day length or heat/cold stress could have influence on the balance between sexuality and apomixis, these do not tend to be either universal (i.e., most day length dependencies would rule out controlled apomixis in the tropics, where its introduction could be of such value), nor controllable (heat shocking a field is not a realistic option).

Yet to maximize the benefits to the agricultural community, it must be possible to introduce apomixis into a wide spectrum of genetic backgrounds, ideally retaining the ability to cross with related germplasm, then apomictically fix any genotype for subsequent tests and multiplication. The ability to suppress conditionally the apomictic trait should also be available to the small farmer and to the resource-limited breeding and crop management community.

To maximize the utility of the trait and its acceptance in the farming community, it should probably have virtually complete penetrance; apomixis must not be a tendency but a strict and obligate feature of the plant's reproduction until, as noted above, that trait is suppressed to make new combinations. There may be situations in which this is not the case, but in most scenarios, it would be most likely essential.

While many apomicts require pollen either to fertilize the central nuclei to form a functional triploid endosperm and/or to stimulate the development of the embryo and endosperm (pseudogamous apomicts), there exist a substantial number of apomictic plants that have absolutely no dependence on pollination whatsoever (autonomous apomicts). These include the well characterized genera *Taraxacum* (dandelions) and *Hieracium*, with their many agamospecies.

The advantages of autonomous apomixis are substantial, most prominently in the avoidance of threats to crop production associated with failures in pollination and/or fertilization. The events of sexual hybridization, male flower development, pollen formation and shedding, pollen transport, and fertilization combine to present a window of vulnerability in cropping systems. This vulnerability often results in catastrophic and unpredictable yield losses in difficult cropping systems and environments, and in some crops leads to suboptimal performance even in benign environments. Heat, cold, or drought stress, water stresses, submergence, pollinator failures (for instance from promiscuous insecticide usage), and other environmental and biotic factors can have enormous effects on pollen formation, viability, spread, and effectiveness. For instance, a fairly brief period of excessive heat and drought during this window of vulnerability can destroy maize silks and prohibit reasonable seed set, even if the crop has been well managed throughout its life. In rice, moderately low temperatures during pollen meiosis can result in drastically reduced fertility. Thus, autonomous apomixis would render many cropping systems robust and reliable whereas under current practice they are fragile and risk prone.

The importance of this feature will clearly differ between crops. In self fertile forage crops, in which harvest is associated with the mature plant and pollination is relatively assured by the structure of the flower, pseudogamous reproduction is quite an acceptable option. In self fertile grains, however, although pollination is relatively reliable, the harvest is principally associated with the endosperm tissue. Variation in the genotype of that tissue is reflected in yield, both with respect to the quality and quantity of grain harvested. This would be of particular concern if apomixis is widely used to capture heterosis from highly heterozygous populations through single plant selection. Autonomous endospermy would allow the composition of the endosperm to be addressed through direct selection, and the improvements made could be retained in a highly heterotic genetic background.

The increasing vertical integration of agricultural production, and the concomitant privatization and centralization of agricultural research, instills legitimate fears that key methods and opportunities could become unavailable through intellectual property restrictions. This concern is voiced by plant breeders in the private and public sectors alike, as well as by agricultural policy makers and the general public. If apomixis technology is patented and restricted to just one entity, rather than widely licensed to all potential users, its application for broad environmental and economic benefit could be greatly reduced. This restriction would be particularly unfortunate, because apomixis, more than any other innovation mooted in agriculture could have a profound decentralizing and diversifying effect.

Components of an Integrated and Effective Research and Development Strategy for Apomixis

Several steps are seen as being necessary to achieve an adequate understanding of the processes that underlie apomixis, and the introduction of an inducible (or suppressible) form of the trait into target species:

(1) Detailed molecular and cellular analysis of a model apomict;

(2) Genetic screens for apomictic mutants in *Arabidopsis thaliana*;

(3) Analysis of plant genes by complementation of meiotic mutants in yeasts;

(4) Development of tools to delimit and control gene expression within target plants; and

(5) Development of tools for field level control of transgene expression.

Detailed Molecular and Cellular Analysis of a Model Apomict

Most studies of apomixis have focused on describing the cellular mechanisms employed, or on the ecological implications of the trait for different species. While the diversity of these data can be valuable, it is often difficult to rationalize disparate pieces of information from widely different systems. There is, therefore, a need to develop a model system to study apomixis. Before proposing any candidates, however, it is helpful to consider the features required, with particular regard to a molecular study of the trait.

Ideally, a model plant should be easily cultivated, both *in vivo* and *in vitro*, and be a perennial that is easily propagated vegetatively, to permit the maintenance of sterile or self-incompatible sexual biotypes. Small stature, a short generation time, abundant seed set and a simple mechanism for hybridization facilitate the analysis of inheritance and the rapid turnover of experimental populations. Both sexual and apomictic biotypes need to be available, preferably employing autonomous endospermy to avoid difficulties associated with pseudogamy. Apomixis needs to be easily assessed, preferably in a format that can be quantified, to facilitate the evaluation of allelic differences and of the additive and epistatic influences of modifier loci. Of the mechanisms of apomixis reported, apospory appears to be the most appropriate for the initial study, although comparative analyses of diplosporous and adventitious embryogenic systems at a later time would help to identify elements common to natural apomictic systems and regions of control that overlap with sexuality. Apospory appears to be relatively simply inherited, at least in the small number of cases studied, which should simplify the molecular analysis. It is also typically a facultative mechanism in which both zygotic and clonal seeds can be harvested from the same plant. Facultative systems have the advantage that the selective inactivation of either the sexual or apomictic developmental pathway through mutation is likely to lead to the exclusive expression of the other. When apospory is conferred by the inheritance of a dominant allele, mutation results in the expression of sexuality, the recessive condition. This provides an invaluable internal control since the formation of sexual seed indicates that megagametogenesis and embryogenesis remain functional.

There are also a number of requirements which would specifically assist the advancement of a molecular research program. A model system for this type of study must be amenable to genetic transformation to permit the introduction of marker genes, mutagenic sequences, and the reintroduction of putative control sequences. It is preferable that it have a small genome and ideally already be characterized with respect to morphological and molecular markers (deletions, translocations, RFLPs, RAPDs, transposons, etc.), to facilitate the localization of critical loci. For commercial reasons it would also be advantageous, although not essential, to use a crop species to facilitate the transfer of research findings into practical outcomes.

Although more than 300 flowering plants from more than 35 families have been described as apomictic no species meets all of the above criteria. Model systems, of course, have been proposed previously. Most have been monocotyledonous species that are either important forage crops such as *Pennesetum* (Ozias-Akins et al, 1993) and *Brachiaria* (Borges do Valle et al, 1994), or wild relatives of grain crop species, such as *Tripsacum* (related to maize), *Elymus* (wheat), and *Panicum* (Millet). Monocotyledons, however, are typically difficult to transform, imposing limitations on the approaches that can be taken to identify and isolate the genetic elements involved. Some dicotyledonous aposporic species have been used as model systems. Rutishauser (1948) studied *Potentilla* (Rosaceae) and Nogler (1984) conducted a series of elegant experiments on *Ranunculus auricomus* (Ranunculaceae). Unfortunately both of these species are pseudogamous, presenting difficulties associated with pollination and the subsequent need to prove clonal seed formation. Autonomous endospermy is not common among aposporous apomicts, but is known in several genera of the *Asteraceae*. Within that family, the taxon that appears to be best suited for use as a model system for a molecular study of apomixis is *Hieracium* subgenus *Pilosella*, a compilation of over 60 apo-species native to Eurasia and North America.

Hieracium species display many of the features listed above. Most are small herbaceous perennials that are easily propagated and maintained in the greenhouse. Yeung (1971) reported that *Hieracium floribundum* can be induced to flower after exposure to five or more days of continuous light. Similar findings have been reported for *H. robustum* (Bergstrom, 1969) and *H. boreale* (Philipson, 1948). It is interesting to note that *Hieracium* does not flower in response to gibberellic acid application, unlike many other Long Day rosette species (Peterson and Yeung, 1972). Using day-length-extension lighting in the greenhouse, plants of *H. pilosella* and *H. aurantiacum* can be encouraged to flower throughout the year. Seed sets within 3–4 months of germination, allowing 3–4 generations per annum.

Hieracium seed develops by facultative apospory coupled to autonomous endospermy. Pollination is therefore not required for the formation of clonal seed, and apomixis can be scored by seed set after the exclusion of pollen. The capitulum of *Hieracium* is a compound inflorescence containing 60 to 120 individual florets. Decapitation of the immature bud removes both anthers and stigmas, which prevents sexual seed formation but not clonal seed set (Ostenfeld, 1906). Gadella (1991) has reported that many of the sexual forms of *H. pilosella* that he has studied were self-incompatible tetraploids. Self-incompatiblity greatly simplifies hybridization in a plant that bears such small flowers. In common with almost all apomictic plants, *Hieracium* species are typically polyploid. Reported examples range from 3 to 8× (Tutin et al, 1976).

The cellular mechanism of apomixis in *Hieracium* is relatively well documented (Skalinska 1971, 1973; Skalinska and Kubien 1972). Until recently, however, very little information has been available on the experimental manipulation of these plants. The apparent suitability of *Hieracium* for use as a model system has stimulated an effort in our group to develop the methods necessary for a molecular analysis of apospory. A range of tissue culture techniques have now been described, including methods for micropropagation and shoot regeneration from leaf tissue (Bicknell, 1994a). An efficient genetic transformation system has also been developed (Bicknell and Borst, 1994b), and activity and inheritance have been demonstrated for chimeric genes conferring resistance to spectinomycin, kanamycin, and hygromycin, susceptibility to 5 fluorocytosine, and for β glucuronidase activity. Introduced activator transposable elements from maize have been demonstrated to move in this system (Bicknell, 1994d). One interesting outcome of this work has been the use of introduced dominant marker genes to quantify segregation and recombination in facultatively apomictic biotypes of *Hieracium*. Chimeric heterologous antibiotic marker genes are dominant, highly penetrant, and can be scored all together at the seedling stage. Inheritance can then be confirmed by Southern blot or polymerase chain reaction (PCR) analyses. Inheritance of a dominant positive marker sequence, such as kanamycin resistance, from a paternal parent can be used to quantify recombination. Similarly, the loss of a negative selection marker, such as 5-fluorocytosine sensitivity, in the maternal parent can be used for a measure of segregation. Current work on *Hieracium* focuses on the development of a two element transposon tagging system in this plant, to facilitate the isolation of the genetic control elements involved in the expression of apomixis (Bicknell, 1994c).

Genetic Screens for Apomictic Mutants in Arabidopsis thaliana

The small crucifer *Arabidopsis thaliana* has now become the most important model for plant genetic research. Its genetics and molecular biology have been developed to a remarkable degree of sophistication, allowing the rapid isolation of mutants in numerous processes and the subsequent cloning of the relevant genes. This has been largely due to the small size of its genome, its short generation time, and the substantial critical mass of research now being conducted using this species.

One consequence has been the often unspoken adoption of the *Arabidopsis* genome as the representative genetic compliment of a flowering plant. Apomixis is a developmental process and therefore will not be fully understood until it is considered within the context of the total developmental program operating in flowering plants. As most of our understanding of plant developmental genetics relates to the biology of *Arabidopsis*, it will probably be necessary to conduct comparative studies with this plant early in the development of any new research strategy.

A.M. Chaudhury and his colleagues at the Commonwealth Scientific and Industrial Research Organization (CSIRO) Division of Plant Industry are conducting experiments to obtain and analyze apomictic mutants of *Arabidopis*. Using well chosen combinations of visible and male-sterile mutations as the parental material in a screen for apomictic characters, Chaudhury has obtained a set of *Arabidopsis* mutations with promising components of apomixis including several fertilization independent seed (*fis*) mutants. Substantial further analysis is underway (Chaudhury, 1995).

Molecular Genetic Analysis of Candidate Genes Affecting Ovule Development

Understanding of the molecular processes that control the formation of flowers and the various floral organs has increased considerably over the past few years. Homeotic genes play an essential role in these developmental processes. Analysis of homeotic mutants in *Arabidopsis* and *Antirrhinum* has resulted in a powerful genetic model for floral organ development (Weigel and Meyerowitz, 1994). Following this paradigm, Angenent and Dons at the Centre for Plant Breeding and Reproduction Research (CPRO-DLO) in the Netherlands are taking an interesting approach to obtaining and understanding genes involved in ovule development in *Petunia.* They have isolated a series of genes belonging to the MADS box homeotic gene family, and they have been analyzing these by ectopic and heterochronic transgene expression. These genes are designated floral binding protein (fbp) genes and have been shown to be involved in the formation of the floral organs. (Angenent et al, 1993).

It is clear that understanding the molecular regulation of ovule development and the formation of the embryo sac might open new ways for the induction and control of apomixis. Recently two MADS box genes (fbp 7 and fbp 11) that determine the identity of ovules within the pistil of *Petunia hybrida* have been characterized (Angenent et al, 1995). Both genes are expressed only in the developing ovules. Knocking out the genes or causing ectopic expression of the genes in transgenic plants clearly reveals that expression of these genes is necessary for proper development of functional ovules.

These results are very encouraging and may provide insights to further unravel the genetic control of processes taking place in ovules, especially the formation of the embryo sac. Since MADS box genes encoding transcription factors control a number genes, it seems likely that isolation of such target genes will bring us closer to genes directly involved in embryo sac formation, and thus the control of apomixis.

Analysis of Plant Genes by Complementation of Meiotic Mutants in Yeasts

Because apomixis is fundamentally a trait affecting the meiotic process, and in particular the points of entry and exit from meiosis, it would be sensible to make use of the wealth of information already obtained from genetic and molecular studies of the yeasts, *Saccharomyces cerevisiae* and *Schizosaccharomyces pombe.*

These systems have allowed the characterization of numerous genes and gene products associated with the decisions to enter and exit meiosis. These include mutants affecting sporulation (*spo*) (Honigberg. and Esposito, 1994), cell division control (*cdc*), meiotic induction (*ime*) (Kassir et al, 1988), and even apomictic strains that produce true diploid spores (Bilinski et al, 1989). Many of the well-characterized effects are mediated by protein kinases operating in cascades (Yoshida et al, 1990).

The degree of conservation in function, and indeed often in primary protein structure, between yeasts and plants is likely to be substantial, given the critical role of meiosis in evolution. Studies conducted by Hirt, Heberle-Bors, and their collaborators in Vienna have clearly shown that plant genes can complement cell cycle defects in yeasts (Hirt et al, 1991, 1992, 1993; Jonak et al, 1993), and other groups are making substantial headway in isolating genes associated with cell-cycle regulation (Hemerly et al, 1992; Kobayashi et al, 1994).

Characterization of plant genes isolated by complementation of yeast meiotic lesions, and yeast genes expressed in plants may be very valuable in illuminating the parallels and differences in the entry and exit points of meiosis. This in turn could lead to new avenues to inhibit entry to or precociously exit from a meiotic phase, the essence of apomixis.

Tools to Delimit and Control Gene Expression within Target Plants—Cell Type Specificity

All transgenic strategies will hinge upon an ability to tightly control the expression of introduced genes within particular target cells and to particular times of development. This will require the isolation and characterization of promoters and other controlling sequences that are specific in their action to nucellar cells, megaspore mother cells, and other cells of the embryo sac. Methodology for isolation and characterization of cDNAs specific to particular tissue types in plants has been well described (Koltunow et al, 1990); however, the difficulties in preparing pure populations of megaspore mother cells are substantial. Whereas even a few years ago this task would have been unmanageable due to the need to isolate large quantities of homogenous material from which to prepare cDNA libraries or probes, the advent of PCR now allows true micro scale amplification of mRNA populations from even single cells. For example, a recent publication describes work in which a large cDNA library has been developed from only about 100 egg cells of maize (Dresselhaus et al, 1994).

Using RNA mediated Arbitrarily Primed PCR (Welsh, et al, 1992), or RAP PCR (also called RNA display) (Liang and Pardee, 1993), it is also possible now to compare extremely small amounts of RNA (in the nanogram range) without library generation and subtraction. This opens possibilities for using single or a few excised nucellar or embryo mother cells at particular stages, or from apomictic and non-apomictic siblings, to establish the differential patterning and allow subsequent gene isolation. Heterochronic or ectopic expression of these candidate genes, or suppression of their activity through antisense or cosuppression approaches, can then be a particularly valuable avenue to explore function.

In recent studies at Cold Spring Harbor Laboratory, Dr Ueli Grossniklaus has developed a large set of *Ds*-mediated enhancer trap mutant lines of *Arabidopsis* in which a promoter-less or promter-deficient GUS gene is inserted in single copies into the genome. After extensive screening of stained and cleared ovules, Grossniklaus has found lines in which GUS is expressed in a variety of cells and stages of the ovule, including some lines in which the insertion causes single cells in the embryo sac to express GUS (U. Grossniklaus, personal communication). This extremely powerful approach, in one step, allows not only mutants to be generated which are defective in particular aspects of female gametophyte gene function, but also the ready analysis of gene expression of the insertion, and a routine cloning to rescue the inserted gene for subsequent analysis. This brute-force approach seems certain to give us a vast suite of necessary tools as well as mutants to investigate and manipulate ovule and female gametophyte development and genetics, and hence to obtain and control apomixis.

Development of Tools for Field Level Control of Transgene Expression and Environmental Limitation of the Trait

Responsible development of apomixis must involve the use of mechanisms to ensure both the environmental and ecological restriction of the trait. This could probably be achieved through concomitant male sterility, engineered as early in the development of the male floral organs as possible. This trait could best be envisioned as being facultative, so that on demand, the trait could be passed to sexually related species through conventional breeding methods.

It is important to ensure that the apomictic trait is expressed as the default condition, but suppressible at the will of the farmer or breeder. Molecular mechanisms need to be developed that will allow breeders to switch the trait off to perform a sexual cross, then release the switch to allow reversion of the F1 to apomictic mode. Thus, systems comprising compounds that can be applied to induce or repress defined, corresponding promoters need to be developed.

Interesting candidates would be compounds that are not present in plants, are not toxic, and that have a well characterized ability to induce transgenes. Such systems, showing varying degrees of success, have been developed using induction by copper ions (Mett et al, 1993), or glucocorticoids (Schena et al, 1992).

A particularly good example of the type of approach showing promise for controlling complex field traits is afforded by the excellent work of Gatz and her colleagues in engineering the bacterial tetracycline repressor/operator interaction in transgenic plants (Gatz and Quail, 1988; Gatz et al, 1992; Roder et al, 1994; Weinman et al, 1994). In these studies, Gatz et al have generated a set of molecular tools by which transgene action can either be induced (in which the tet repressor functions as a negative regulator) or repressed (in which the tet repressor is fused to a transcriptional activator, and is hence a positive effector) through the action of exogenously applied tetracycline, with a dynamic range of up to 1000 in some cases. While the tetracycline system is by far the best developed, there is no reason to expect that other systems cannot be developed to provide field level control of transgene action as well.

Some Challenging Questions about Apomixis

The Physical Nature of the Lesion

Is there a chance that apomixis is not encoded by a gene *per se*? What if apomixis is the result of structural features of a chromosome(s) that cause inappropriate or precocious entry into a pathway of megaspore mother cell formation and exiting of meiosis? What if it is a structure that interferes with the initiation of meiosis? For instance, one can envision a binding site for a key protein that has too high an affinity and thus titrates the protein before it can build to a sufficient level to initiate the meiotic process. Is it conceivable that there is not any gene expression in a traditional sense associated causally with apomixis. Variants on this model could also explain the tendency to find apomixis preferentially in polyploids and its tendency to be a dominant trait. If this were so, then differential gene expression screens would probably not show a causative mRNA expressed, and hence would not be productive.

Misleading Single locus Behaviour

Apomixis has seemingly occured in many diverse plant taxa independently, and because of the diversity of mechanisms and phenotypes (autonomous or pseudogamous, aposporous or diplosporous, facultative or obligate etc), it seems very possible that this trait is not produced by a simple mutation of a single target gene. However, in the majority of the few well-studied apomictic systems, apomixis does seem to segregate as a single dominant locus.

It is very possible that apomixis is a result of several different mutations that independently produce deleterious effects but which, in combination, would produce the requisite failure of the meiotic process with a concommitant ability to precociously enter embryogenesis. Because of the independent deleterious effects of either of these traits alone, one would anticipate that propagation of apomixis would often, if not always, behave genetically as a single locus. Segregation of either lethal or sublethal trait would guarantee the elimination of the alleles from any sexual population, placing strong selection pressure to ensure cosegregation or pseudolinkage. This could occur by any of a variety of mechanisms such as association on a balancer chromosome that is recombinationally deficient due to its structure. Thus, the appearance of apomixis as a single mutation, locus, or linkage group is almost certainly to be expected no matter how many actual genes are involved in the process. This caveat could prove very troublesome to approaches relying on gene disruption and isolation through transposon insertion, for example.

Summary

No technological intervention in such a fundamental activity as agriculture can ever be socially or environmentally neutral. The form that the development of apomixis takes will greatly influence who stands to benefit from the innovation. When it is achieved, however, the degree to which agriculture and society will change will be almost beyond comprehension. The advent of apomixis will make the advances and the problems of the Green Revolution seem tractable and easy to predict. Because of the changes in agricultural practice that could ensue, the magnitude, diversity, and types of socio-economic impacts must be considered soon. It is only through foresight that we can encourage a responsible shaping of the research and development strategies. The development of policies and infrastructures suitable to cope with an innovation of this magitude in anticipation of its arrival, rather than *ex post facto* is essential and will greatly enhance its utility and minimize its adverse effects.

While it is clear that apomixis has the potential to have substantial impacts on agriculture, it is important to consider the optimum criteria for the development of the trait, to focus research efforts, and target the most appropriate crops. Social, agricultural, and environmental benefits will not necessarily flow directly from the simple invention or discovery of an apomixis gene. A concerted and ethical effort should be made to ensure that maximum social and environmental good occurs from the innovation.

REFERENCES

Angenent GC, Franken J, Busscher M, Colombo L, van Tunen AJ (1993): Petal and stamen formation in petunia is regulated by the homeotic gene fbp1. *Plant J* 4:101–112

Angenent GC, Franken J, Busscher M, van Dijken A, van Went JL, Dons JJM, van Tunen AJ (1995): Ovule identity in petunia is determined by a novel class of MADS box genes (in press)

Asker SE, Jerling L (1992): *Apomixis in plants*. Boca Raton, FL: CRC Press

Bergstrom G (1969): Influence of temperature, light and resting stage on morphology, meiosis, pollen formation and seed fertility in apomictic Hieracium robustum. *Hereditas* 62:429–433

Bicknell RA (1994a): Micropropagation of *Hieracium aurantiacum*. *Plant Cell Tissue Organ Cult* 37:197–199

Bicknell RA, Borst NK (1994b): Agrobacterium-mediated transformation of Hieracium aurantiacum. *Int J Plant Sci* 155(4):467–470

Bicknell RA (1994c): Hieracium; A model system for studying the molecular genetics of apomixis. *ANL* 7:8–10

Bicknell RA (1994d): Evidence for the transposition of the Ac transposable element from maize, in the facultative apomict, *Hieracium aurantiacum*. In: *Proceedings of the Queenstown Molecular Biology Meeting*. Queenstown, New Zealand.

Bilinski CA, Marmiroli N, Miller JJ (1989): Apomixis in Saccharomyces cerevisiae and other eukaryotic micro-organisms. *Adv Microb Physiol* 30:23–52

Borges do Valle C, Glienke C, Leguizamon GOC (1994): Inheritance of apomixis in Brachiaria, a tropical forage grass. *ANL* 7:42–43

Chaudhury AM (1995): personal communication

Dresselhaus T, Lorz H, Kranz E (1994): Representative cDNA libraries from few plant cells. *Plant J* 5(4):605–610

Gadella TWJ (1991): Variation, hybridization and reproductive biology of Hieracium pilosella *L Proc Kon Ned Akad V Wetensch* 94(4):455–488

Gatz C, Quail PH (1988): Tn10–encoded tet repressor can regulate an operator-containing plant promoter. *Proc Natl Acad Sci U S A* 85:1394–7

Gatz C, Frohberg C, Wendenburg R (1992): Stringent repression and homogeneous de-repression by tetracycline of a modified CaMV 35S promoter in intact transgenic tobacco plants. *Plant J* 2:397–404

Gustafsson A (1946): *Apomixis in Higher Plants.* Svalof, Lund

Hemerly A, Bergounioux C, Van Montagu M, Inze D, Ferreira P (1992): Genes regulating the plant cell cycle: isolation of a mitotic-like cyclin from Arabidopsis thaliana. *Proc Natl Acad Sci USA* 89:3295–9

Hirt H, Pay A, Bogre L, Meskiene I, Heberle-Bors E (1993): cdc2MsB, a cognate cdc2 gene from alfalfa, complements the G1/S but not the G2/M transition of budding yeast cdc28 mutants. *Plant J* 4:61–9

Hirt H, Mink M, Pfosser M, Bogre L, Gyorgyey J, Jonak C, Gartner A, Dudits D, Heberle-Bors E (1992): Alfalfa cyclins: differential expression during the cell cycle and in plant organs. *Plant Cell* 4:1531–8

Hirt H, Pay A, Gyorgyey J, Bako L, Nemeth K, Bogre L, Schweyen RJ, Heberle-Bors E, Dudits D (1991): Complementation of a yeast cell cycle mutant by an alfalfa cDNA encoding a protein kinase homologous to p34cdc2 *Proc Natl Acad Sci U S A* 88:1636–40

Honigberg SM, Esposito RE (1994): Reversal of cell determination in yeast meiosis: postcommitment arrest allows return to mitotic growth. *Proc Natl Acad Sci U S A* 91:6559–6563

Jefferson RA (1992): Strategic development of apomixis as a general tool for agriculture. In: Wilson KJ, ed. *Proceedings of the International Workshop on Apomixis in Rice*; 1992 Jan; Changsha, China. New York: Rockefeller Foundation

Jonak C, Pay A, Bogre L, Hirt H, Heberle-Bors E. (1993): The plant homologue of MAP kinase is expressed in a cell cycle-dependent and organ-specific manner. *Plant J* 3:611–7

Kassir Y, Granot D, Simchen G (1988): IME1, a positive regulator gene of meiosis in S. cerevisiae. *Cell* 52:853–62

Kobayashi T, Hotta Y, Tabata S (1993): Isolation and characterization of a yeast gene that is homologous with a meiosis-specific cDNA from a plant. *Mol Gen Genet* 237:225–32

Koltunow AM, Truettner J, Cox KH, Wallroth M, Goldberg RB (1990): Different temporal and spatial gene expression patterns occur during anther development. *Plant Cell* 2:1201–1224

Liang P, Pardee AB (1992): Differential display of eukaryotic messenger RNA by means of the Polymerase Chain Reaction. *Science* 257:967–971

Mariani C, Goldberg RB, Leemans J (1991): Engineered male sterility in plants. *Symp Soc Exp Biol* 45:275–9

Mendel GJ (1866): Versuche ber Pflanzenhybriden. *Verh Naturforsch Ver Brunn* 4:3

Mett VL, Lochhead LP, Reynolds PH (1993): Copper-controllable gene expression system for whole plants. *Proc Natl Acad Sci U S A* 90:4567–71

Nogler GA (1975): Genetics of apospory in Ranunculus auricomus. VI. Embyology of F3 and F4 backcross offspring. *Phytomorph* 25:485–490

Ostenfeld CH (1906): Experimental and cytological studies in the Hieracia. I. Castration and hybridization experiments with some species of Hieracia. *Bot Tids* 27:225–248

Ozias-Akins P, Lubbers EL, Hanna WW, McNay JW. (1993): Transmission of the apomictic mode of reproduction in Pennisetum: co-inheritance of the trait and molecular markers. *Theor Appl Genet* 1993(85):632–638

Peterson RL, Yeung EC (1972): Effect of two gibberellins on species of the rosette plant *Hieracium. Bot Gaz* 133:190–198

Philipson WR (1948): Studies in the development of inflorescence. IV. The capitula of Hieracium boreale Fries and Dahlia gracilis Ortg. *Ann Bot* 12:65–75

Richards AJ (1986): *Plant Breeding Systems.* London: George Allen & Unwin

Roder FT, Schmulling T, Gatz C (1994): Efficiency of the tetracycline-dependent gene expression system: complete suppression and efficient induction of the rolB phenotype in transgenic plants. *Mol Gen Genet* 243:32–8

Rutishauser A (1948): Pseudogamie und Polymorphie in der Gattung Potentilla. *Arch Julius-klaus-Stiftung f Vererb-Forsch* 23:267–424

Schena M, Lloyd AM, Davis RW (1991): A steroid-inducible gene expression system for plant cells. *Proc Natl Acad Sci USA* 88:10421–5

Skalinska M (1971): Experimental and embryological studies in *Hieracium aurantiacum* L. *Acta Biol Crac ser Bot* 14:139–155

Skalinska M (1973): Further studies in facultative apomixis of *Hieracium aurantiacum* L. *Acta Biol Crac ser Bot* 16:121–137

Skalinska M, Kubien E (1972): Cytological and embryological studies in *Hieraceum pratense* Tausch. *Acta Biol Crac ser Bot* 15:39–50

Staiger CJ, Cande WZ (1992): Ameiotic, a gene that controls meiotic chromosome and cytoskeletal behavior in maize. *Dev Biol* 154:226–30

Tutin TG, Heywood VH, Burges NA, Moore DM, Valentine DH, Walters SM, Webb DA (1976): *Flora Europea.* Cambridge: Cambridge University Press

Weinmann P, Gossen M, Hillen W, Bujard H, Gatz C (1994): A chimeric transactivator allows tetracycline-responsive gene expression in whole plants. *Plant J* 5:559–69

Yeung E C, Peterson RL (1971): Studies on the rosette plant Hieracium floribundum. I. Observations related to flowering and axillary bud development. *Can J Bot* 50:73–78

Yoshida M, Kawaguchi H, Sakata Y, Kominami K, Hirano M, Shima H, Akada R, Yamashita I (1990): Initiation of meiosis and sporulation in Saccharomyces cerevisiae requires a novel protein kinase homologue. *Mol Gen Genet* 221: 176–86

Weigel D, Meyerowitz EM (1994): The ABCs of floral homeotic genes. *Cell* 78: 203–209

Welsh J, Chada K, Dalal SS, Cheng R, Ralph D, McClelland M (1992): Arbitrarily primed PCR fingerprinting of RNA. *Nucleic Acids Research* 20:4965–4970

6

The Role of Meiotic Recombination in Generating Novel Genetic Variability

PATRICK S. SCHNABLE, XIAOJIE XU, LAURA CIVARDI, YIJI XIA, AN-PING HSIA, LEI ZHANG, BASIL J. NIKOLAU

Long-term selection experiments have revealed the unexpected plasticity of the maize genome. One such series of experiments, begun at the University of Illinois in 1896, has concentrated on increasing or decreasing the percentage of oil or protein in maize kernels. Even though these experiments have been conducted on closed populations, i.e., without the introduction of additional genetic variability (and the initial population had a narrow genetic base), selection continues to provide significant genetic gains even after 95 cycles (Dudley and Lambert, 1992). Of even more significance, it has been possible effectively to reverse selection subsequent to significant gains. Another example comes from the continued gains in grain yields and other agronomic traits obtained in the BSSS population which has been under reciprocal recurrent selection at Iowa State University for fourteen cycles (Keeratinijakal and Lamkey, 1993; Schnicker and Lamkey, 1993; Lamkey, 1995). Again, even though the initial population had a relatively narrow genetic base (a intermated group of sixteen inbreds), selection does not appear to have exhausted the available genetic diversity within this closed population. Of broader significance is the continued world-wide gain in grain yields that can be attributed to genetic improvement. Corn breeders have been able to increase grain yields via genetic improvement an average of approximately 2% per year (Russell, 1993) over the last several decades. Again, these increases have occurred even in the absence of significant amounts of introgression of novel genetic material.

The explanations for these observations are either that vast amounts of genetic diversity have been present in the founding populations and that these reservoirs have not been exhausted, or that novel genetic diversity is being generated during the course of the selection experiments. The first possible explanation is unlikely given the relatively narrow genetic base from which these populations have been created. In the following discussion, we will suggest that recombina-

The Impact of Plant Molecular Genetics
BWS Sobral, Editor

tion per se contributes to the creation of genetic diversity upon which selection (natural or artificial) can act. Namely, we provide support for the view that a large proportion of recombination occurs within genic sequences (Thuriaux, 1977) and demonstrate that these events can create novel alleles.

It has long been known that large physical intervals of the maize genome do not exhibit uniform rates of recombination per physical distance. For example, although the BA translocation breakpoint 3Sb (a cytological marker) resides at the midpoint of the short arm of chromosome 3, it maps very close to the centromere on genetic maps (Coe, 1993). This indicates that most recombination occurs in the distal half of chromosome 3s.

To address the question of whether recombination rates per kb are nonuniform within smaller physical intervals, the interval between the *a1* and *sh2* loci on the long arm of chromosome 3 was selected as a model. Test crosses were used to determine the genetic distance associated with this interval. Mutations at the *a1* and *sh2* loci confer phenotypes that readily allow the identification and isolation of meiotic recombinants; kernels homozygous for mutant alleles of the *a1* and *sh2* loci are colorless (due to *a1*) and shrunken (because of *sh2*). Genetic recombinants were isolated from a testcross (Cross 1) by virtue of their non-parental phenotypes (i.e., colorless, round and colored, shrunken kernels).

Cross 1: *A1 Sh2/a1::rdt sh2* X *a1::rdt sh2/a1::rdt sh2*

Two experiments were conducted. In the first experiment (conducted in 1991), the *A1 Sh2* chromosome in Cross 1 was derived from Line C (a color-converted W22 inbred line). In second experiment (conducted in 1993), the *A1 Sh2* donor was the inbred line LH82. The validity of each of the putative recombinants was confirmed via genetic crosses and restriction fragment length polymorphisms (RFLP) analyses using *a1*- and *sh2*-specific probes. Based upon the rates at which nonparental gametes were recovered (Table 1), the genetic distance between the *a1* and *sh2* loci was determined to be 0.086 cM in the Line C experiment and 0.106 cM in the LH82 experiment. Because these experiments were conducted in different years, the statistically significant difference in recombination rates between the two experiments may reflect environmental influences. Alternatively, this difference may be attributable to either *cis* or *trans* genetic effects. Most likely each of the these factors can contribute to differences in rates of recombination.

The *a1-sh2* physical interval was cloned on a 470-kb YAC. Subsequent molecular analyses using yeast chromosome fragmentation technology have established that these two loci are separated by 140 kb (Civardi et al, 1994). Hence, the relationship between genetic and physical distance (which we have termed ρ) is 1630 and 1320 kb/cM in the Line C and LH82-derived populations, respectively. This compares to the genome average of 1500 kb/cM.

A number of studies have established that the value of ρ within defined maize loci is one to two orders of magnitude smaller than the genome average value of ρ (Dooner et al, 1985; Dooner, 1986; Dooner et al, 1991; Nelson, 1968; Wessler

Table 1. The Relationship Between Genetic and Physical Distances in the *a1-sh2* Interval

Interval	No. X-overs	Pop'l Size	Interval Size (kb)	Interval Size (cM)	ρ (kb/cM)
Genome	N/A*	N/A	3×10^6	2061	1456
A1-Sh2 (Line C, 1991)	58	67,000	n.t.**	0.086	1628†
A1-Sh2 (LH82, 1993)	195	184,000	140	0.106	1321

* N/A indicates not applicable.

** n.t. indicates not tested.

† Assumes that the *a1-sh2* interval from Line C has a physical distance of 140 kb.

and Varagona, 1985; Freeling, 1977; Sachs et al, 1986). Since smaller values of ρ indicate more recombination per kb, plant genes therefore serve as recombination hot spots. However, it is unclear whether genes are hot-spots for recombination per se, or whether genes cluster in regions of the genome that are themselves recombinationally hyperactive.

To address this question and the related question of whether all recombination hot spots are genes, the molecular positions of a series of meiotic recombination breakpoints that occurred within the 140-kb *a1-sh2* interval (from Table 1) are being mapped. To accomplish this, DNA from each recombinant plant is being subjected to RFLP analyses using as probes single-copy DNA fragments isolated from a cosmid contig that spans the cloned 140-kb interval. Based on these analyses, it has been possible to demonstrate that 1/3 (16/58) of all of the recombination breakpoints occurred in a 6-kb interval that includes the *a1* gene. This result suggests that the *a1* locus is recombinationally hyperactive, a view that is supported by intragenic recombination rates at this locus (Brown and Sundaresan, 1991; Civardi et al, 1994).

To further characterize this recombination hot spot, intragenic recombinants have been isolated that arise via recombination events within a 1.2-kb interval of the *a1* locus (Civardi et al, 1994). Rare intragenic *a1* recombination events have been selected from cross 2.

Cross 2: *a1-mum2/a1::rdt* X *a1::rdt /a1::rdt*

The recessive *a1-mum2* allele contains a 1.4-kb *Mu1* transposon insertion at nucleotide −97 in the *a1* gene (O'Reilly et al, 1985; Shepherd et al, 1988), and the recessive *a1::rdt* allele contains a 0.7-kb *rdt* transposon insertion at nucleotide +1,083 (Brown et al, 1989, see Figure 1). Each of these mutants conditions a colorless phenotype as a consequence of its associated transposon insertion. Most kernels isolated from cross 2 would be expected to have genotypes of *a1-mum2/a1::rdt* or *a1::rdt/a1::rdt* and to therefore exhibit a colorless phenotype. However, rare intragenic recombination events at the *a1* locus which occur within the 1.2-kb interval between the two transposon insertion sites can generate chimeric *A1'* alleles that condition colored kernels that are readily distinguishable from the colorless phenotype conditioned by *a1-mum2* and *a1::rdt*..

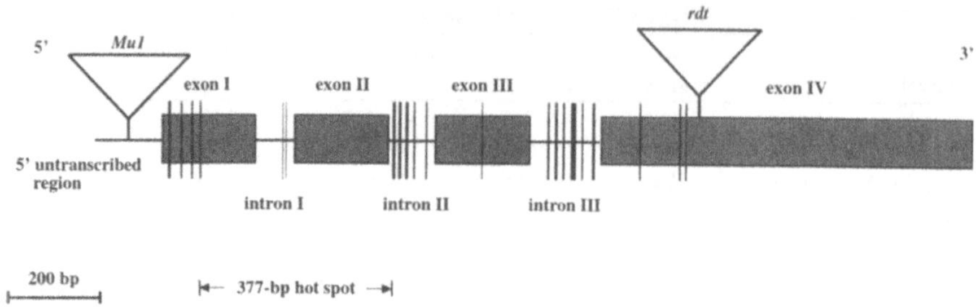

Figure 1. Identification of the *a1* recombination hot spot by intragenic recombination between the *a1::rdt* and *a1-mum2* alleles. The figure illustrates the molecular structure of the *a1::rdt* and *a1-mum2* alleles, which contain *rdt* and *Mu1* transposon insertions at the indicated locations. In addition, the *a1::rdt* and *a1-mum2* alleles show sequence polymorphisms (vertical lines, the width of which indicate the number of base pairs involved in the polymorphism). These polymorphisms were used to define intervals to which recombination breakpoints of *A1'* alleles were mapped. Fourteen out of 15 recombination breakpoints mapped to the 377-bp interval labeled as a recombination hot spot.

Twenty-four such putative recombinants have been isolated from a total population of 742,100 kernels, and 15 of these have been analyzed. The validity of each of these putative recombinants has been confirmed via genetic crosses and DNA gel blotting experiments. In this recombination experiment, the genetic distance associated with the 1.2-kb interval of the *a1* locus that was assayed is 0.0065 cM (+/−0.0009 cM). Therefore the value of ρ for this interval is 185 kb/cM. Thus, within this interval the rate of recombination per kb is eight-fold higher than the genome average of 1500 kb/cM. Significantly, the *a1* locus is more recombinationally active than the neighboring 140-kb interval between the *a1* and *sh2* loci (Table 1). These findings are consistent with the hypothesis that recombination occurs at higher rates in genes *per se*, rather than because genes are most likely to be found in large, recombinationally active chromosomal segments.

To determine whether recombination events resolve randomly across the 1.2-kb recombination hotspot within the *a1* gene, the 15 *A1'* alleles generated via intragenic recombination have been PCR (polymerase chain reaction) amplified and sequenced. Because the *a1-mum2* and *a1::rdt* alleles exhibit abundant DNA sequence polymorphisms relative to each other (Figure 2A), it has been possible to precisely map the positions of recombination breakpoints associated with each of these 15 *A1'* alleles. These analyses demonstrated that recombination breakpoints do not resolve uniformly across the previously defined 1.2-kb recombination hotspot, rather they cluster within a 377-bp stretch of the *a1* gene; fourteen of the fifteen recombination breakpoints are resolved within this stretch, which exhibits a value of ρ equal to 62 kb/cM. Thus, this 377-bp recombination hotspot is 23 times more recombinagenic than the genome as a whole. However, it could be argued that the presence of the *Mu1* transposon in the 5′ end of the *a1-mum2* allele used in the experiment to isolate the *A1'* alleles is affecting the rate and specificity of recombination within the *a1* gene. This is not the case. In the earlier experiment using the wild-type *A1* allele extracted from the inbred

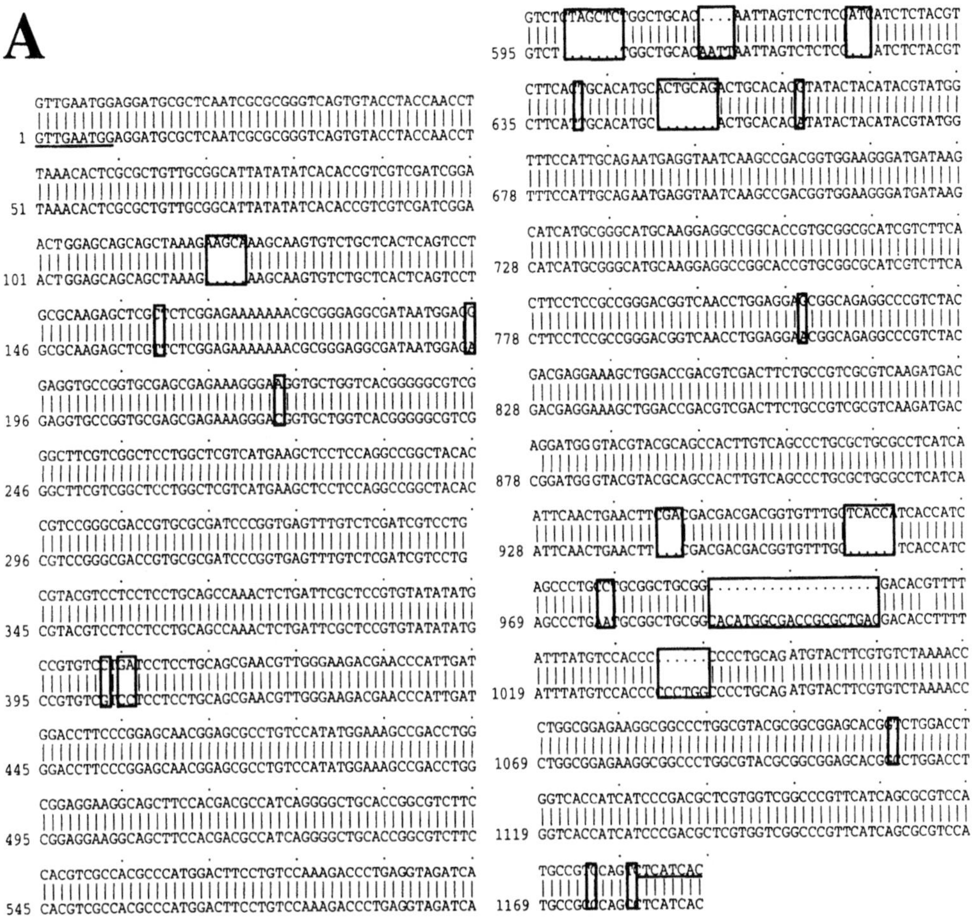

Figure 2. Polymorphisms among *a1* alleles. **A:** Polymorphisms between the *a1::rdt* (upper sequence) and *a1-mum2* alleles are boxed. The sequences between the *Mu1* and *rdt* inserts in these alleles are shown. The target site duplications associated with each transposon insertion are underlined. **B:** Sequences of *a1::rdt*, *A1′-278* and *a1-mum2* are compared from positions 1 to 216 and from 390 to 684 (according the numbering scheme used in Figure 2A). *A1′-278* DNA sequences which exhibit polymorphisms relative to either (or both) the *a1::rdt* and *a1-mum2* alleles are indicated. At these polymorphic sites, like sequences are boxed. *A1-278* carries *a1::rdt*-specific polymorphisms at its 5′-end and *a1-mum2*-specific polymorphisms towards its 3′-end. It is therefore probable that *A1-278* arose via a recombination event between two *A1* alleles related to the progenitors of *a1::rdt* and *a1-mum2*. Based upon the sequence of *A1-278*, this recombination event would have occurred within the 377-bp recombination hot spot. The *A1-278*-specific polymorphisms represent mutations (or conversions) that occurred independently of the recombination event.

Line C, 16 of 58 crossovers between the *a1* and *sh2* loci mapped to the 6-kb interval that includes the *a1* locus. Eight of these crossovers have been PCR amplified and sequenced; four arose via cross-overs within the 377-bp recombination hot spot, creating novel *A1′* alleles (the remainder map to a region 5′ of the

B

```
al::rdt   .............................................GTTGAATG
A1-278    .............................................GTTGAATG
a1-mum2   .............................................GTTGAATG

al::rdt   GAGGATGCGC TCAATCGCGC GGGTCAGTGT ACCTACCAAC CTTAAACACT
A1-278    GAGGATGCGC TCAATCGCGC GGGTCAGTGT ACCTACCAAC CTTAAACACT
a1-mum2   GAGGATGCGC TCAATCGCGC GGGTCAGTGT ACCTACCAAC CTTAAACACT

al::rdt   CGCGCTGTTG CGGCATTATA TATCACACCG TCGTCGATCG GAACTGGAGC
A1-278    CGCGCTGTTG CGGCATTATA TATCACACCG TCGTCGATCG GAACTGGAGC
a1-mum2   CGCGCTGTTG CGGCATTATA TATCACACCG TCGTCGATCG GAACTGGAGC

al::rdt   AGCAGCTAAA GAAGCAAAGC AAGTGTCTGC TCACTCAGTC CTGCGCAAGA
A1-278    AGCAGCTAAA GAAGCAAAGC AAGTGTCTGC TCACTCAGTC CTGCGCAAGA
a1-mum2   AGCAGCTAAA G.....AAGC AAGTGTCTGC TCACTCAGTC CTGCGCAAGA

al::rdt   GCTCGCTCTC GGAGAAAAAA ACGCGGGAGG CGATAATGGA GCGAGGTGCC
A1-278    GCTCGCTCTC GGAGAAAAAA ACGCGGGAGG CGATAATGGA GCGAGGTTCC
a1-mum2   GCTCGTTCTC GGAGAAAAAA ACGCGGGAGG CGATAATGGA GAGAGGTGCC

al::rdt   GGTGCGAGCG AGA                  ATA TGCCGTGTCC
A1-278    GGTGCGAGCG AGA   ——//——        ATA TACCGTGTCG
a1-mum2   GGTGCGAGCG AGA                  ATA TGCCGTGTCG

al::rdt   TGATCCTCCT GCAGCGAACG TTGGGAAGAC GAACCCATTG ATGGACCTTC
A1-278    TGATCGTCCT GCAGCGAACG TTGGGAAGAC GAAGCCATTG ATGGACCTTC
a1-mum2   TCCTCCTCCT GCAGCGAACG TTGGGAAGAC GAAGCCATTG ATGGACCTTC

al::rdt   CCGGAGCAAC GGAGCGCCTG TCCATATGGA AAGCCGACCT GGCGGAGGAA
A1-278    CCGGAGCAAC GGAGCGCCTG TCCATATGGA AAGCCGACCT GGCGGAGGAA
a1-mum2   CCGGAGCAAC GGAGCGCCTG TCCATATGGA AAGCCGACCT GGCGGAGGAA

al::rdt   GGCAGCTTCC ACGACGCCAT CAGGGGCTGC ACCGGCGTCT TCCACGTCGC
A1-278    GGCAGCTTCC ACGACGCCAT CAGGGGCTGC ACCGGCGTCT TCCACGTCGC
a1-mum2   GGCAGCTTCC ACGACGCCAT CAGGGGCTGC ACCGGCGTCT TCCACGTCGC

al::rdt   CACGCCCATG GACTTCCTGT CCAAAGACCC TGAGGTAGAT CAGTCTCTAG
A1-278    CACGCCCATG GACTTCCTGT CCAAAGACCC TGAGGTAGAT CAGTCT....
a1-mum2   CACGCCCATG GACTTCCTGT CCAAAGACCC TGAGGTAGAT CAGTCT....

al::rdt   CTCTGGCTGC AC....AATT AGTCTCTCGA TCATCTCTAC GTCTT..CAG
A1-278    ...TGGCTGC ACAATTAATT AGTCTCTCG. ..ATCTCTAC GTCTTCACAG
a1-mum2   ...TGGCTGC ACAATTAATT AGTCTCTCG. ..ATCTCTAC GTCTT..CAT

al::rdt   TGCACATGCA CTGCAGACTG CACACGTATA CTACATACGT ATGGTTTCCAT
A1-278    TGCACATGC. ......ACTG CACACGTATA CTACAGACGT ATGGTTTCCAG
a1-mum2   TGCACATGC. ......ACTG CACACATATA CTACATACGT ATGGTTTCCAT
```

Figure 2. Continued.

a1 locus). Hence, even in the absence of the *Mu1* transposon, the *a1* locus is behaving as a recombination hot spot.

Assuming that the estimate of the value of ρ within the entire *a1* gene (185 kb/cM) reflects the average value of ρ within all maize genes, and assuming the 2000 cM maize genome (Coe, personal communication, cited in Civardi et al, 1994) contains 80,000 genes (Okamuro and Goldberg, 1989; Bird, 1995) which average 2-kb in size, intragenic recombination could account for half of all recombination in the genome. In addition, the value of 185 kb/cM for ρ at the *a1* locus is higher than most determinations of this ratio for other maize genes (Dooner et al, 1985; Dooner, 1986; Dooner et al, 1991; Nelson, 1968; Wessler and Varagona, 1985; Freeling, 1977; Sachs et al, 1986). Therefore, intragenic recombination could conceivably account for a very large proportion of the total recombination that occurs in the maize genome, even though genic sequences may represent as little as 5% of the total genome. This view that recombination occurs preferentially within genes is consistent with the finding that the rate of recombination per gene appears uniform across species (Bart and Bell, 1987).

Such a finding would suggest that there is an evolutionary advantage to an organism in having most, if not all, of its recombination occur within genes. One such advantage would be if intragenic recombination serves as a mechanism for the creation of novel alleles that can contribute to genetic diversity. Indeed the recombination studies presented here have resulted in the generation of novel *A1'* alleles. Further, a wild-type *A1* allele has been identified whose chimeric structure could most simply be explained if it originated via intragenic recombination between two progenitor *A1* alleles (Figure 2B). Hence, meiotic recombination may be one of the mechanisms that generates the vast reservoir of genetic diversity available to maize breeders.

Given that intragenic recombination events occur at frequencies that can be readily detected, this phenomenon could potentially impact the generation of DNA sequence-based phylogenetic trees. In the generation of such trees each sequence polymorphism is treated as an independent evolutionary event. The number of such polymorphisms is typically used as a measure of the evolutionary distance between two alleles. However, if intragenic cross-overs are shuffling genic polymorphisms at high rates, two alleles could display multiple polymorphisms relative to each other, even though they differ by only one evolutionary event (a recombination event). This bias will only be introduced if the two alleles in question have the potential to undergo meiotic recombination, i.e., is they are carried by interfertile, intercrossing species. Hence, current phylogenetic approaches have the potential to overestimate short evolutionary distances.

ACKNOWLEDGMENTS

This manuscript is Journal Paper No. J-16520 of the Iowa Agriculture and Home Economics Experiment Station, Ames, Iowa, Project No. 3125.

REFERENCES

Bart A, Bell G (1987): Mammalian chiasma frequencies as a test of two theories of recombination. *Nature* 326:803–805

Bird AP (1995): Gene number, noise reduction and biological complexity. *Tr Genet* 11:94–100

Brown J, Sundaresan V (1991): A recombination hotspot in the maize intragenic region. *Theor Appl Genet* 81:185–188

Brown JJ, Mattes MG, O'Reilly C, Shepherd NS (1989): Molecular characterization of *rDt*, a maize transposon of the *"Dotted"* controlling element system. *Mol Gen Genet* 215:239–244

Civardi L, Xia YJ, Edwards K, Schnable PS, Nikolau BJ (1994): The relationship between the genetic and physical distances of the cloned *a1-sh2* interval of the *Zea mays* L. genome. *Proc Natl Acad Sci* 91:8268–8272

Coe EH (1993): Gene list and working maps. *Maize Genet Coop Newsl* 67: 133–166

Dooner HK (1986): Genetic fine structure of the *bronze* locus in maize. *Genetics* 113:1021–1036

Dooner HK, Keller J, Harper E, Ralston E (1991): Variable patterns of transposition of the maize element Activator in tobacco. *Plant Cell* 3:473–482

Dooner HK, Weck E, Adams S, Ralston E, Favreau M, English J (1985): A molecular genetic analysis of insertions in the bronze locus in maize. *Mol Gen Genet* 200:240–246

Dudley JW, Lambert RJ (1992): Ninety generations of selection for oil and protein in maize. *Maydica* 37:81–87

Freeling M (1978): Allelic variation at the level of intragenic recombination. *Genetics* 89:211–224

Keeratinijakal V, Lamkey KR (1993): Responses to reciprocal recurrent selection in BSSS and BSCB1 maize populations. *Crop Sci* 33:73–77

Lamkey KR (1995): personal communication

Nelson OE, (1968): The *waxy* locus in maize. II. The location of the controlling element alleles. *Genetics* 60:507–524

Okamuro JK, Goldberg RB (1989): Regulation of plant gene expression: General principles. In: *The Biochemistry of Plants: A Comprehensive Treatise, Vol 15*, Stumpf PK, Conn EE, eds. New York: Academic Press

O'Reilly C, Shepherd NS, Pereira A, Schwarz-Sommer Z, Bertram I, Robertson DS, Peterson PA, Saedler H (1985): Molecular cloning of the *a1* locus of *Zea mays* using the transposable elements *En* and *Mu1*. *EMBO J* 4:877–882

Russell WA (1993): Achievements of maize breeders in North America. In: *International Crop Science I*, Frey KJ et al, eds. Madison, Wisconsin: Crop Science Society of America, Inc

Sachs M, Dennis E, Gerlach W, Peacock WJ (1986): Two alleles of *Adh1* have 3′ structural and poly (A) addition polymorphisms. *Genetics* 113:449–467

Schnicker BJ, Lamkey KR (1993): Interpopulation genetic variance after reciprocal recurrent selection in BSSS and BSCB1 maize populations. *Crop Sci* 33:90–95

Shepherd NS, Sheridan WF, Mattes MG, Deno G (1988): The use of *Mutator* for gene tagging: Cross-referencing between transposable element systems. In: *Plant Transposable Elements*, Nelson O, ed. New York: Plenum Press

Thuriaux P (1977): Is recombination confined to structural genes on the eukaryotic genome? *Nature* 268:460–462

Wessler SR, Varagona MJ (1985): Molecular basis of mutations at the waxy locus of maize: Correlation with the fine structure genetic map. *Proc Natl Acad Sci* 82:4177–4181

Part II

Evolution and Phylogenetics

7

Molecular Markers in Plant Conservation Genetics

WILLIAM J. HAHN AND FRANCESCA T. GRIFO

Introduction

One of the most unfortunate consequences of expanding human populations is the demise and extinction of other species. Although many of these losses are due only indirectly to human activities, the current rate of extinction is unprecedented in the geologic record, and our own pervasiveness has left little room for recovery in the wild of those species that do manage to survive. Recognition of this destruction has led to a desire to conserve what is left, but choices among alternate strategies are often clouded by a lack of objective criteria and baseline information. Because of these deficiencies, the need for a rational and scientifically sound basis to any conservation strategy has become increasingly evident.

In response to the call for a scientific approach to conservation, workers in a variety of academic and applied fields have directed their attention to the problem of extinctions, with the collective effort referred to as conservation biology. Among the various approaches involved in this field, the specific application of genetic analysis has been one of the most influential as well as one of the most controversial. Borrowing from the most practical to the most theoretical aspects of population and molecular genetics, conservation genetics (like forensic genetics) represents an interface in which legal action or management decisions often rely heavily on the quality and type of science involved. Different sampling strategies, types of genetic data gathered, method of data analysis, and interpretation of results may each contribute to a decision and strategy that will determine the future of a particular species or population.

Of the different types of genetic data available, none has so heavily influenced our understanding of the evolutionary process as have molecular markers. Beginning with the initial observations of protein diversity by workers in the 1960s and leading to today's understanding of variation at the nucleotide level, the development of new molecular markers has almost always resulted in theoretical

The Impact of Plant Molecular Genetics
BWS Sobral, Editor

and conceptual advances, many of which concern conservation issues. Furthermore, the proliferation of new techniques has allowed for a more comprehensive view of the evolution and biology of the organisms in question. In this chapter, we survey some of the molecular markers employed in plant conservation genetics and outline the role and influence of genetics in conservation biology.

Types of Molecular Markers

Despite the greater number of endangered plant species, molecular conservation genetic studies are fewer in number and much less sophisticated for plant groups than for their vertebrate counterparts. A variety of techniques are routinely used in mammalian and avian conservation genetics, largely due to the importance attached to individual species in these groups and to the greater public appeal they maintain (Avise, 1989, 1994; Dowling, et al, 1992). Many of these vertebrate groups have been studied from a number of molecular perspectives thus providing independent measures of genetic diversity and a more comprehensive understanding of its origin and maintenance (O'Brien, 1994, 1995). For plants, the majority of conservation studies have used the single technique of isozyme analysis in spite of the potential limitations of the technique and the availability of numerous alternatives (Schaal, et al, 1991a). Appropriate models for the evaluation of these other techniques can be found in the crop breeding literature and in comparison to studies of vertebrate systems. Short reviews of the relative merits and costs of these markers can be found in Rafalski and Tingey (1993) and Whitkus et al, (1994).

As summarized by Avise (1994), two basic analytic approaches have been taken with conservation genetic data. The first is a more traditional population genetics perspective which is founded on estimates of heterozygosity (Allendorf and Leary, 1986; Antonovics, 1984; Clegg, 1988; Ellstrand and Elam, 1993; Guerrant, 1986; Huenneke, 1991). For maximum statistical power, the molecular methods used in this approach must be able to distinguish and recover alternate alleles within and among individuals. This perspective has been the most common in plant conservation studies and serves as the basis for many conservation and management strategies.

An alternative and complementary perspective is that of systematics in which the population is typically the smallest unit considered, and the objective of the study is an estimate of the historical relations between different populations or species (Avise et al, 1987; Avise and Ball, 1990). In this approach, both discrete and continuous characters (including frequency data) may be used, but greater resolving power is possible with the former. Relatively few studies of plant conservation genetics have employed this approach although many molecular systematic studies have included rare and endangered species.

Isozymes

Isozymes are a class of enzymes that share a common substrate but exhibit different electrophoretic mobilities (Markert and Moller, 1959). When different electromorphs are identified by genetic analysis as allelic, they are called allozymes (Prakash et al, 1969). The ease with which different loci and alleles can be scored and the relatively low costs have made allozymes the marker of choice for most plant population and conservation studies. The codominant nature of allozymes allows for a rich array of statistical analyses centering on heterozygosity estimates (Gottlieb, 1981; Nei, 1987; Weir, 1990), and a considerable degree of correlation has been noted between life history traits and the degree and pattern of allozyme variation (Hamrick and Godt, 1989; Hamrick et al, 1991; Mitton, 1989).

A number of limitations are found with isozymes, however. Factors such as tissue specific and developmentally controlled expression as well as environmentally induced expression can cause difficulty in detecting activity (Markert and Moller, 1959; Wendel and Weeden, 1989). Gene duplications (Gottlieb, 1982) can complicate both interpretation of goals and data analysis, in part because the means by which duplications occur are not always the same (Weeden and Wendel, 1989). Additionally, because allozymes are often under functional or selective constraints, they may not be the most appropriate markers for the estimation of genetic diversity in conservation studies (Hedrick et al, 1986; Mitton, 1994). Perhaps the most important shortcomings of allozymes, however, are the limited number of resolvable loci (about 30–50 in most taxa) and the amount of variation detectable with standard techniques. The usual method of detection for allozymes, staining of protein variants on a starch gel following electrophoresis, may not detect all of the actual nucleotide variation present among electromorphs (Aquadro and Avise, 1982), and comparison of allozyme data with other types of information (see below) often shows the failure of allozymes to accurately estimate genomic variation.

Despite these limitations, isozyme studies constitute the majority of genetic data for plant and animal conservation studies and will continue to provide invaluable markers for heterozygosity estimates and correlation with life history traits (Mitton, 1989). Isozymic studies of rare plants generally demonstrate low levels of genetic diversity in comparison to widespread congeners (Hamrick and Godt, 1989; Karron, 1991; Barret and Kohn, 1991; Soltis and Soltis, 1991). Extreme examples are found in *Bensoniella oregona* (Saxifragaceae: Soltis et al, 1992), *Pedicularis furbishiae* (Scrophulariaceae: Waller et al, 1987), *Howellia aquaticus* (Campanulaceae: Lesica et al, 1988), *Lactoris fernandeziana* (Lactoridaceae: Crawford et al, 1994), and *Trifolium reflexum* (Leguminosae: Hickey et al, 1991) in which a complete absence of variation was noted in all loci examined. Low levels of genetic diversity are also commonly recorded in island taxa even though they may be relatively common. Examples of this are found in Hawaii where ferns (Ranker, 1992) and members of the genus *Bidens* (Asteraceae: Helenurm and Ganders, 1985) show much lower levels of within and between population diversity in comparison to mainland relatives. Although these types of observations are frequent, some rare species are known to exhibit significant levels of protein variation (*Cypripedium calceolus* (Orchidaceae): Case, 1993).

Correlations with life history traits often help explain the various patterns of genetic diversity. For example, many rare plant species are self-pollinating which tends to decrease overall genetic diversity but more strongly differentiates separate populations. Because of this, the variation present in rare species is usually distributed among populations rather than within (e.g. *Limnanthes floccosa* subsp. *californica* (Limnanthaceae): Dole and Sun, 1992) whereas widespread species tend to distribute diversity more evenly across their range (Hamrick et al, 1991). In contrast, Richter et al, (1994) detected relatively high levels of allozyme variation in the narrow endemic *Delphinium viridescens* (Ranunculaceae) attributing it to the species' outbreeding mode of reproduction.

Finally, isozymes are useful in detecting hybrids because of the biparental mode of inheritance and their codominant phenotypic expression (Rieseberg, 1991). Many studies have used allozymes as markers of hybridization, which is particularly widespread in plants. From a conservation perspective, hybridization may result in outbreeding depression or the loss of genetic diversity through assimilation of smaller populations into larger ones (Rieseberg, 1991; Templeton, 1986).

These types of observations have provided much of the empirical genetic evidence on which generalizations about rare and endangered species are based. Given the number of exceptions, however, measures of protein diversity must be put into a proper context before final conclusions are reached. It is estimated that only about 0.5% of the genome of most organisms is needed to code for all proteins (10,000–50,000) in the average eukaryote (Flavell, 1980) of which isozymes are a very small fraction. In light of the small number of loci available and the above mentioned limitations, many extrapolations of allozyme data to explain evolutionary phenomena are probably inaccurate (Mitton, 1994). Clearly, additional types of evidence are needed to properly assess the relevance and validity of allozyme diversity measures. In this review, we reference some of the general results obtained with allozymes in comparison to the various DNA markers discussed below. For more details on allozymes in plants, readers are referred to the reviews found in Soltis and Soltis (1989), Tanksley and Orten (1983), and Moss (1982).

Restriction Fragment Length Polymorphisms

The ability of a class of enzymes known as restriction endonucleases to cleave DNA at specific nucleotide sequences or recognition sites has greatly enhanced our ability to examine structure and variation of the primary genetic material. The loss or gain of a restriction site between two related taxa is generally assumed to be caused by a single nucleotide mutation at the recognition site and allows for the estimation of percent nucleotide difference between the two taxa. Other forms of mutation including indels, inversions, and translocations can disrupt a recognition site, but the patterns then produced are generally quite distinctive. The resultant restriction fragment polymorphisms (RFLPs) can provide broad coverage of the genome under consideration when a sufficiently large array of probes or primer pairs are used to generate and visualize the fragments. Increased character sampling can then be accomplished by using more enzymes or selecting more frequently cutting enzymes (Helentjaris et al, 1985; Holsinger and Jansen, 1993; Olmstead and Palmer, 1994).

Most RFLP studies have used either digestion of isolated organellar DNA with visualization of the restriction fragments directly on agarose gels, or they have used filter hybridization techniques with digestion of total DNA and visualization of resultant fragments via homologous or heterologous DNA probes. Although the techniques of filter hybridization are more thorough in recovering all fragments and assuring homology, they can be laborious, and visualization of fragments may involve either radioactivity or toxic chemicals. An alternate approach, that eliminates these cumbersome steps uses restriction endonucleases to survey variation in PCR amplified fragments. The resultant PCR-RFLPs or amplified fragment length polymorphisms (AFLPs) have proved useful in studies where individual regions are sufficiently characterized to allow primer construction. The ease and speed with which these regions can be screened with PCR technology permits much faster sampling of a large number of individuals (Sobral and Honeycutt, 1994). The trade-offs between these two approaches are of broader coverage of the genome (filter hybridization) versus larger samples of individuals (PCR-RFLPs). Several different classes of DNA have been surveyed for RFLPs and will be discussed individually.

Chloroplast DNA

For plant systematics chloroplast DNA (cpDNA) has been the molecule of choice for phylogeny estimation (Palmer et al, 1988; Olmstead and Palmer, 1994; Sytsma and Hahn, 1994). Most surveys have considered only restriction site variation across the genome although studies of gene order and structural rearrangements are also possible (Downie and Palmer, 1992). The mostly uniparental inheritance plastid and the discrete (presence versus absence) character type renders cpDNA RFLP data particularly amenable to phylogenetic analysis (Holsinger and Jansen, 1993). No confirmed reports of intermolecular recombination are known for the molecule, but a few studies have reported heteroplamsy or within-individual polymorphisms, perhaps due to biparental inheritance or somatic mutations of the plastid genome. Because of the essentially homogeneous population of cpDNA within an individual, measures of individual heterozygosity are not applicable to cpDNA RFLP data.

In general, plastid RFLP data generated with filter hybridization techniques are valuable among species and very closely related genera (Palmer et al, 1988) although intraspecific variation is known in a number of taxa (Soltis et al, 1992). PCR-RFLP techniques are often more appropriate for measures of population-level variation, particularly in the variable intergenic spacer regions (MacCauley, 1994). Some of the intraspecific variation reported for chloroplast RFLPs (Soltis et al, 1992) is partitioned along geographic lines although relatively few studies have specifically addressed this issue (Liston et al, 1992).

While few cpDNA studies have been directed specifically at rare plants, a number of endangered taxa have been included in broader surveys. In these studies, relationships of rare taxa are often more clearly defined, and some measure of distinctiveness can be made. Frequently, the analysis of additional populations uncovers intraspecific genetic structure and indicates a more complicated evolutionary history than seen with other types of data. Like allozyme data, cpDNA RFLP variation is often associated with life history traits (Fenster and Ritland, 1992; Olmstead, 1990) but because of the uniparental mode of chloroplast inheritance, cpDNA variation is usually more strongly partitioned than allozyme diversity (MacCauley, 1994).

Chloroplast DNA RFLP studies have also proved useful in documenting cases of hybridization. Rieseberg and Wendel (1993) have surveyed the literature, and they report that the chloroplast is far more likely to introgress than are nuclear markers. This type of information is particularly useful in verifying the validity of gene trees as actually species trees and detecting the introgression of foreign genetic material in rare and endangered taxa (see above).

Although RFLP studies have been declared outdated by some, the large body of comparative systematic data and the prospect of faster and cleaner techniques will most certainly keep the general technique alive.

MITOCHONDRIAL DNA

The preferred molecule for many systematic zoologists and population geneticists has been mitochondrial DNA (mtDNA). The highly conserved structure but significant sequence variability of the animal mitochondrial genome have rendered it particularly useful for studies at a variety of taxonomic levels (Avise et al, 1987; Harrison, 1989; Moritz et al, 1987) and a number of studies have had application to conservation issues (Moritz, 1994). Additionally, like the chloroplast, the uniparental pattern of inheritance and non-recombining nature of the genome has allowed for a number of evolutionary insights. Unlike the animal mitochondrion, however, the plant mitochondrial genome is characterized by a highly volatile genome structure but relatively low levels of sequence divergence (Palmer, 1992). Although the molecule is presently under study in a number of labs, current techniques limit application of this genome in evolutionary studies.

SINGLE COPY NUCLEAR DNA

Although much of the plant genome is composed of repetitive DNA (Lapitan, 1992), a significant proportion consists of low or single copy nuclear DNA (scnDNA) which has considerable potential for population-level RFLP studies. Probes for these regions are obtained either through random cloning of total genomic DNA (gDNA) or through the cloning of complementary DNA (cDNA) from isolated total RNA. Probes generated for one species usually show sufficient homology to detect variation in closely related taxa, but the limits of utility vary from group to group. A major advantage to this system is that a broad sampling of the genome can be accomplished given a sufficiently large clone library. Phenotypically, scnDNA markers are codominant which allows for genetic analyses similar to those used for allozymes. Levels of genetic diversity detected with gDNA probes are roughly equivalent to those estimated with allozyme markers whereas that seen with cDNA probes is significantly higher. In *Brassica campestris*, for example, McGrath and Quiros (1992) detected levels of cDNA polymorphism three times that found with gDNA probes with the differences in polymorphism attributed to intron and flanking regions detected by the cDNA probes but not present in regions homologous to gDNA probes.

scnDNA probes have been used in phylogenetic studies (e.g., *Gossypium*: Brubaker and Wendel, 1994) and in the study of hybridization (e.g., *Populus*: Keim et al, 1989). This technique has not been applied specifically to plant conservation problems, but several vertebrate studies have used these types of markers (Karl et al, 1991). Although the success of this technique in crop improvement and population-level studies is encouraging, the costs of development are significant, and the potential for application to endangered plant species may be limited to economically important taxa or genera in which a number of endangered species are found (e.g., many Hawaiian genera).

Repetitive DNA

Plant genomes are characterized by wide variance in size (Lapitan, 1992) ranging from 0.07 pg per nucleus in *Arabidopsis thaliana* to 17 pg in *Triticum aestivum* among well-characterized taxa. Some of these differences can be related to the large number of polyploids in the plant kingdom, but a considerable percentage of this variation is attributable to varying amounts of repetitive DNA. As a generalization, species of plants with relatively low genome sizes are characterized by a low number of repetitive sequence. For example, *Arabidopsis* has only 14% repeated sequence whereas grasses contain as much as 60%–80% repetitive DNA. Variation due to repetitive DNA can result in differences of several orders of magnitude even among species within a single family (Bennett et al, 1982).

The nature of repetitive DNA ranges from simple gene duplications of protein coding loci (i.e., gene families) to the highly repeated tandem arrays of ribosomal genes (rDNA) to dispersed repeats of simple sequences (VNTRs). Duplicated protein loci constitute a relatively small amount of total repetitive DNA and are discussed here in the context of allozymes.

The next largest fraction of repetitive DNA is the high copy number nuclear ribosomal DNA (nrDNA) which has been used extensively in evolutionary studies. Ribosomal repeats are normally homogenized by various forms of concerted evolution resulting in similar if not identical sequences among all copies (Dvorak, 1988). This feature eliminates many of the homology problems associated with duplicated protein loci and provides a readily detectable set of markers given the large number of copies present in most plants. RFLP studies of nrDNA have been used to document intraspecific variation in a number of plant groups (Brauner et al, 1993; Sytsma and Schaal, 1985). Several cases of within-individual variation are known (Learn and Schaal, 1987; Schaal and Learn, 1988). Additional uses are seen in studies of hybrids and introgression (Rieseberg et al, 1991).

Of the remainder of repetitive DNA found in most plant species, the majority is present as tandem repeats of short repeated sequences (Tautz and Renz, 1984) referred to as satellite DNA or variable number tandem repeats (VNTR). Several other names are in the literature, but these two are the most common. Smaller repeat units (>10bp) are known as minisatellites whereas very short repeat units (<10bp) are referred to as microsatellites with the separation between these two classes arbitrary and not always consistent. The evolutionary relationships between the two classes have been examined by Wright (1994). VNTR loci are readily screened using either filter hybridization or PCR techniques (Rogstad, 1993) and are generally thought to be inherited in a codominant Mendelian fashion (Jeffreys et al, 1985; Akkaya et al, 1992; Morgante and Olivieri, 1993). These features, coupled with the ability to resolve slight differences in repeat length, have provided powerful new markers for the description of some basic population genetic parameters (Bruford and Wayne, 1994).

The first class of repeated DNA used in population studies, minisatellites or DNA fingerprints, were first described in plant species by Dallas (1988) and Rogstad et al (1988). Since then, over 400 studies have been published, principally in agronomically important plant groups (Weisling et al, 1995). Minisatellite markers typically show levels of variation higher than that found with allozymes or nrDNA RFLPs and form the core of modern forensic DNA technology. They are being used with increasing frequency in studies of rare and endangered animal species (Fleischer et al, 1993) but for plants, studies are still somewhat limited.

Alberte et al (1994) have described high amounts of minisatellite diversity in the seagrass *Zostera marina* and have demonstrated restricted gene flow among relatively close populations. Furthermore, they have demonstrated a lower level of genetic diversity in disturbed populations relative to undisturbed populations. Among Californian populations of *Zostera marina* surveyed, those in highly disturbed populations have much higher within-population similarity coefficients than those from relatively undisturbed populations. Previous studies by these authors (Fain et al, 1992) have found significant interpopulation differentiation but little variation within populations among nrDNA RFLPs.

DNA fingerprints have also been used to compare the effects of different breeding systems. Wolff et al (1994) has surveyed three species of *Plantago* for minisatellite variation and has found that the highly-selfing species *P. major* is low in diversity within populations but relatively high between populations. The outcrossing species *P. lanceolata* possesses higher variability within populations and only moderate differentiation between populations. *Plantago coronopus*, with a mixed mating system, shows intermediate measures of genetic diversity. In all cases, levels of variation within and between populations correspond to those found in an earlier study using isozyme markers (Wolff et al, 1991) but which were only weakly correlated.

Microsatellite markers, also known as simple sequence repeats (SSRs: Jacob et al, 1991) or short tandom repeats (STRs; Edwards et al, 1991), have been characterized in a number of crop plant groups, particularly in the development of linkage maps (Wang et al, 1994). A good example of this is the integration of 30 microsatellite markers into the densely mapped genome of *Arabidopsis* (Bell and Ecker, 1994). Microsatellite markers are under intensive study in many animal systems for conservation purposes although the relatively long development time currently limits application as with scnDNA markers discussed above. Among the first uses in wild plant population genetics has been the survey of several topical tree species by Condit and Hubbell (1991). Possible applications suggested by these authors are the examination of parentage, quantification of gene flow, and study of the nature of genetic diversity within and among populations. A review of PCR-mediated approaches to plant microsatellite variation is found in Morgante and Olivieri (1993).

Microsatellites are usually several times more variable than allozymes. For example, in the wild yam species, *Dioscorea tokoro*, Teruachi and Konuma (1994) have found that the number of alleles and levels of microsatellite heterozygosity is more than twice as high as that observed by allozyme analysis (Teruachi, 1990).

RAPDs and AP-PCR

Randomly Amplified Polymorphic DNA (RAPD) is a popular technique that uses the polymerase chain reaction (PCR) but with short nonspecific primers under conditions of modest stringency (Williams et al, 1990). Typically these are 9 or 10-mer oligonucleotides that are random in sequence but often biased in nucleotide content. For arbitrarily primed PCR (AP-PCR), the primers are longer, but low annealing stringencies are used for the first few rounds of amplification (Welsh and McClelland, 1990). The net effect in each of these techniques is to amplify up to several regions of the genome each of which are flanked by the specified priming sites. Although the number of regions flanked by the same priming sites is very large, the specific reaction conditions employed determine the size range of possible products thereby limiting the total number of fragments or amplicons obtained. The modest stringencies allow for some base mismatch, but most variation is assumed to be derived from single base pair mutations in the priming sites. Many other factors may influence the production of a given fragment, but the basic assumption is that comigrating fragments between accessions represent homologous regions of the genome (see below).

RAPD phenotypes are inherited in a dominant fashion (as with cpDNA RFLPs) and therefore do not allow direct estimates of heterozygosity (Tingey and DelTufo, 1992). Estimates of population genetic structure are nonetheless possible, but sampling intensities of two to ten times that needed for codominant markers are required for equivalent statistical power (Lynch and Milligan, 1994).

Features that make RAPDs attractive are their relatively low cost and the ease and speed with which variants can be screened (Rafalski and Tingey, 1993).

At a first glance, RAPDs offer the potential to distinguish among accessions at several levels of differentiation. Useful polymorphisms have been found at a variety of taxonomic levels ranging from varieties and cultivars to subspecies and species. At and above the species level, however, homology between comigrating bands becomes questionable. A large number of factors can contribute to incorrect homology assessments with many of these difficulties inherent in the random nature of the procedure (Ellsworth et al, 1993; Hedrick, 1992). Absence of phenotypes (bands) may arise due to insertion/deletion events at the primer site(s), sufficient base pair mismatch due to point mutations at the primer site(s), complete absence of corresponding loci (or at least one or both of the primer sites), and biased synthesis of alternate loci in the same reaction (Smith et al, 1994).

A comparison of random nuclear RFLP and RAPD markers in the genus *Brassica* reveals similar patterns of intraspecific diversity but very different patterns between closely related species (Thormann et al, 1994). The root cause is nonspecificity of the RAPD primers which produce nonhomologous comigrating bands among species. Using a filter hybridization check for homology, the authors have found that three RAPD probes out of the fifteen chosen do not hybridize to all comigrating bands. The three that failed involve interspecific comparisons suggesting that about 20% of the interspecific comparisons result in inaccurate homology calls. As a further control, these authors have examined both high and low amplification intensity effects for twelve of the RAPD bands scored. Fully one third of these probes hybridized to high-copy-number sequences for both high and low intensity RAPD bands. These results suggest that genomic copy does not influence the relative intensity of a band but that primer-template homology and primer competition among fragments might. Similar results have been obtained in *Xanthomonas* in which the authors specifically address the problems of phylogenetic analysis using RAPDs (Smith et al, 1994). The experience from these studies suggests that RAPD approaches are quite appropriate for population studies, but that extreme caution must be used for interspecific comparisons.

Although most within-species comparisons have demonstrated consistent results with RAPDs showing resolving power equivalent to that seen with scnDNA RFLPs (Halldén et al, 1994), significant deviations from Mendelian ratios have been noted. Echt et al (1992) working with alfalfa and Reiter et al (1993) working with *Arabidopsis* have found that only 76% and 57% respectively of the RAPD fragments examined segregate in a dominant Mendelian fashion. Additional deviations from expected dominant inheritance have been observed by Heun and Helentjaris (1993) in maize F_1 hybrids. Other studies on primates have noticed an excess of nonparental bands in known pedigrees (Riedy et al, 1992).

Despite these limitations, RAPDs have been used by several authors to examine genetic diversity in rare plant species. Variation in RAPD banding patterns has been uncovered in the Jaun Fernandez Island endemic *Lactoris fernandeziana* (Lactoridaceae) by Crawford et al (1994) as has been variation in rDNA intergenic spacers (Brauner et al, 1992) even though no allozyme variation has been detected. Additionally, RAPDs have been used as nuclear markers in the study of hybridization. Crawford et al (1993) have examined the hybrid origin of *Margyracena skottsbergii* (Rosaceae) on the Juan Fernandez Islands and Arnold et al (1992) have confirmed the hybrid origin of *Iris nelsonii* relative to the two putative parents *I. fulva* and *I. hexagona* and estimated the extent of gene flow among these taxa.

DNA Sequence Approaches

Sequence studies in conservation genetics are essentially nonexistent with only a few currently available genes readily applicable to the task. The nuclear ribosomal internal transcribed spacer (nrDNA ITS) has been utilized in a number of plant groups (Baldwin et al, 1995) with several of the taxa surveyed qualifying as rare and endangered. The power to differentiate among species is limited for ITS, but some inference was possible for the origins and status of *Clarkia franciscana* (Hahn et al, 1993) and *Streptanthus glandulosus* (Mayer and Soltis, 1994; Mayer et al, 1994). For insects, Vogler and DeSalle (1994a, b) have examined ITS-1 sequence variation in several populations of eastern North American Tiger Beetles and have specifically discussed some of the conservation implications of their results.

Chloroplast DNA sequence studies have only recently been attempted at the species level. The gene *mat*K has been examined by Steele and Vilgalys (1994) and Johnson and Soltis (1994) demonstrating that interspecific variation is sufficient for phylogenetic resolution. Specific application to conservation studies might be warranted as the value of plastid markers in such studies using RFLPs has been amply demonstrated (see above).

Other Types of Genetic Data

Although the data types described above are fairly diverse, many are in the early stages of application and further evaluation is still needed. A number of additional approaches have been described (Arnheim et al, 1990; Dweikat et al, 1993; Lessa, 1993; Newbury and Ford-Lloyd, 1993) which might warrant further examination. Of the various types of molecular genetic data discussed in this review, however, each has utility only at the level of genetic markers which are used to estimate historical relationships and estimate current diversity. Because the important components of species survival revolve around the genotype-phenotype-environment interface and not simple genetic markers (Lande, 1988), more relevant information would concern the genetic basis of those traits directly involved in fitness, reproduction, phenotypic plasticity, and adaptability in rare plants. A comprehensive discussion of this issue is outside the scope of this review, but a few brief comments seem appropriate.

A major component of the genome that is rarely considered in conservation studies is that which contributes to metric traits. These quantitative trait loci (QTL) have been under intensive study in many crop plants and have contributed enormously to improved yield, disease resistance, and other economically important qualities (Bulmer, 1991; Falconer, 1981; Knapp, 1994). In contrast to the typical single locus isozyme gene or DNA marker, polygenic QTLs are much more mutable with spontaneous rates of mutation per generation relative to environmental variance roughly 1000 times higher than that of single locus genes (Lande, 1976, 1977, 1980). The additive effect of many small genes increases the odds of mutation, and the resultant phenotypes show more continuous variation than the quantum mutations exhibited by single locus traits. Because of this, our estimates of low genetic diversity in rare plants may be of little importance in regard to estimating the odds of survival for a given species. Only limited studies of wild species have been conducted using QTLs (Bijlsma et al, 1994; van Houten et al, 1994), but the importance of these traits in crop improvement suggests that considerable insight might be gained from such studies on endangered species.

Specific physiologic systems involving nutrition, water relations, heat and drought tolerance, antiherbivory mechanisms, and heavy metal tolerance, among others, are all reasonably well characterized for many crop or model-organism species and could provide considerable insight into the nature of adaptation and survival for highly specialized species (Bennetzen et al, 1993; Kellogg and Birchler, 1994). For animal conservation genetics, an active area of study concerns the major histocompatibility complex (MHC) in which diversity is thought to be maintained in order to meet the evolving challenges of new diseases (Yuhki and O'Brien, 1990). Roughly equivalent studies in plants might concern the phytoalexin system (Fritz and Sims, 1992) or the biosynthetic pathways that control secondary compound antiherbivore or antipathogen production (Harborne, 1988). Finally, these observations also point to the need for studies of the molecular basis of adaptation including controlling regions, transposons, and regulation of expression.

What Are the Needs of Conservation Biology?

The chief concern of conservation biology is the study and conservation of endangered species. A first question to ask might be about the distinctiveness of the entities of concern because the issue of what level of diversity we should actually try to maintain is still under debate (Walker, 1992; Woodruff, 1989). Some authors argue that highly unique entities should be given preference over equally rare taxa with close relatives of abundant distribution (Vane-Wright et al, 1991) while others argue that the evolutionary potential is highest in species-rich groups since the ability to adapt is seemingly greater (Erwin, 1991). At a smaller scale, the details of species concepts are brought into question (Crothers, 1992; Rojas, 1992; Vogler & DeSalle, 1994b), and the importance of species versus subspecies, hybrids, and populations has generated considerable debate about the scientific legitimacy of legal conservation units (O'Brien and Mayr, 1991). Therefore, the first measures to be taken with molecular methods are taxon specific markers and estimation of the degree of differentiation between units.

While most threatened species are rare, the converse is not necessarily true as rarity itself for a given species might be normal (Rabinowitz, 1981). When, however, rarity is due to human influence or when naturally rare species are sensitive to human activities, the threat of extinction becomes more of a concern. In dealing with the causes of rarity, one must ask to what degree is it of human origin and to what extent is it determined by the life history traits of the species in question (Kunin and Gaston, 1993; Weller, 1994). These traits, such as breeding system and morpho-physiological features, are the link between the genotype and the environment and are those that are involved in the day to day establishment, growth, survival, and reproduction of an organism. In many cases, life history traits will be the limiting factor for a species' likelihood of success whereas genetic diversity (as traditionally measured) is often inconsequential or is only a byproduct of a particular type of breeding system (Lande, 1988; Milligan et al, 1994). The relationship between patterns of genetic diversity and life history traits has been the focus of many molecular population genetic surveys and assumptions, but the need to specifically focus on the relevant forms of diversity is not always addressed. As mentioned above, we need to know more about the genetic basis of life history traits and its relevance to adaptation and survival.

An important controversy in conservation biology is the relative importance of *in situ* versus *ex situ* management (Falk and Holsinger, 1991; Shoenwald-Cox et al, 1983). While strict preservation of habitat is usually the best measure to prevent extinction, many species require intervention such as ex situ cultivation, captive breeding, and reintroduction. Seed banking (Hamilton, 1994) and plant reintroduction programs (Fenster and Dudash, 1994) are often heavily dependent upon genetic data yet the assumptions behind these practices are frequently not met by the type of genetic data employed.

The nature of the specific genetic marker used can also strongly bias estimates of diversity or understanding of the evolutionary process. As previously discussed, only a very small percentage of the genome is described by traditional allozyme studies yet these data form of majority of all genetic information for rare and threatened plant species. The need for additional perspectives on molecular diversity is clearly needed. Comparison of the techniques discussed earlier strongly suggests that different data types are giving different pictures of the evolutionary process and that a combination of data types is the most appropriate for a complete understanding.

The importance of multiple independent datasets has been shown in animal conservation genetics O'Brien, 1994). For plant conservation genetics, however, relatively few groups have been studied with more than one technique. One example concerns the Californian serpentine endemic *Clarkia franciscana* (Onagraceae) which is known from only two small populations: one in the Presidio of San Francisco and the other in the foothills above Oakland. This rare species is known to vary widely in population size from year to year and is almost completely homozygous at all allozyme loci surveyed (Gottlieb, 1973). Original morphological, cytological, and ecological evidence suggests a close relationship between *C. franciscana* and *C. rubicunda* (Lewis and Raven, 1958), and these authors propose that a combination of chromosomal rearrangements and the evolution of inbreeding accompanied the rapid differentiation of *C. franciscana* from a population of *C. rubicunda*. Allozyme studies fail to support this interpretation but instead place *C. franciscana* equidistant from *C. rubicunda* and *C. amoena* (Gottlieb, 1974).

When a second population of *C. franciscana* was discovered in Oakland, morphological examination suggested that it represents a human-mediated introduction and that it does not represent a unique population. An electrophoretic study of the second population indicates, however, that the allozyme differences between the two populations of *C. franciscana* are considerable with fixed differences at five of the 31 loci examined (Gottlieb and Edwards, 1992). Additional molecular studies indicate an even more complicated situation. Chloroplast DNA RFLP data indicate a closer relationship between *C. franciscana* and *C. amoena* (Sytsma and Smith, 1992) whereas a nrDNA ITS study (Hahn et al, 1993) suggests a closer affinity between *C. franciscana* and *C. rubicunda*. Consideration of all available data suggests that an early and rapid differentiation of all three species from each other occurred leaving the actual pattern of speciation obscured. From a conservation perspective, these results indicate that not only is *C. franciscana* a distinct species relative to *C. rubicunda* and *C. amoena*, but that the two populations of *C. franciscana* are quite divergent and thus equally deserving of protection. Furthermore, the highly inbred nature of both populations of *C. franciscana* might merely be the normal outcome of habitat specialization and an adaptive gene complex and not a consequence of population decline.

Conclusions

Conservation biology has been called the reunion of applied and theoretical biology. Early biologists were very much concerned about the application of their work to specific industrial problems, but the middle part of this century saw a separation of the more theoretically inclined from their production-oriented counterparts. The recognition of our effect on the abundance of wild species, including those of no direct economic value, has highlighted the need for a more integrative approach to conservation. Genetics has always been within the domain of both sides of this spectrum and has participated fully in a reconciled conservation biology. In particular, recent advances in molecular and population genetics have had an enormous impact on conservation biology and will undoubtedly continue to do so.

The reassessment of basic questions and needs within the field and the integration of new types of data and methods of analysis will continue to advance our understanding of the evolutionary process and how we might best preserve some of the diversity that we seem so adept at exterminating.

ACKNOWLEDGMENTS

The authors thank T. Glenn and S. Weller for helpful comments and discussion. This project was supported in part by NSF grant DEB 9303266 to WJH.

REFERENCES

Akkaya MS, Bhagwat AA, Cregan PB (1992): Length polymorphisms of simple sequence repeat DNA in soybean. *Genetics* 132:1131–1139

Allendorf FW, Leary RF (1986): Heterozygosity and fitness in natural populations of animals. In: *Conservation Biology*, Soule ME, ed. Sunderland, MA: Sinauer Associates

Alberte RS, Suba GK, Procaccini G, Zimmerman RC, Fain SR (1994): Assessment of genetic diversity of seagrass populations using DNA fingerprinting: Implications for population stability and management. *Proc Natl Acad Sci USA* 91:1049–1053

Antonovics J (1984): Genetic variation within populations. In: *Perspectives on Plant Population Biology*, Dirzo R, Sarukhan J, eds. Sunderland, MA: Sinauer Associates

Aquadro CF, Avise JC (1982): An assessment of "hidden" heterogeneity within electromorphs at three loci in deer mice. *Genetics* 102:269–284

Arnheim N, White T, Rainey WE (1990): Application of PCR; organismal and population biology. *BioScience* 40(3):174–182

Arnold ML, Buckner CM, Robinson JJ (1991): Pollen mediated introgression and hybrid speciation in Louisiana irises. *Proc Natl Acad Sci USA* 88:1398–1402

Avise JC (1994): *Molecular Markers, Natural History and Evolution*. New York: Chapman and Hall

Avise JC (1989): A role for molecular genetics in the recognition and conservation of endangered species. *Tr Ecol Evol* 4:279–281

Avise JC, Ball RM Jr (1990): Principles of genealogical concordance in species concepts and biological taxonomy In: *Oxford Surveys in Evolutionary Biology*, Futuyama D, Antonovics J, eds. New York: Oxford University Press

Avise JC, Arnold J, Ball RM, Bermingham E, Lamb T, Neigel JE, Reeb CA, Saunders NC (1987): Intraspecific phylogeography: the mitochondrial DNA bridge between populations genetics and systematics. *Ann Rev Ecol Syst* 18:489–522

Baldwin BG, Sanderson MJ, Porter JM, Wojciechowski MF, Cambell CS, Donoghue MJ (1995): The ITS region of nuclear ribosomal DNA: a valuable source of evidence on angiosperm phylogenyu. *Ann Missouri Bot Gard*: in press

Barret SCH, Kohn JR (1991): Genetic and evolutionary consequences of small population size in plants: implications for conservation. In: *Genetics and Conservation of Rare Plants*, Falk DA, Holsinger KE, eds. New York: Oxford University

Bell CJ, Ecker JR (1994): Assignment of 30 microsatellite loci to the linkage map of *Arabidopsis. Genomics* 19:137–144

Bennett MD, Smith JB, Heslop-Harrison JS (1982): Nuclear DNA amounts in angiosperms. *Phil Trans R Soc Lond, B.* 216:179–182

Bennetzen JL, Freeling M (1993): Grasses as a single genetic system: Genome composition, collinearity and compatibility. *Tr Gen* 9:259–261

Bijlsma R, Ouborg NJ, van Treuren, R (1994): On genetic erosion and population extinction in plants: A case study in *Scabiosia columbaria* and *Salvia pratensis*. In: *Conservation Genetics*, Loeschcke V, Tomiuk J, Jain SK, eds. Basel: Birkhäuser Verlag

Brauner S, Crawford DJ, Stuessy TF (1992): Ribosomal and RAPD variation in the rare plant family Lactoridaceae. *Amer J Bot* 79:1436–1439

Brubaker CL, Wendel JF (1994): Reevaluating the origin of domesticated cotton (*Gossypium hirsutum*; Malvaceae) using nuclear restriction fragment length polymorphisms (RFLPs). *Amer J Bot* 81(10):1309–1326

Bruford MW, Wayne RK (1994): Microsatellites and their application to population genetic studies. *Curr Opin Gen Dev* 3:939–943

Bulmer M (1991): *The Mathematical Theory of Quantitative Genetics*. Oxford: Oxford University Press

Case, MA (1993): High levels of allozyme variation within *Cypripedium calceolus* (Orchidaceae) and low levels of divergence among its varieties. *Syst Bot* 18(4):663–677

Clegg, MT (1988): Molecular diversity in plant populations. In: *Plant Population Genetics, Breeding, and Genetic Resources*, Brown AHD, Clegg MT, Kahler AL, Weir BS, eds. Sunderland, MA: Sinauer Associates

Condit R, Hubbell SP (1991): Abundance and DNA sequence of two-base repeat regions in tropical tree genomes. *Genome* 34:66–71

Crawford DJ, Brauner S, Cosner MB, Stuessy TF (1993): Use of RAPD markers to document the origin of the intergeneric hybrid *Margyracena skottsbergii* (Rosaceae) on the Juan Fernandez Islands. *Amer J Bot* 80(1):89–92

Crawford DJ, Stuessy TF, Cosner MB, Haines DW, Wiens D, Penaillo P (1994): *Lactoris fernandeziana* (Lactoridaceae) on the Juan Fernandez Islands: allozyme uniformity and field observations. *Cons Biol* 8(1):277–280

Crothers BI (1992): Genetic characters, species concepts, and conservation biology. *Cons Biol* 6:314

Dallas JF (1988): Detection of DNA "fingerprints" of cultivated rice by hybridization with a human minisatellite DNA probe. *Proc Natl Acad Sci USA* 85:6831–6835

Dole JA, Sun M (1992): Field and genetic survey of the endangered Butte County meadowfoam—*Limnanthes floccosa* susp. *californica* (Limnanthes). *Cons Biol* 6:549–558

Dowling TE, Minckley WL, Douglas ME, Marsh PC, Demarais BD (1992): Response to Wayne, Novak and Henry: use of molecular characters in conservation biology. *Cons Biol* 6:600–603

Downie SR, Palmer JD (1992): Use of chloroplast DNA rearrangements in reconstructing plant phylogeny. In: *Molecular Systematics of Plants*, Soltis PS, Soltis DE, Doyle JJ, eds. New York: Chapman and Hall

Dvorak J (1988): Evolution of multigene families: the ribosomal RNA loci of wheat and related species: In: *Plant Population Genetics, Breeding, and Genetic Resources*, Brown AHD, Clegg MT, Kahler AL, Weir BS, eds. Sunderland, MA: Sinauer Associates

Dweikat I, Mackenzie S, Levy M, Ohm H (1993): Pedigree assessment using RAPD-DGGE in cereal crop species. *Theor Appl Genet* 85:497–505

Echt CS, Erdahl LA, McCoy TJ (1992): Genetic segregation of random amplified polymorphic DNA in diploid cultivated alfalfa. *Genome* 35:84–87

Edwards A, Civitello H, Hammond HA, Caskey CT (1991): DNA typing and genetic mapping with trimeric and tetrameric tandem repeats. *Am J Hum Genet* 49:746–756

Ellstrand NC, Elam DR (1993): Population genetic consequences of small population size: implications for plant conservation. *Ann Rev Ecol Syst* 24:217–242

Ellsworth DL, Rittenhouse KD, Honeycutt RL (1993): Artifactual variation in randomly amplified polymorphic DNA banding patterns. *BioTechniques* 14:214–217

Erwin T (1991): An evolutionary basis for conservation strategies. *Science* 253:750–752

Fain SR, DeTomaso A, Alberte RS (1992): Characterization of disjunct populations of *Zostera marina* (eelgrass) from California: genetic differences resolved by restriction-fragment length polymorphisms. *Mar Biol* 112: 683–689

Falconer DS (1981): *Introduction to Quantitative Genetics*. New York: Longman

Falk DA, Holsinger KE (1991): *Genetics and Conservation of Rare Plants*. New York: Oxford University Press

Fenster CB, Dudash MR (1994): Genetic considerations for plant population restoration and conservation. In: *Restoration of Endangered Species*, Bowles ML, Whelan CJ, eds. Cambridge: Cambridge University Press

Fenster CB, Ritland K (1992): Chloroplast DNA and isozyme diversity in two *Mimulus* species (Scrophulariaceae) with contrasting mating systems. *Amer J Bot* 79(12):1440–1447

Flavell R (1980): The molecular characterization and organization of plant chromosomal DNA sequences. *Ann Rev Pl Physiol* 31:569–596

Fleisher RC, Tarr CL, Pratt TK (1994): Genetic structure and mating system in the palila, and endangered Hawaiian honeycreeper, as assessed by DNA fingerprinting. *Mol Ecol* 3:383–392

Fritz RS, Simms EL, eds. (1992): *Plant Resistance to Herbivores and Pathogens. Ecology, Evolution, and Genetics*. Chicago: University of Chicago Press

Gottlieb LD (1973): Enzyme differentiation and phylogeny in *Clarkia franciscana, C. rubicunda*, and *C. amoena*. *Evolution* 27:205–214

Gottlieb LD (1981): Electrophoretic evidence and plant populations. *Prog Phytochem* 7:1–46

Gottlieb L (1982): Conservation and duplication of isozymes in plants. *Science* 216:373–380

Gottlieb LD, Edwards SW (1992): An electrophoretic test of the genetic independence of a newly discovered population of *Clarkia franciscana. Madroño* 39(1):1–7

Guerrant EO Jr (1986): Genetic and demographic considerations in the sampling and reintroduction of rare plants. In: *Conservation Biology. The Theory and Practice of Nature Conservation, Preservation and Management*, Fiedler PL, Jain SK, eds. New York: Chapman and Hall

Hahn WJ, Karol K, Sytsma KJ (1993): Nuclear ribosomal internal transcribed spacer phylogenetics of the genus *Clarkia* (Onagraceae). *Amer J Bot* 80(suppl):152

Halldén C, Nilsson N-O, Rading IM, Säll T (1994): Evaluation of RFLP and RAPD markers in a comparison of *Brassica napus* breeding lines. *Theor Appl Genet* 88:123–128

Hamilton MB (1994): *Ex situ* conservation of wild plant species: time to reassess the genetic assumptions and implications of seed banks. *Cons Biol* 8(1):39–49

Hamrick JL, Godt MJW (1989): Allozyme diversity in plant species. In: *Plant Population Genetics, Breeding, and Genetic Resources*, Brown AHD, Clegg MT, Kahler AL, Weir BS, eds. Sunderland, MA: Sinauer Associates

Hamrick JL, Godt MJW, Murawski DA, Loveless MD (1991): Correlations between species traits and allozyme diversity: Implications for conservation biology. In: *Genetics and Conservation of Rare Plants*. Falk DA, Holsinger KE, eds. New York: Oxford University Press

Harborne JB (1988): *Introduction to Ecological Biochemistry*, 3rd ed. London: Academic Press

Harrison RG (1989): Animal mitochondrial DNA as a genetic marker in population and evolutionary biology. *Tr Ecol Evol* 4:6–11

Hedrick P (1992): Shooting the RAPD's. *Nature* 355:679–680

Hedrick PW, Brussard PF, Allendorf FW, Beardmore JA, Orzack S (1986): Protein variation, fitness, and captive propagation. *Zoo Biol* 5:91–99

Helentjaris T, King G, Slocum M, Siedenstrang C, Wegman S (1985): Restriction fragment length polymorphisms as probes for plant diversity and their development as tools for applied plant breeding. *Pl Mol Biol* 5:109–118

Helenurm K, Ganders FR (1985): Adaptive radiation and genetic differentiation in Hawaiian *Bidens. Evolution* 39:753–765

Heun M, Helentjaris T (1993) Inheritance of RAPDs in F1 hybrids of corn. *Theor Appl Genet* 85:961–968

Hickey RJ, Vincent MA, Guttman SI (1991): Genetic variation in running buffalo clover (*Trifolium stoloniferum*, Fabaceae). *Cons Biol* 5:309–316

Holsinger KE, Jansen RK (1993): Phylogenetic analysis of restriction site data. In: *Molecular Evolution: Producing the Biochemical Data. Methods in Enzymology*, Vol 224, Zimmer EA, White TJ, Cann RL, Wilson AC, eds. New York: Academic Press van Houten W, van Raamsdonk L, Bachman K (1994): Intraspecific evolution of *Microseris pygmaea* (Asteraceae, Lactuceae) analyzed by cosegregation of phenotypic characters (QTLs) and molecular markers. *Pl Syst Evol* 190:49–67

Huenneke LF (1991): Ecological implications of genetic variation in plant populations. In: *Genetics and Conservation of Rare Plants*, Falk DA, Holsinger KE, eds. New York: Oxford University Press

Jacob HJ, Lindpainter K, Lincoln SE, Kusumi K, Bunker RK, Mao Y-P, Genten D, Dzau VJ, Lander ES (1991): Genetic mapping of a gene causing hypertention in the stroke-prone spontaneously hypertensive rat. *Cell* 67:213–224

Jeffreys AJ, Wilson V, Thein SL (1985): Hypervariable 'minisatellite' regions in human DNA. *Nature* 314:67–73

Johnson LA, Soltis DE (1994): matK DNA sequences and phylogenetic reconstruction in Saxifragaceae s. str. *Syst Bot* 19(1):143–156

Karl SA, Bowen BW, Avise JC (1992): Global population genetic structure and male-mediated gene flow in the green turtle (*Chelonia mydas*): RFLP analyses of anonymous nuclear loci. *Genetics* 131:163–173

Karron JD (1991): Patterns of genetic variation and breeding systems in rare plant species. In: *Genetics and Conservation of Rare Plants*. Falk DA, Holsinger KE, eds. New York: Oxford University Press

Keim P, Paige KW, Whitham TG, Lark KG (1989): Genetic analysis of an interspecific hybrid swarm of *Populus*: Occurrence of unidirectional introgression. *Genetics* 123:557–565

Kellogg EA, Birchler JA (1994): Linking phylogeny and genetics: *Zea mays* as a tool for phylogenetic studies. *Syst Biol* 42(4):415–439

Knapp SJ (1994): Mapping quantitative trait loci. In: *DNA-Based Markers in Plants*, Phillips RL, Vasil IK, eds. Dordrecht: Kluwer Academic Publishers

Kunin WE, Gaston KJ (1993): The biology of rarity: Patterns, causes, and consequences. *Tr Ecol Evol* 8(8):298–301

Lande R (1976): The maintenance of genetic variability by mutation in a polygenic character with linked loci. *Genet Res* 26:221–235

Lande R (1977): The influence of the mating system on the maintenance of genetic variability in polygenic characters. *Genetics* 86:485–498

Lande R (1980): Genetic variation and phenotypic evolution during allopatric speciation. *Am Nat* 116:463–479

Lande R (1988): Genetics and demography in biological conservation. *Science* 241:1455–1460

Lapitan NLV (1992): Organization and evolution of higher plant nuclear genomes. *Genome* 35:171–181

Learn GH, Schaal BA (2987): Population subdivision for ribosomal DNA repeat variants in *Clematis fremontii*. *Evolution* 41:433–438

Lesica P, Leary RF, Allendorf FWA, Bilderback DE (1988): Lack of genic diversity within and among populations of an endangered plant, *Howellia aquatilis*. *Cons Biol* 2:275–282

Lessa EP (1993): Analysis of DNA sequence variation at population level by polymerase chain reaction and denaturing gradient gel electrophoresis. In:

Molecular Evolution: Producing the Biochemical Data. Methods in Enzymology, Vol 224, Zimmer EA, White TJ, Cann RL, Wilson AC, eds. New York: Academic Press

Lewis H, Raven PH (1958): Rapid evolution in *Clarkia. Evolution* 12:319–336

Liston A, Rieseberg LH, Hanson MA (1992): Geographic partitioning of chloroplast DNA varation in the genus *Datisca* (Datiscaceae). *Pl Syst Evol* 181:121–132

Lynch M, Milligan BG (1994): Analysis of population genetic structure with RAPD markers. *Mol Ecol* 3:91–99

MacCauley DE (1994): Contrasting the distribution of chloroplast DNA and allozyme polymorphism among local populations of *Silene alba*: Implications for studies of gene flow among plants. *Proc Natl Acad Sci USA* 91:8127–8131

Markert CL, Moller F (1959): Multiple forms of enzymes: tissue, ontogenetic and species specific pattern. *Proc Natl Acad Sci USA* 45:753–763

Mayer MS, Soltis PS (1994): The evolution of serpentine endemics: A chloroplast DNA phylogeny of the *Streptanthus glandulosus* complex (Cruciferae). *Syst Bot* 19(4):557–574

Mayer MS, Soltis PS, Soltis DE (1994): The evolution of the *Streptanthus glandulosus* complex (Cruciferae): Genetic divergence and gene flow in serpentine endemics. *Amer J Bot* 81:1288–1299

McGrath JM, Quiros CF (1992): Genetic diversity at isozyme and RFLP loci in *Brassica campestris* as related to crop type and geographical origin. *Theor Appl Genet* 83:783–790

Miller JC, Tanksley SD (1993): Effect of different restriction enzymes, probe source, and probe length in detecting restriction fragment length polymorphism in tomato. *Theor Appl Genet* 80:385–389

Milligan BG, Leebens-Mack J, Strand AE (1994): Conservation genetics: beyond the maintenance of marker diversity. *Mol Ecol* 3:423–435

Mitton JB (1989): Physiological and demographic variation associated with allozyme variation. In: *Isozymes in Plant Biology*. Soltis DE, Soltis PS, eds. Portland, OR: Dioscorides Press

Mitton JB (1994): Molecular approaches to population biology. *Ann Rev Ecol Syst* 25:45–69

Morgante M, Olivieri AM (1993): PCR-amplified microsatellites as markers in plant genetics *Plant J* 3:175–182

Moritz C (1994): Application of mitochondrial DNA analysis in conservation: a critical review. *Mol Ecol* 3:401–411

Moritz C, Dowling TE, Brown WM (1987): Evolution of animal mitochondrial DNA: relevance for population biology and systematics. *Ann Rev Ecol Syst* 18:269–292

Moss DW (1982): *Isoenzymes*. New York: Chapman and Hall

Nei M (1987): *Molecular Evolutionary Genetics*. New York: Columbia University Press

Newbury HJ, Ford-Lloyd BV (1993): The use of RAPD for assessing variation in plants. *Pl Growth Regul* 12:43–51

O'Brien SJ (1994): A role for molecular genetics in biological conservation. *Proc Natl Acad Sci USA* 91:5748–5755

O'Brien SJ (1994): Genetic and phylogenetic analyses of endangered species. *Ann Rev Genet*: in press

O'Brien SJ, Mayr E (1991): Bureaucratic mischief: Recognizing endangered species and subspecies. *Science* 251:1187–1188

O'Brien SJ, Roelke ME, Marker L, Newman A, Winkler CA, Meltzer D, Colly L, Evermann JF, Bush M, Wildt DE (1985): Genetic basis for species vulnerability in the cheetah. *Science* 227:1428–1434

Olmstead RG (1990): Biological and historical factors influencing genetic diversity in the *Scutellaria angustifolia* complex (Labiatae). *Evolution* 44:54–70

Olmstead RG, Palmer JD (1994): Chloroplast DNA systematics: A review of methods and data analysis. *Amer J Bot* 81(9):1205–1224

Palmer JD (1992): Mitochondrial DNA in plant systematics: applications and limitations. In: *Molecular Systematics of Plants*, Soltis PS, Soltis DE, Doyle JJ, eds. New York: Chapman and Hall

Palmer JD, Jansen RK, Michaels HJ, Chase MW, Manhart JR (1988): Chloroplast DNA variation and plant phylogeny. *Ann Missouri Bot Gard* 75:1180– 1208

Prakash S, Lewontin SC, Hubby JL (1969): A molecular approach to the study of genic heterozygosity in natural populations. IV. Patterns of genic variation in central, marginal and isolated populations in *Drosophila pseudobscura. Genetics* 61:841–858

Rabinowitz D (1981): Seven forms of rarity. In: *The Biological Aspects of Rare Plant Conservation*, Synge H, ed. New York: Wiley & Sons

Rafalski JA, Tingey SV (1993): Genetic diagnostics in plant breeding: RSPDs, microsatellites and machines *Tr Gen.* 9(8):275–280

Ranker TA (1992): Genetic diversity of endemic Hawaiian epiphytic ferns: implications for conservation. *Selbyana* 13:131–137

Reiter RS, Williams JGK, Feldman KA, Rafalski JA, Tingey SV, Scolnik PA (1992): Global and local genome mapping in *Arabidopsis thaliana* by using recombinant inbred lines and random amplified polymorphic DNAs. *Proc Natl Acad Sci USA* 89:1477–1481

Richter TS, Soltis PS, Soltis DE (1994): Genetic variation within and among populations of the narrow endemic *Delphinium viridescens* (Ranunculaceae). *Amer J Bot* 81(8):1070–1076

Riedy MF, Hamilton WJ, Aquadro CF (1992): Escess of non-parental bands in offspring from known primiate pedigrees assayed using RAPD PCR. *Nucl Acids Res* 20:918

Rieseberg LH (1991): Hybridization in rare plants: Insights from case studies in *Cercocarpus* and *Helianthus* In: *Genetics and Conservation of Rare Plants*, Falk DA, Holsinger KE, eds. New York: Oxford University Press

Rieseberg LH, Wendel JF (1993): Introgression and its consequences in plants. In: *Hybrid Zones and the Evolutionary Process*, Harrison RG, ed. New York: Oxford University Press

Rogstad SH, Patton JC, Schaal BA (1988): M13 repeat probe detects DNA minisatellite-like sequences in gymnosperms and angiosperms. *Proc Nat Acad Sci USA* 85:9176–9178

Rogstad SH (1993): Surveying plant genomes for variable number of tanden repeat loci. In: *Molecular Evolution: Producing the Biochemical Data. Methods in Enzymology*, Vol. 224, Zimmer EA, White TJ, Cann RL, Wilson AC, eds. New York: Academic Press

Rojas M (1992): The species problem and conservation: What are we protecting? *Cons Biol* 6:170–178

Schaal BA, Learn GH, Jr (1988): Ribosomal DNA variation within and among plant populations. *Ann Missouri Bot Gard* 75:1207–1216

Schaal BA, Leverich WL, Rogstad SH (1991a): Comparison of methods for assessing genetic variation in plant conservation biology. In: *Genetics and Conservation of Rare Plants*, Falk DA, Holsinger KE, eds. New York: Oxford University Press

Schaal BA, O'Kane SL, Rogstad SH (1991b): DNA variation in plant populations. *Tr Ecol Evol* 6:329–333

Schonewald-Cox CM, Chambers SM, MacBryde B, Thomas L (1983): *Genetics and Conservation. A Reference for Managing Wild Animal and Plant Populations*. Menlo Park, CA: Benjamin/Cummings

Smith JJ, Scott-Craig JS, Leadbetter JR, Bush GL, Roberts DL, Fulbright DW (1994): Characterization of random amplified polymorphic DNA (RAPD) products from *Xanthomonas campestris* and some comments on the use of RAPD products in phylogenetic analysis. *Mol Phyl Evol* 3(2):135–145

Sobral BWS, Honeycutt RJ (1994): Genetics, plants, and the polymerase chain reaction. In: *The Polymerase Chain Reaction*, Mullis KB, Ferré F, Gibbs RA, eds. Boston: Birkhauser.

Soltis DE, Soltis PS (1989): *Isozymes in Plant Biology*. Portland OR: Dioscorides Press

Soltis DE, Soltis PS, Milligan B (1992): Intraspecific chloroplast DNA variation: systematic and phylogenetic implications. In: *Plant Molecular Systematics*, Soltis PE, Soltis DE, Doyle JJ, eds. New York: Chapman and Hall

Soltis PS, Soltis DE (1991): Genetic variation in endemic and widespread plant species: Examples from Saxifragaceae and *Polystichum* (Dryopteridaceae). *Aliso* 13:215–223

Soltis PS, Soltis DE, Tucker TL, Lang FA (1992): Allozyme variability is absent in the narrow endemic *Bensoniella oregona* (Saxifragaceae). *Cons Biol* 6:131–134

Steele KP, Vilgalys R (1994): Phylogenetic analyses of Polemoniaceae using nucleotide sequences of the plastid gene *mat*K. *Syst Bot* 19(1):126–142

Systma KJ, Hahn WJ (1994): Molecular Systematics: 1991–1993. pp. 307–333 In: *Progess in Botany*. vol. 55, Behnke H-D, Lüttge U, Esser K, Kadereit JW, Runge M, eds. Heidelberg: Springer Verlag

Sytsma KJ, Smith JF (1992): Molecular systematics of Onagraceae: Examples from *Clarkis* and *Fuchsia*. In: *Plant Molecular Systematics*, Soltis PE, Soltis DE, Doyle JJ, eds. New York: Chapman and Hall

Sytsma KJ, Schaal BA (1985): Phylogenetics of the *Lisianthus skinneri* (Gentianaceae) species complex in Panama utilizing DNA restriction site fragment analysis. *Evolution* 39:594–608

Tautz D, Renz M (1984): Simple sequences are ubiquitous repetitive components of eukaryotic genomes. *Nucl Acids Res* 12:4127–4138

Tanksley SD, Orton TJ (1983): *Isozymes in Plant Breeding and Genetics*. Amsterdam: Elsevier Science Publishers

Templeton AR (1986): Coadaptation and outbreeding depression. In: *Conservation Biology*, Soule M, ed. Sunderland MA: Sinauer Associates

Terauchi R (1990): Genetic diversity and population structure of Dioscorea todoro Makino, a dioecious climber. *Pl Species Biol* 5:243–253

Terauchi R, Konuma A (1994): Microsatellite polymorphism in *Dioscorea tokoro*, a wild yam species. *Genome* 37:794–801

Tingey SV, DelTufo JP (1992): Genetic analysis with Random Amplified Polymorphic DNA markers. *Pl Physiol* 101:349–352

Thormann CE, Ferreira ME, Camargo LEA, Tivang JG, Osborn TC (1994): Comparison of RFLP and RAPD markers to estimating genetic relationships within and among cruciferous species. *Theor Appl Genet* 88:973–980

Vane-Wright RI, Humphries DJ, Williams PH (1991): What to protect? Systematics and the agony of choice. *Biol Cons* 55:235–254

Vogler AP, DeSalle R (1994a): Evolution and phylogenetic information content of the ITS-1 region in the Tiger Beetle *Cicindela dorsalis*. *Mol Biol Evol* 11(3):393–405

Vogler AP, DeSalle R (1994b): Diagnosing units of conservation management. *Cons Biol* 8(2):354–363

Walker RK (1992): Biodiversity and ecological redundancy. *Cons Biol* 6:18–23

Waller DM, O'Malley DM, Gawler SC (1987): Genetic variation in the extreme endemic *Pedicularis furbishiae* (Scrophulariaceae). *Cons Biol* 1:335–340

Wang Z, Weber JL, Zhong G, Tanksley SD (1994): Survey of plant short tandem DNA repeats. *Thor Appl Genet* 88:1–6

Weeden NF, Wendel JF (1989): Genetics of plant isozymes. In: *Isozymes in Plant Biology*, Soltis DE, Soltis PS, eds. Portland OR: Dioscorides Press

Weir BS (1990): *Genetic Data Analysis*. Sunderland MA: Sinauer Associates

Weisling K, Wolff K, Nybom H, Meyer W (1995): *DNA Fingerprinting in Plants and Fungi*. Boca Raton FL: CRC Press

Weller SJ (1994): The relationship of rarity to plant reproductive biology. In: *Restoration of Endangered Species*, Bowles ML, Whelan CJ, eds. Cambridge: Cambridge University Press

Welsh J, McClelland M (1990): Fingerprinting genomes using PCR with arbitrary primers. *Nucl Acids Res* 18:7213–7218

Wendel JF, Weeden NF (1989): Visualization and interpretation of plant isozymes. In: *Isozymes in Plant Biology*, Soltis DE, Soltis PS, eds. Portland OR: Dioscorides Press

Whitkus R, Doebley J, Wendel JF (1994): Nuclear DNA markers in systematics and evolution. In: *DNA-Based Markers in Plants*, Phillips RL, Vasil IK, eds. Dordrecht: Kluwer Academic Publishers

Williams JGK, Kubelik AR, Livak KJ, Rafalski JA, Tingey SV (1990): DNA polymorphisms amplified by arbitrary primers are useful as genetic markers. *Nucl Acids Res* 18:6531

Wolff K, Rogstad SH, Schaal BA (1994). Population and species variation of minisatellite DNA in *Plantago. Theor Appl Genet* 87:733–740

Woodruff DS (1989): The problems of conserving genes and species. In: *Conservation for the Twenty-First Century*, Western D and Pearl M, eds. New York: Oxford University Press

Wright JM (1994): Mutation at VNTRs: Are minisatellites the evolutionary progeny of microsatellites? *Genome* 37:45–347

Yuhki N, O'Brien SJ (1990): DNA variation of the mammalian major histocompatibility complex reflects genomic diversity and population history. *Proc Natl Acad Sci USA* 87:836–840

8

Identifying Links Between Genotype and Phenotype Using Marker Loci and Candidate Genes

KEITH A. CRANDALL

Introduction

The elucidation of the causal links between variation at the genetic level and the emergent phenotype has been at the heart of theoretical and empirical studies in quantitative and population genetics since the rediscovery of Mendel's work and the synthesis of Mendelism, Darwinism, and Biometry through population genetics (Provine, 1971). Wright (1980, 1982) has summarized the historical emergence of four predominant interpretations to the relationship between genotype and phenotype (Figure 1). The first relationship (Figure 1A) represents a one to one mapping of genotype to phenotype. This relationship applies to genes of major effect, the traditional Mendelian view of inheritance (Haldane, 1932). Kimura's (1983) neutral theory of evolution is depicted in relationship (Figure 1C), where variation at the genotype has no relationship to variation seen at the phenotypic level. Figures 1B and 1D represent the predominant views of the relationships between genotype and phenotype. Relationship (1B) represents multiple genes with minor additive effects on phenotypic variation, a view forcefully put forth by Fisher (1930). Relationship (1D) represents genic relationships with pleiotropic effects producing a maze of interaction influencing the phenotype (Wright, 1931; Wright, 1932). This chapter outlines the uses of molecular techniques to identify loci, genes, alleles, and mutations associated with variation at the phenotypic level.

One major approach to the study of the genotype/phenotype relationship has been the unmeasured genotype approach. With this approach, the locus or loci that affect the phenotype are unknown, and the researcher attempts to infer links

The Impact of Plant Molecular Genetics
BWS Sobral, Editor

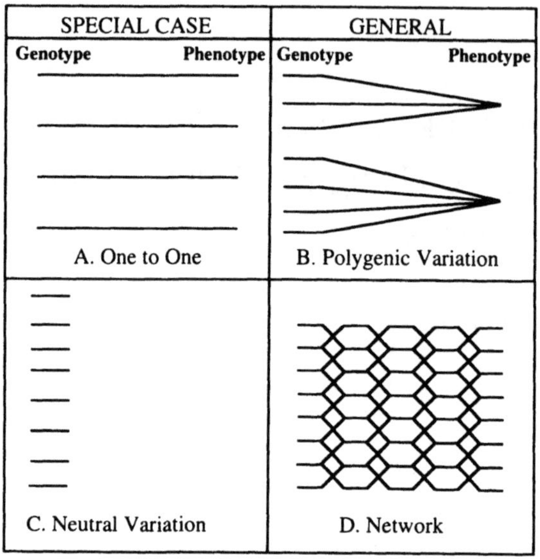

Figure 1. Four main interpretations of the relationship between genotype and phenotype (Wright, 1980; Wright, 1982). Reprinted with permission of *Evolution*.

via correlation between relatives, a response to selection, or by the use of hybridization and controlled crosses. Basic to this approach is the assumption that phenotypes are normally distributed; therefore, they can be completely characterized by a mean and variance. Thus, the relationship between phenotype and genotype is defined as;

$$P_{ij} = \mu + g_i + E_j \qquad [1]$$

where P_{ij} is the phenotype of the ith genotype in the jth environment, μ is the population mean, g_i is the genotypic deviation and E_j is the environmental deviation (Falconer, 1989). The genotypic value (G_i) is then the average phenotype of genotype i or $G_i = \mu + g_i$. This relationship describes the association of phenotype and genotype for an individual; however, in a sexually mating population, genes, rather than genotypes, are passed on to the next generation. Thus, the genotypic value must be placed within a population context, referring to genes, and not to genotypes. The average effect of an allele accomplishes this by taking into account the genotypic values as well as gene frequencies. For the single locus, two allele model; $\alpha = a + d(q - p)$, where α is the average effect, a is the genotypic value (or additive effect), d is the dominance deviation, q is the frequency of one allele, and p is the frequency of the other allele. The relationship among these values is depicted in Table 1.

When the locus or loci affecting a given phenotype is unknown, the unmeasured genotype approach offers three basic modes of analysis to gain further information to add to the basic quantitative genetic model given in equation [1]: (1) correlation between relatives; (2) response to selection; and (3) hybridization and controlled crosses (Fisher, 1918; Wright, 1921; Falconer, 1989). While insightful in many ways, the unmeasured genotype approach has a number of limitations (Boerwinkle et al, 1986). The major limitation to this approach is that it offers little insight into the genetic architecture underlying the studied phenotype (Figure 2). The genetic architecture of a quantitative trait refers to the number of genes involved in the manifestation of a given phenotype, the number of functional alleles at each gene and their relative frequencies, the arrangement of these alleles into genotypes, and the impact of alleles and genotypes on the trait of interest and other traits (Boerwinkle et al, 1986; Sing et al, 1988; Sing et al, 1992a; Haviland et al, 1995a). These questions can be addressed using the measured genotype approach to study quantitative trait loci (QTL) (Boerwinkle et al, 1986; Sing et al, 1988).

The Measured Genotype Approach

There are two basic subapproaches within the measured genotype approach used to identify causal links between the QTL and phenotypic trait of interest, (1) the marker locus approach, and (2) the candidate gene approach. These two methods detect associations at two distinct levels of the genetical hierarchy (Figure 2). The marker locus approach identifies genetic markers linked to gene regions associated with a phenotypic change in the trait of interest. Here, the genotypes measured are just markers and have no effect on the phenotype of interest. The candidate gene approach, on the other hand, identifies, a priori, a gene region with presumed functional relationship to the phenotype under consideration. This gene is then surveyed for genetic variation to identify functional alleles whose average effects can be calculated. In addition to the complementarity of application in the genetical hierarchy, the two methods are also complementary from practical standpoints (Cheverud and Routman, 1993; Routman and Cheverud, 1994). The candidate gene approach is applicable to natural populations without the requirement of interspecific crosses or intraspecific crosses of very divergent phenotypes, as is the case with the marker locus approach. The marker locus approach allows for the identification of new loci affecting a quantitative trait, whereas the candidate gene approach restricts inference to the physiologically relevant gene under consideration. Thus, these two approaches are complemen-

Table 1. The Relationship Between Average Effects and Genotypic Values for a Single Locus Two Allele Model

Genotype	AA	Aa	aa
G_i	a	d	$-a$
frequency	p^2	$2pq$	q^2
Average Effect (α_A)		$q[a + d(q - p)]$	
α_a		$-p(a + d(q - p)]$	

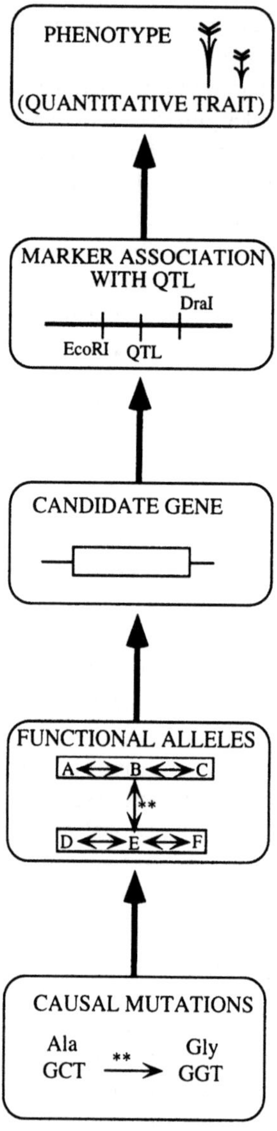

Figure 2. The hierarchical relationship of the genetic architecture underlying a phenotypic (quantitative) trait.

tary at many levels with the limitations of one method being the strengths of the alternative (Cheverud and Routman, 1993; Routman and Cheverud, 1994).

The Marker Locus Approach

The underlying principle of the marker locus approach is to saturate the genome with markers in hopes that a marker will be tightly linked to a gene region responsible for variation in the quantitative trait (Tanksley, 1993). This linkage disequilibrium results in a correlation between the marker genotypes and the quantitative trait values. The effectiveness of the marker locus approach depends on the resolution of the linkage map. With recent advances in molecular technologies, it has become feasible to saturate genomes with molecular markers and thereby provide high resolution linkage maps for a variety of organisms (Table 2). For example, Tanksley et al (1992) recently published molecular linkage maps for potato and tomato with an average spacing between markers of 1.2 cM (~900kb). A variety of molecular techniques have been used to generate linkage maps for genomes or specific gene regions, including: restriction fragment length polymorphisms (RFLPs) (Causse et al, 1994); simple sequence repeat polymorphisms (SSRPs or microsatellites) (Ellegren et al, 1994); random amplified polymorphic DNA (RAPD) (Philipp et al, 1994); and restriction landmark genomic scanning (RLGS) (Hayashizaki et al, 1994). These methods and their relative strengths and weaknesses have recently been reviewed by Routman and Cheverud (1994).

STATISTICAL MODELS

Once a linkage map has been constructed and a quantitative trait of interest identified, statistical procedures are used to indicate significant correlations between maker loci and the QTL. The analytical procedures fall into two categories: (1) linear regression models (Soller et al, 1976; Cowen, 1989; Haley and Knott, 1992; Martinez and Curnow, 1992; Moreno-Gonzalez, 1992; Jansen, 1993; Rodolphe and Lefort, 1993; Zeng, 1993; Haley et al, 1994; Jansen and Stam, 1994; Zeng, 1994); and (2) maximum likelihood models (Weller, 1986; Weller, 1987; Lander and Botstein, 1989; Knapp et al, 1990; Knott and Haley, 1992a, 1992b). The traditional methods of detecting QTL via linear regression models (Soller et al, 1976), while meeting with limited success, had many difficulties, including: (1) under-estimation of phenotypic effect of the QTL due to recombination; (2) lack of statistical power to detect the QTL with small samples; (3) lack of definitive positioning of the QTL; and (4) no correction for multiple comparisons, i.e., the type I error rate is inflated (Lander and Botstein, 1989). Because of these difficulties Lander and Botstein (1986, 1989) developed interval mapping based on a maximum likelihood model. Let A and B be inbred strains differing for a quantitative trait and let B_1 be the backcross performed with A as the recurrent parent. Then, the following linear model was used to test for a QTL located on an interval of markers for a backcross population:

$$P_j = a + bg_j + e_j \quad \text{for } j = 1,2, \ldots ,n \qquad [2]$$

where P_j is the trait value (phenotype) of the jth individual in the population, g_j is the indicator variable (0, 1) equal to the number of B alleles, e_j is a random residual variable for the jth individual with mean 0 and variance σ^2, a is the population mean (a parameter), and b is the effect of the putative QTL expressed as a difference in effects between the homozygote and heterozygote (Lander and Botstein, 1989). The likelihood estimation procedure then maximizes the probability $L(a,b,\sigma^2)$. The null hypothesis of no QTL linked to the marker locus (H_0:

Table 2. Linkage Maps in Plants Covering Organismal Genomes

Organism	Common Name	Linkage Groups	Markers	Reference*
Phaseolus vulgaris	bean	11	RFLP, isozyme, seed protein, color	1
Lycopersicon esculentum x L. pennellii	tomato	12	RFLP, isozyme, morphology	2
Solanum tuberosum x S. berthaultii	potato	12	RFLP, isozyme, morphology	2
Arabidopsis thaliana	—		RAPD	3
Lactuca sativa	lettuce	13	RFLP, RAPD, isozyme, resistance, morphology	4
Secale cereale	rye	8	RFLP, RAPD, isozyme, morphology, physiology	5
Pinus taeda	loblolly pine	20	RFLP	6
Cucumis sativus	cucumber	10	RFLP, RAPD, isozyme, resistance, morphology	7
Hordeum vulgare	barley	7	RFLP, isozyme, morphology	8
Zea mays	corn	11	RFLP, isozyme	9
Brassica oleracea	mustard	11	RFLP, isozyme, morphology	10
Pinus pinaster	maritime pine	17	Proteins	11
Pisum sativum	pea	7	RFLP, morphology	12
Brassica napus	mustard	11	RFLP	13
Musa acuminata	banana	15	RFLP, RAPD	14
Arachis	peanut	11	RFLP	15
Pinus elliottii	slash pine	13	RAPD	16
Glycine max	soybean	31	RFLP, isozyme, morphology	17
Medicago sativa	alfalfa	10	RFLP	18
Vicia faba	faba bean	11	RFLP, RAPD, isozyme	19
Apium graveoleus	celery	8	RFLP, isozyme, morphology	20
Citrus	citrus	11	RFLP, isozyme	21
Beta vulgaris	sugar beet	9	RFLP, isozyme, morphology	22
Cuphea lanceolata	—	6	RFLP, allozyme	23
Oryza sativa	rice	12	RFLP, RAPD, isozyme, morphology	24
Saccharum spontaneum	sugarcane	64	RFLP, AP-PCR	25

* 1. (Vallejos et al, 1992); 2. (Tanksley et al, 1992); 3. (Reiter et al, 1992); 4. (Kesseli et al, 1994); 5. (Philipp et al, 1994); 6. (Devey et al, 1994); 7. (Kennard et al, 1994); 8. (Kleinhofs et al, 1993); 9. (Gardiner et al, 1993); 10. (Kianian and Quiros 1992); 11. (Gerber et al, 1993); 12. (Ellis et al, 1992); 13. (Landry et al, 1991); 14. (Faure et al, 1993); 15. (Halward et al, 1993); 16. (Nelson et al, 1993); 17. (Lark et al, 1993); 18. (Brummer et al, 1993); 19. (Torres et al, 1993); 20. (Huestis et al, 1993); 21. (Durham et al, 1992); 22. (Pillen et al, 1992); 23. (Webb et al, 1992); 24. (Causse et al, 1994); 25. (Da Silva et al, 1995).

$b = 0$) is then tested against the alternative of linkage (H_a: $b \neq 0$) using the *LOD* score with a predefined threshold value (the difficulties in determining threshold values are discussed below):

$$LOD = \log_{10}\left[\frac{L(\hat{a},\hat{b},\hat{\sigma}^2)}{L(\hat{\mu}_A,0,\hat{\sigma}^2_{B1})}\right]. \quad [3]$$

where $\hat{\mu}_A$ is the estimated mean of the phenotype in strain A and $\hat{\sigma}^2_{B1}$ is the estimated variance of the phenotype in the backcross population. Thus the maximum likelihood estimate, $L(\hat{a},\hat{b},\hat{\sigma}^2)$, is compared to the constrained maximum likelihood estimate, $L(\hat{\mu}_A,0,\hat{\sigma}^2_{B1})$ under the assumption of $b = 0$. When the *LOD* score exceeds the predetermined threshold value, the null hypothesis (H_0: $b = 0$) is rejected, and the existence of a QTL is suggested.

Interval mapping has a number of advantages over the traditional regression methods, including: (1) the method provides a quantitative assessment of possible QTL at various points along the genome; (2) the inferred phenotypic effects are asymptotically unbiased, provided there is only a single QTL on a chromosome; (3) the probable position of the QTL is given by support intervals; and (4) fewer individuals are required to detect QTL (Lander and Botstein, 1989). This method has been successfully applied in a number organisms, especially in agriculture, e.g., Paterson et al (1988) identified six QTL affecting tomato fruit weight, four QTL affecting soluble solids, and five affecting fruit pH.

Recently, however, a number of limitations to the interval mapping techniques have been identified, including: (1) the test procedures become biased if multiple QTL occur on a chromosome (Knott and Haley, 1992a; Martinez and Curnow, 1992); (2) multiple QTL can lead to inappropriate intervals being identified (Zeng, 1994); and (3) the use of only two markers at a time to test for QTL is inefficient (Knapp, 1991; Haley and Knott, 1992; Jansen and Stam, 1994; Zeng, 1994). These difficulties have been addressed in similar ways independently by two research groups (Jansen, 1993; Zeng, 1993; Jansen and Stam, 1994; Zeng, 1994). Their approach is to combine the interval mapping technique with a multiple regression model allowing for simultaneous analysis of multiple QTL while partitioning individual QTL effects. This is accomplished by summing the b effects in equation [2] over the number of ordered markers (t). Thus, the multiple regression model becomes:

$$P_j = a + \sum_{i=1}^{t} b_i g_{ij} + e_j \quad \text{for } j = 1,2,\ldots,n \quad [4]$$

where g_{ij} is the ith marker in the jth individual (Zeng, 1994). This multiple regression model (equation [4]) is then combined with the interval mapping model (equation [2]) to yield a more accurate and efficient mapping method:

$$P_j = a + bg_j + \sum_{k \neq i,i+1} b_k g_{kj} + e_j \quad \text{for } j = 1,2,\ldots,n \quad [5]$$

where b_k is the partial regression coefficient of the phenotype P on the kth marker and g_{kj} is a known coefficient for the kth marker in the jth individual, taking a value 1 or 0 depending on whether the marker type is homozygote or heterozygote (Zeng, 1994). Rejection of the null hypothesis is assessed using a likelihood ratio test similar to that shown in equation [3]. The combination of multiple linear regression with interval mapping has a number of strengths: (1) mapping precision is improved by conditioning on linked markers; (2) the method preserves the likelihood profile to assess relative evidence of QTL at various positions along the genome; and (3) the method can confine the test to one region at a time reducing the dimensionality of the search, making it more efficient and accurate (Zeng, 1994).

DETERMINING THRESHOLDS

While the combination of multiple regression procedures with interval mapping has alleviated some problems associated with the detection of QTL, certain difficulties still remain. These difficulties center around the establishment of threshold values for the acceptance or rejection of the null hypothesis of no linkage based on equation [3]. In statistical hypothesis testing, two types of error can occur: type I error, the rejection of the null hypothesis when it is, in fact, true (i.e., QTL are identified where none exists, false positives) and type II error, the acceptance of the null hypothesis when the alternative is true (i.e., an undetected QTL). There are two major difficulties in applying traditional hypothesis testing to the detection of QTL (Lander and Botstein, 1989). The first is the distribution of the test statistic (equation [3]) is unknown because the conditions that ensure an asymptotic chi-square distribution for the test statistic are not satisfied (Churchill and Doerge, 1994). This effect is compounded by small sample sizes, unknown distributional properties of the quantitative trait, the composition of the genome, and the genetic map density (Darvasi et al, 1993; Churchill and Doerge, 1994). The second major difficulty in establishing thresholds is the problem of multiple comparisons implicit in genome searchers for QTL (Churchill and Doerge 1994; Haley et al, 1994; Jansen and Stam, 1994; Zeng, 1994).

Researchers have taken three approaches in attempting to improve the reliability of threshold values (i.e., to reduce type I and type II errors) in QTL studies. The first is to apply an alternative parametric test statistic. Gerber and Rodolphe (1994), among others (e.g., Carbonell et al, 1992), have suggested the use of a chi-square based test statistic and demonstrate its performance relative to *LOD* scores on a data set from maritime pine. They demonstrate that in this case, the *LOD* scores give conservative threshold values (increasing type II error) and conservative estimates of recombination relative to the chi-square tests. The problem of multiple comparisons on the threshold value is not addressed by Gerber and Rodolphe (1994). Also, it is unclear whether their results are generally applicable to diverse data sets.

The second approach, taken in an attempt to gain more accurate threshold values, has been to use additional markers as cofactors in a multiple regression model (Jansen, 1994; Rebai et al, 1994; Zeng, 1994). This approach shows promise in establishing accurate threshold values when explored via computer simulation. However, they are subject to difficulties pertaining to the number of markers relative to sample size (too many markers with small samples can inflate the threshold level) (Zeng, 1994).

Finally, Churchill and Doerge (1994) have offered a nonparametric permutation based test to determine empirically based threshold levels. The advantage of this method is the lack of assumptions about the underlying distribution of the data. Furthermore, threshold values obtained by this method are limited specifically to the data set on which the permutations are performed, thereby taking into account the specifics of the particular experiment (Churchill and Doerge, 1994).

Currently, there is no clear picture of which method will provide the best thresholds for the diverse experimental situations experienced by researchers. The field of determining appropriate thresholds, and identifying QTL in general, is extremely active, with improved techniques and additional difficulties being discovered regularly. Despite these difficulties, marker locus studies have been very successful in identifying gene regions that associate with quantitative genetic variation (Paterson et al, 1988; Edwards et al, 1992).

The Candidate Gene Approach

While the marker locus approach offers significant insight into the first component of the genetic architecture of phenotypic variation, i.e., the number of genes involved, little insight is gained about the remaining elements. In order to gain knowledge of the number of functional alleles at a gene and the impact of alleles on the trait of interest, an alternative approach is needed; the candidate gene approach. While candidate gene studies are not altogether new (Lusis, 1988), their applicability has been limited to genes associated with well known biochemical pathways, mainly in human genetic studies (Sing and Davignon, 1985; Boerwinkle et al, 1987; Sing et al, 1988; Sing et al, 1992b). Three factors have led to the applicability of the candidate gene approach over a broader range of study systems. The first is the ability to survey specific gene regions for restriction site or nucleotide sequence variation. These molecular techniques offer high-resolution information on variation underlying candidate genes and allow the establishment of cladistic relationships independent of phenotype. Secondly, the accumulation of information on the biochemical and physiological pathways of phenotypes of interest allows for the identification of a broader spectrum of candidate genes. For example, in studies of genetic variability in serum cholesterol levels, more than 30 genes have been identified as candidate genes worthy of genetic analyses (Sing et al, 1988). As mapping studies progress, candidate genes associated with a host of phenotypes in a variety of organisms are being identified (Bonierbale et al, 1994; Kilian et al, 1994; Veldboom et al, 1994). Finally, the analytical tools have been developed that allow the identification and localization of genetic variation associated with phenotypic changes (Templeton et al, 1987, 1988, 1992; Templeton and Sing, 1993).

IDENTIFYING GENOTYPE/PHENOTYPE ASSOCIATIONS

The candidate gene approach begins with the identification of a gene known to influence a quantitative trait by virtue of its role in the biochemical or physiological pathway associated with the phenotype. Alternatively, marker locus studies can be used to identify candidate genes affecting a given phenotype. Once a candidate gene is identified, it is surveyed for genetic variation using RFLP analyses or nucleotide sequencing. This variation is then used to establish evolutionary relationships among haplotypes. Because levels of variation within species are typically low for a given gene region, traditional procedures for estimating phylogenetic relationships perform poorly (Crandall, 1994). Intraspecific data can be subject to a number of phenomena typically ignored by traditional methods of phylogeny reconstruction, e.g., recombination (Crandall et al, 1994). Templeton et al (1992) developed a cladogram estimation procedure based on a statistical assessment of the parsimony criterion which accounts for these intraspecific phenomena, including recombination. This method has its statistical power when few differences separate haplotypes because the method utilizes both shared sites and differences in establishing mutational connections (Crandall, 1994). Once the evolutionary relationships among haplotypes have been estimated, the resulting networks of alternative cladograms provide a statistical framework for testing associations between genotype and phenotype.

Templeton et al (1987) and Templeton and Sing (1993) have developed statistical procedures for detecting significant associations between phenotype and genotype within this cladogram framework. Their procedures use the cladogram structure from the above estimation procedure to define a nested statistical design, thereby allowing the clustering of individuals based on genotype rather than phenotype. The statistical analysis allows ambiguity in the cladogram estimation and is compatible with either quantitative or categorical phenotypes. The central assumption behind this method is that if a mutation causing a phenotypic effect has occurred in the evolutionary history of the population, it would be embedded within the same historical structure represented by the cladogram (Templeton et al, 1987, 1988; Templeton and Sing, 1993). The nesting procedure consists of nesting n-step clades within $(n + 1)$-step clades, where n refers to the number of transitional steps used to define the clade. By definition, each haplotype is a 0-step clade. The $(n + 1)$-step clades are formed by the union of all n-step clades that can be joined together by $n + 1$ mutational steps. The nesting procedure begins with tip clades, i.e., those clades with a single mutational connection and proceeds to interior clades. The nesting procedure results in hierarchical nests with nesting level directly correlated to evolutionary time, i.e., the lower the nesting level the more recent the evolutionary events relative to higher nesting levels. The nesting design can then be used to test for significant associations of phenotype and genotype by either a nested analysis of variance (NANOVA) for continuous data (Templeton et al, 1987) or a permutation chi-squared contingency test for categorical data (Roff and Bentzen, 1989; Templeton and Sing, 1993).

The candidate gene approach provides a statistical framework for exploring the associations of phenotypic variation and the underlying genetic variation in natural populations. Unlike the marker locus approach, interspecific crosses or crosses between phenotypically divergent populations are not necessary to detect the underlying genetic associations. Furthermore, the associations detected are physiologically or biochemically relevant to the phenotypic trait under consideration, yielding direct estimates of genotypic effect at the QTL itself. The disadvantages of this approach are twofold. First, a substantial knowledge of the biochemical or physiological pathway of the phenotype is required to identify appropriate candidate loci. Second, only the genotypic effects of the candidate genes are explored, preventing the discovery of new loci involved in the phenotypic expression of the trait.

This approach has been used most extensively in human population genetics. For example, Haviland et al (1995a) used the cladistic approach to examine associations among variation at the candidate gene low density lipoprotein receptor (LDLR) and variation in three phenotypes, plasma lipid, lipoprotein, and apolipoprotein levels. Their analysis identified three haplotypes with small effects on different plasma lipid traits. These effects were identified as gender specific pleiotropic effects, indicating that pooling by gender can obfuscate underlying associations (Haviland et al, 1995a). Thus the cladistic analysis allows the mapping of multiple haplotypes to multiple traits. Such a strategy recognizes the complexity of the relationship between genotype and phenotype envisioned by Wright (Figure 1D).

The application of this cladistic approach to candidate gene studies has been expanded recently in two ways. First, Templeton (1995) has extended the method to include the analysis of case/control data, a common sampling design in genetic/disease association studies. He demonstrates the method by showing associations between the candidate locus for apoproteins E, CI, and CII and the phenotype of sporadic early and late-onset forms of Alzheimer's disease. Second, Hallman et al (1994) have developed a likelihood based approach to the cladistic analysis for employing family data in the test for association between candidate genes and quantitative phenotypes. This method is demonstrated by identifying effects of the apolipoprotein B (Apo B) gene on total-, low-density-lipoprotein, and high-density-lipoprotein (HDL)-cholesterol, triglyceride, and Apo B levels using haplotypes from 121 French nuclear families. They concluded that 10% of the genetic variance and 5% of the total variance in the HDL-cholesterol and triglyceride levels were associated with haplotype effects at the Apo B locus (Hallman et al, 1994).

AN EXAMPLE FROM THE AMY LOCUS IN DROSOPHILA MELANOGASTER

Using restriction site data from 49 lines of *D. melanogaster* covering a 15kb region encompassing the duplicated *Amy* locus (Langley et al, 1988), Templeton and Sing (1993) examined genetic variation relative to amylase activity. The cladogram estimation procedure indicated that the gene region should be subdivided into three regions, each with no evidence of internal recombination, but with recombination between subregions (Templeton et al, 1992). The resulting cladograms for the three subregions associated nesting designs are given in Figure 3a, 3b, and 3c. The first subregion, shown in Figure 3a, represents approximately 6kb 5′ of the left duplicated *Amy* locus (see restriction map in Langley et al, 1988). The nesting procedure for this region produces groupings up to the two-step level. A nested analysis of variance of amylase activity in this subregion indicates no significant associations (Templeton and Sing, 1993). Figure 3c represents the 3′ subregion of the amylase locus consisting of approximately 6.5kb. The NANOVA of amylase activity in this subregion also failed to detect associations between the underlying genetic variation and variation in amylase activity. Figure 3b represents cladistic relationships for haplotypes within the middle 2.5kb of the restriction map, spanning the area between the duplicated genes. Within this subregion, the cladogram is so simple that all haplotypes would be placed within a single

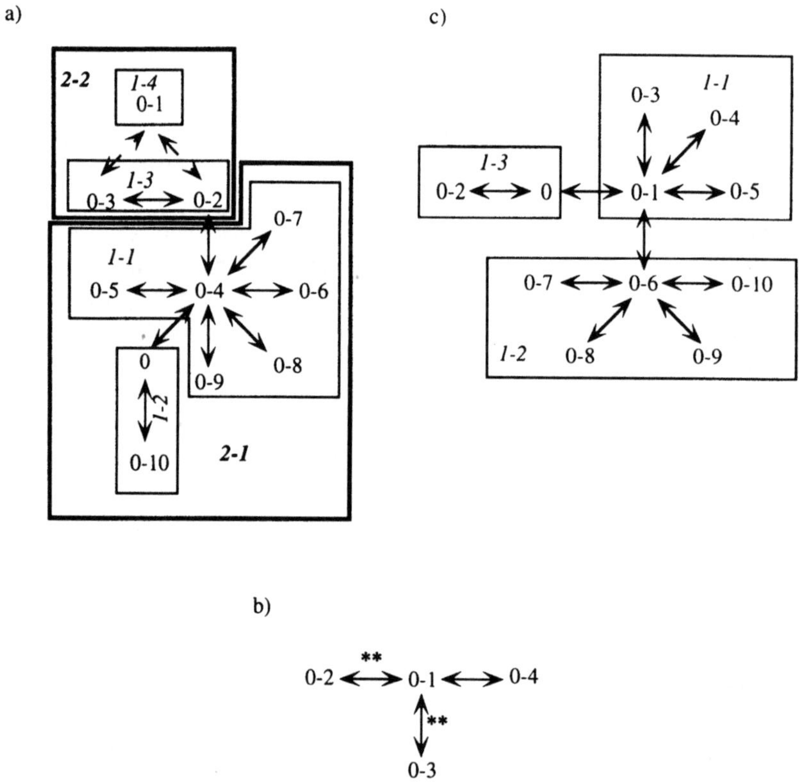

Figure 3. Cladograms and nested analyses for the *Amy* locus from Templeton and Sing (1993). Haplotypes are designated by nesting level then haplotype number. Zero haplotypes represent missing intermediates. Arrows represent mutational connections among haplotypes, with dashed arrows indicating ambiguous relationships. Haplotypes have been renamed for simplification. Cladograms for the left, middle, and right subregions of the *Amy* locus are given in figures 3a, 3b, and 3c, respectively.

Table 3A. Analysis of Variance of Amylase Activity in the Middle Subregion of the *Amy* Locus*

Source	Sum of Squares	Degrees of Freedom	Mean Square	F-statistic
0-step clades	0.38987	3	0.12996	8.09***
Error	0.72326	45	0.01607	

Table 3B. Bonferroni Multiple Comparison Tests of the Evolutionarily Relevant Contrasts Among the Haplotypes Found in the Middle Subregion of the *Amy* Locus Defined by the Cladogram Structure of Figure 3b*

Haplotype Contrast	Bonferroni Significance
1 vs. 2	<1%
1 vs. 3	<1%
1 vs. 4	>5%

* (Templeton and Sing, 1993)

nesting category, so no nesting is needed. Thus a standard analysis of variance is performed indicating a significant association (Table 3A). The cladogram structure not only defines the nesting categories, but also the evolutionarily relevant contrasts to perform when significant associations are indicated. By the cladogram structure for this subregion (Figure 3b), three relevant contrasts are indicated; 1 versus 2, 1 versus 3, and 1 versus 4. When these contrasts are made (Table 3B), two significant associations are revealed. Thus, statistically significant phenotypic associations are localized within this small subregion of the *Amy* locus using the evolutionarily relevant contrasts. This is contrary to a standard ANOVA, which indicates a significant line effect, but does not localize where the effect may be (Templeton and Sing, 1993). Additionally, the cladistic approach has the ability to detect phenotypically convergent, but evolutionarily independent, mutations and the ability to narrow the possibilities of which mutations are actually responsible for the phenotypic effects both to a small subset of haplotypes as well as to a subregion of DNA. In this example, the causal mutations fall within the 2.5kb region (narrowed from 15kb) and are between haplotypes 1–2, and 1–3.

Subsequent analyses to define these causal mutations can then concentrate sequencing efforts on these contrasts over this relatively small region of the *Amy* locus. Once sequence data is obtained from these two haplotypes for the 2.5kb region, a similar nested analysis based on the genealogical regionships of sequenced haplotypes can be performed to identify specific causal mutations for a shift in phenotype. The ultimate test for causal relationships could be accomplished by performing site-directed mutagenesis and observing effects of the phenotype of interest. Such an approach has been used in the study of amino acid replacements and their effects on wavelength absorption in vertebrate visual pigments (Yokoyama, 1995). Yokoyama and Yokoyama (1990) have shown that the evolutionary relationships of amino acid sequences from vertebrate visual pigments suggest three amino acid changes important in modifying the green-sensitive visual pigment into red-sensitive visual pigment. These changes have been subsequently tested by site-directed mutagenesis and shown to produce a wavelength shift by about 30 nm (Chan et al, 1992). Thus the evolutionary approach has been a powerful predictor of the causal relationships between genotypic variation and phenotypic variation.

Phenotypes Are Complex

While the marker locus approach and candidate gene approach provide complementary insights into the underlying genetic architecture of phenotypic traits, they are simplified views of this architecture. The candidate gene approach comes closest to providing insights into epistatic interactions and pleiotropic effects among loci, while current marker locus approaches ignore such interactions. In fact, complex genetic architectures greatly hinder the effectiveness of QTL searches using the current marker locus approaches (Eaves, 1994). Another facet of the genotype/phenotype relationship all but ignored in the marker locus approach is the influence of environment. It is clear that in studying the etiology of complex diseases, such interactions cannot be ignored (Sing and Reilly, 1993; Haviland et al, 1995b). Studies in human genetics attempting to account for environmental influence indicate that such influence can be significant (Boerwinkle and Hallman, 1993). Likewise, in studies that explore variation in marker loci across diverse ecogeographic areas, significant amounts of genetic variation within species are found. For example, Zhang et al (1993) has analyzed variation in RFLPs from wild barley in diverse areas and has found up to 30% of the loci differed among individuals within populations, and about 50% of the loci differed among plants in different populations.

While the marker locus approaches to date have ignored the biological realities of epistasis, pleiotropy, and environmental influence, this is more a reflection of the stage of development of these methods rather than limitations to this approach. For example, Jiang and Zeng (1995) have recently expanded the model represented in equation [5] to incorporate the simultaneous mapping of multiple traits. Given a sample of n individuals from an F_2 population of two inbred lines with observations on m quantitative traits, we can denote the value of each marker as 2, 1, and 0 for the homozygote in one parental line, heterozygote, and homozygote in the other parental line, respectively. Then let P_{jk} denote the value of the kth trait in the jth individual. The model in equation [5] can then be extended to test for a QTL on a marker interval $(i, i + 1)$ as:

$$\begin{aligned}
P_{j1} &= b_{01} + b_1^* x_j^* + d_1^* z_j^* + \sum_l^t (b_{l1} x_{jl} + d_{l1} z_{jl}) + e_{j1} \\
P_{j2} &= b_{02} + b_2^* x_j^* + d_2^* z_j^* + \sum_l^t (b_{l2} x_{jl} + d_{l2} z_{jl}) + e_{j2} \\
&\vdots \\
P_{jm} &= b_{0m} + b_m^* x_j^* + d_m^* z_j^* + \sum_l^t (b_{lm} x_{jl} + d_{lm} z_{jl}) + e_{jm} \\
j &= 1, \ldots, n
\end{aligned} \qquad [6]$$

where b_{0k} represents the mean effect of the model for trait k, b_k^* is the additive effect of the putative QTL on trait k, x_j^* counts the number of the allele at the putative QTL from one of the two parents, d_k^* is the dominance effect of the putative QTL on trait k, z_j^* is the indicator variable of the heterozygosity at the QTL, x_{jl} and z_{jl} are corresponding variables for marker l other than i and $i + 1$ in individual j with regression coefficients b_{lk} and d_{lk} on trait k, and e_{jk} is the residual effect on trait k for individual j (Jiang and Zeng, 1995). This model allows for the mapping of multiple traits jointly. The advantages to the joint mapping of multiple traits is threefold: (1) increase in statistical power of detecting QTL; (2) improvement of precision of parameter estimation; and, most importantly, (3) provision of an hypothesis testing framework for the interaction of multiple traits (Jiang and Zeng, 1995). For example, Jiang and Zeng (1995) have used the model in equation [6] to develop formal statistical tests for pleiotropy, QTL by environment interaction, and pleiotropy versus close linkage. They then demonstrate the power of this approach through computer simulation.

The evolutionary approach to candidate gene studies (Hallman et al, 1994) and the marker locus approach (Jiang and Zeng, 1995) offer analytical frameworks for the consideration and partitioning of genetic diversity of the QTL and their effects on phenotype. Future studies and advances in analytical approaches to the study of quantitative trait loci will incorporate further complexities of the populations examined, including epistasis, pleiotropy, and environmental influences on genetic diversity. As these models become more realistic by the incorporation of these phenomena, our ability becomes enhanced to ask diverse and interesting biological questions relating to the interplay among genotype, phenotype, and the environment.

ACKNOWLEDGMENTS

I thank Matt Brauer, Eric Routman, Charlie Sing, and Zhao-Bang Zeng for helpful comments on earlier drafts of this chapter. This work was supported by grant number DEB-9303258 from the National Science Foundation and the Alfred P. Sloan Foundation.

REFERENCES

Boerwinkle E, Hallman DM (1993): Genotype-by-environment interaction: It's a face of life. In: *Genetics of Cellular, Individual, Family, and Population Variability*, Sing CF, Hanis CL, eds. Oxford: Oxford University Press

Boerwinkle E, Chakraborty R, Sing CF (1986): The use of measured genotype information in the analysis of quantitative phenotypes in man. *Ann Hum Genet* 50:181–194

Boerwinkle E, Visvikis S, Welsh D, Steinmetz J, Hanash SM, Sing CF (1987): The use of measured genotype information in the analysis of quantitative phenotypes in man. II. The role of the apolipoprotein E polymorphism in determining levels, variability and covariability of cholesterol, betalipoprotein and triglycerides in a sample of unrelated individuals. *Am J Med Genet* 27: 567–582

Bonierbale MW, Plaisted RL, Pineda O, Tanksley SD (1994): QTL analysis of trichome-mediated insect resistance in potato. *Theor Appl Genet* 87: 973–987

Brummer CC, Bouton JH, Kochert G (1993): Development of an RFLP map in diploid alfalfa. *Theor Appl Genet* 86:329–332

Carbonell EA, Gerig TM, Balansard E, Asins MJ (1992): Interval mapping in the analysis of nonadditive quantitative trait loci. *Biometrics* 48:305–315

Causse MA, Fulton TM, Cho YG, Ahn SN, Chunwongse J, Wu K, Xiao J, Yu Z, Ronald PC, Harrington SE, Second G, McCouch SR, Tanksley SD (1994): Saturated molecular map of the rice genome based on an interspecific backcross population. *Genetics* 138:1251–1274

Chan T, Lee M, Sakmar TP (1992): Introduction of hydroxyl-bearing amino acids causes bathochromic spectral shifts in rhodopsin. *J Biol Chem* 267: 9478–9480

Cheverud JM, Routman E (1993): Quantitative trait loci: individual gene effects on quantitative characters. *J Evol Biol* 6:463–480

Churchill GA, Doerge RW (1994): Empirical threshold values for quantitative trait mapping. *Genetics* 138:963–971

Cowen NM (1989): Multiple linear regression analysis of RFLP data sets used in mapping QTLs. In: *Development and Application of Molecular Markers to Problems in Plant Genetics*, Helentjaris T, Burr B, eds. Cold Spring Harbor, NY: Cold Spring Harbor Laboratory

Crandall KA (1994): Intraspecific cladogram estimation: Accuracy at higher levels of divergence. *Syst Biol* 43:222–235

Crandall KA, Templeton AR, Sing CF (1994): Intraspecific phylogenetics: Problems and solutions. In: *Models in phylogeny reconstruction*, Scotland RW, Siebert DJ, Williams DM, eds. Oxford, England: Clarendon Press

Darvasi A, Weinreb A, Minke V, Weller JI, Soller M (1993): Detecting marker-QTL linkage and estimating QTL gene effect and map location using a saturated genetic map. *Genetics* 134:943–951

Da Silva JAG, Honeycutt RJ, Burnquist W, Al-Janabi SM, Sorrells ME, Tanksley SD, Sobral BWS (1995): *Saccharum spontaneum* L. 'SES 208' genetic linkage map combining RFLP- and PCR-based markers. *Mol Breed* 1:165–179

Devey ME, Fiddler TA, Liu B-H, Knapp SJ, Neale DB (1994): An RFLP linkage map for loblolly pine based on a three-generation outbred pedigree. *Theor Appl Genet* 88:273–278

Durham RE, Liou PC, Gmitter FG Jr, Moore GA (1992): Linkage of restriction fragment length polymorphisms and isozymes in *Citrus*. *Theor Appl Genet* 84:39–48

Eaves LJ (1994): Effect of genetic architecture on the power of human linkage studies to resolve the contribution of quantitative trait loci. *Heredity* 72:175–192

Edwards MD, Helentjaris T, Wright S, Stuber CW (1992): Molecular-marker-facilitated investigations of quantitative trait loci in maize 4. Analysis based on genome saturation with isozyme and restriction fragment length polymorphism markers. *Theor Appl Genet* 83:765–774

Ellegren H, Chowdhary BP, Johansson M, Marklund L, Fredholm M, Gustavsson I, Andersson L (1994): A primary linkage map of the porcine genome reveals a low rate of genetic recombination. *Genetics* 137:1089–1100

Ellis THN, Turner L, Hellens RP, Lee D, Harker CL, Enard C, Domoney C, Davies DR (1992): Linkage maps in pea. *Genetics* 130:649–663

Falconer DS (1989): *Introduction to Quantitative Genetics*. New York: John Wiley

Faure S, Noyer JL, Horry JP, Bakry F, Canaud C (1993): A molecular marker-based linkage map of diploid bananas (*Musa acuminata*). *Theor Appl Genet* 87:517–526

Fisher RA (1918): The correlations between relatives on the supposition of Mendelian inheritance. *Trans Roy Soc Edinb* 52:399–433

Fisher RA (1930): *The Genetical Theory of Natural Selection*. Oxford: Oxford University Press

Gardiner JM, Coe EH, Melia-Hancock S, Hoisington DA, Chao S (1993): Development of a core RFLP map in maize using an immortalized F2 population. *Genetics* 134:917–930

Gerber S, Rodolphe F (1994): Estimation and test for linkage between markers: a comparison of lod score and chi-square test in a linkage study of maritime pine (*Pinus pinaster* Ait.). *Theor Appl Genet* 88:293–297

Gerber S, Rodolphe F, Bahrman N, Baradat P (1993): Seed-protein variation in maritime pine (*Pinus pinaster* Ait.) revealed by two-dimensional electrophoresis: genetic determinism and construction of a linkage map. *Theor Appl Genet* 85:521–528

Haldane JBS (1932): *The Causes of Evolution*. London: Longmans, Green

Haley CS, Knott SA (1992): A simple regression method for mapping quantitative trait loci in line crosses using flanking markers. *Heredity* 69:315–324

Haley CS, Knott SA, Elsen J-M (1994): Mapping quantitative trait loci in crosses between outbred lines using least squares. *Genetics* 136:1195–1207

Hallman DM, Visvikis S, Steinmetz J, Boerwinkle E (1994): The effect of variation in the apolipoprotein B gene on plasma lipid and apolipoprotein B levels. I. A likelihood-based approach to cladistic analysis. *Ann Hum Genet* 58: 35–64

Halward T, Stalker HT, Kochert G (1993): Development of an RFLP linkage map in diploid peanut species. *Theor Appl Genet* 87:379–384

Haviland MB, Ferrell RE and Sing CF (1995a): A cladistic analysis of the relationship between variation in the low density lipoprotein receptor gene region and interindividual variation in plasma lipid, lipoprotein and apolipoprotein levels. *Am J Hum Genet*: in press

Haviland MB, Kessling AM, Davignon J, Sing CF (1995b): Cladistic analysis of the apolipoprotein AI-CIII-AIV gene cluster using a healthy French Canadian sample. I. Haploid analysis. *Ann Hum Genet*: 59:211-231

Hayashizaki Y, Hirotsune S, Okazaki Y, Shibata H, Akasako A, Muramatsu M, Kawai J, Hirasawa T, Watanabe S, Shiroishi T, Moriwaki K, Taylor BA, Matsuda Y, Elliott RW, Manly KF, Chapman VM (1994): A genetic linkage map of the mouse using restriction landmark genomic scanning (RLGS). *Genetics* 138:1207–1238

Huestis GM, McGrath JM, Quiros CF (1993): Development of genetic markers in celery based on restriction fragment length polymorphisms. *Theor Appl Genet* 85:889–896

Jansen RC (1993): Interval mapping of multiple quantitative trait loci. *Genetics* 135:205–211

Jansen RC (1994): Controlling the type I and type II errors in mapping quantitative trait loci. *Genetics* 138:871–881

Jansen RC, Stam P (1994): High resolution of quantitative traits into multiple loci via interval mapping. *Genetics* 136:1447–1455

Jiang C, Zeng Z-B (1995): Multiple trait analysis of genetic mapping for quantitative trait loci. *Genetics* 140:1111–1127

Kennard WC, Poetter K, Dijkhuizen A, Meglic V, Staub JE, Havey MJ (1994): Linkages among RFLP, RAPD, isozyme, disease-resistance, and morphological markers in narrow and wide crosses of cucumber. *Theor Appl Genet* 89: 42–48

Kesseli RV, Paran I, Michelmore RW (1994): Analysis of a detailed genetic linkage map of *Lactuca sativa* (lettuce) constructed from RFLP and RAPD markers. *Genetics* 136:1435–1446

Kianian SF, Quiros CF (1992): Generation of a *Brassica oleracea* composite RFLP map: linkage arrangements among various populations and evolutionary implications. *Theor Appl Genet* 84:544–554

Kilian A, Kleinhofs A, Villand P, Thorbjornsen T, Olsen O-A, Kleczkowski LA (1994): Mapping of the ADP-glucose pyrophosphorylase genes in barley. *Theor Appl Genet* 87:869–871

Kimura M (1983): *The Neutral Theory of Molecular Evolution*. Cambridge: Cambridge University Press

Kleinhofs A, Kilian A, Maroof SMA, Biyashev RM, Hayes P, Chen FQ, Lapitan N, Fenwick A, Blake TK, Kanazin V, Ananiev E, Dahleen L, Kudrna D, Bollinger J, Knapp SJ, Liu B, Sorrells M, Heun M, Franckowiak JD, Hoffman D, Skadsen R, Steffenson BJ (1993): A molecular, isozyme and morphological map of the barley (*Hordeum vulgare*) genome. *Theor Appl Genet* 86:705–712

Knapp SJ (1991): Using molecular markers to map multiple quantitative trait loci: models for backcross, recombinant inbred, and doubled haploid progeny. *Theor Appl Genet* 81:333–338

Knapp SJ, Bridges WC Jr, Birkes D (1990): Mapping quantitative trait loci using molecular marker linkage maps. *Theor Appl Genet* 79:583–592

Knott SA, Haley CS (1992a): Aspects of maximum likelihood methods for the mapping of quantitative trait loci in line crosses. *Genet Res* 60:139–151

Knott SA, Haley CS (1992b): Maximum likelihood mapping of quantitative trait loci using full-sib families. *Genetics* 132:1211–1222

Lander ES, Botstein D (1986): Strategies for studying heterogeneous genetic traits in humans by using a linkage map of restriction fragment length polymorphisms. *Proc Natl Acad Sci USA* 83:7353–7357

Lander ES, Botstein D (1989): Mapping Mendelian factors underlying quantitative traits using RFLP linkage maps. *Genetics* 121:185–199

Landry BS, Hubert N, Etoh T, Harada JJ, Lincoln SE (1991): A genetic map for *Brassica napus* based on restriction fragment polymorphisms detected with expressed DNA sequences. *Genome* 34:543–552

Langley CH, Shrimpton AE, Yamazaki T, Miyashita N, Matsuo Y, Aquadro CF (1988): Naturally occurring variation in the restriction map of the *Amy* region of *Drosophila melanogaster*. *Genetics* 119:619–629

Lark KG, Weisemann JM, Matthews BF, Palmer R, Chase K, Macalma T (1993): A genetic map of soybean (*Glycine max* L.) using an intraspecific cross of two cultivars: 'Minosy' and 'Noir1'. *Theor Appl Genet* 86:901–906

Lusis AJ (1988): Genetic factors affecting blood lipoproteins: the candidate gene approach. *J Lipid Res* 29:397–429

Martinez O, Curnow RN (1992): Estimating the locations and the sizes of the effects of quantitative trait loci using flanking markers. *Theor Appl Genet* 85: 480–488

Moreno-Gonzalez J (1992): Genetic models to estimate additive and non-additive effects of marker-associated QTL using multiple regression techniques. *Theor Appl Genet* 85:435–444

Nelson CD, Nance WL, Doudrick RL (1993): A partial genetic linkage map of slash pine (*Pinus elliottii* Engelm. var. *elliottii*) based on random amplified polymorphic DNAs. *Theor Appl Genet* 87:145–151

Paterson AH, Lander ES, Hewitt JD, Peterson S, Lincoln SE, Tanksley SD (1988): Resolution of quantitative traits into Mendelian factors using a complete linkage map of restriction fragment length polymorphisms. *Nature* 335: 721–726

Philipp U, Wehling P, Wricke G (1994): A linkage map of rye. *Theor Appl Genet* 88:243–248

Pillen K, Steinrucken G, Wricke G, Herrman RG, Jung C (1992): A linkage map of sugar beet (*Beta vulgaris* L.). *Theor Appl Genet* 84:129–135

Provine WB (1971): *The Origins of Theoretical Population Genetics*. Chicago: University of Chicago Press

Rebai A, Goffinet B, Mangin B (1994): Approximate thresholds of interval mapping tests for QTL detection. *Genetics* 138:235–240

Reiter RS, Williams JG, Feldman KA, Rafalsk JA, Tingey SV, Scolnik PA (1992): Global and local genome mapping in *Arabidopsis thaliana* by using recombinant inbred lines and random amplified polymorphic DNAs. *Proc Natl Acad Sci USA* 89:1477–1481

Rodolphe F, Lefort M (1993): A multi-marker model for detecting chromosomal segments displaying QTL activity. *Genetics* 134:1277–1288

Roff DA, Bentzen P (1989): The statistical analysis of mitochondrial DNA polymorphisms: Chi-square and the problem of small samples. *Mol Biol Evol* 6: 539–545

Routman E, Cheverud JM (1994): Individual genes underlying quantitative traits: Molecular and analytical methods. In: *Molecular Ecology and Evolution: Approaches and Applications*, Schierwater B, Strait B, Wagner GP, DeSalle R, eds. Basel, Switzerland: Birkhauser Verlag

Sing CF, Reilly SL (1993): Genetics of common diseases that aggregate, but do not segregate, in families. In: *Genetics of cellular, individual, family, and population variability*, Sing CF, Hanis CL, eds. Oxford: Oxford University Press

Sing CF, Boerwinkle E, Moll PP, Templeton AR (1988): Characterization of genes affecting quantitative traits in humans. In: *Proceedings of the Second International Conference on Quantitative Genetics*, Weir BS, Eisen EJ, Goodman MM & Namkoong G, eds. Sunderland, MA: Sinauer Associates

Sing CF, Davignon J (1985): Role of the apolipoprotein E polymorphism in determining normal plasma lipid and lipoprotein variation. *Am J Hum Genet* 37: 268–285

Sing CF, Haviland MB, Templeton AR, Zerba KE, Reilly SL (1992a): Biological complexity and strategies for finding DNA variations responsible for inter-individual variation in risk of a common chronic disease, coronary artery disease. *Ann Med* 24:539–547

Sing CF, Haviland MB, Zerba KE, Templeton AR (1992b): Application of cladistics to the analysis of genotype-phenotype relationships. *Eur J Epidemiol* 8: 3–9

Soller M, Brody T, Genizi A (1976): On the power of experimental design for the detection of linkage between marker loci and quantitative loci in crosses between inbred lines. *Theor Appl Genet* 47:35–39

Tanksley SD (1993): Mapping polygenes. *Ann Rev Genet* 27:205–233

Tanksley SD, Ganal MW, Prince JP, de Vicente MC, Bonierbale MW, Broun P, Fulton TM, Giovannoni JJ, Grandillo S, Martin GB, Messeguer R, Miller JC, Miller L, Paterson AH, Pineda O, Roder MS, Wing RA, Wu W, Young ND (1992): High density molecular linkage maps of the tomato and potato genomes. *Genetics* 132:1141–1160

Templeton AR (1995): A cladistic analysis of phenotypic associations with haplotypes inferred from restriction endonuclease mapping or DNA sequencing. V. Analysis of case/control sampling designs: Alzheimer's disease and the apoprotein E locus. *Genetics* 140:403–409

Templeton AR, Sing CF (1993): A cladistic analysis of phenotypic associations with haplotypes inferred from restriction endonuclease mapping. IV. Nested analyses with cladogram uncertainty and recombination. *Genetics* 134: 659–669

Templeton AR, Boerwinkle E, Sing CF (1987): A cladistic analysis of phenotypic associations with haplotypes inferred from restriction endonuclease mapping. I. Basic theory and an analysis of alcohol dehydrogenase activity in Drosophila. *Genetics* 117:343–351

Templeton AR, Crandall KA, Sing CF (1992): A cladistic analysis of phenotypic associations with haplotypes inferred from restriction endonuclease mapping and DNA sequence data. III. Cladogram estimation. *Genetics* 132:619–633

Templeton AR, Sing CF, Kessling A, Humphries S (1988): A cladistic analysis of phenotypic associations with haplotypes inferred from restriction endonuclease mapping. II. The analysis of natural populations. *Genetics* 120: 1145–1154

Torres AM, Weeden NF, Martin A (1993): Linkage among isozyme, RFLP and RAPD markers in *Vicia faba. Theor Appl Genet* 85:937–945

Vallejos CE, Sakiyama NS, Chase CD (1992): A molecular marker-based linkage map of *Phaseolus vulgaris* L. *Genetics* 131:733–740

Veldboom LR, Lee M, Woodman WL (1994): Molecular marker-facilitated studies in an elite maize population: I. Linkage analysis and determination of QTL for morphological traits. *Theor Appl Genet* 88:7–16

Webb DM, Knapp SJ, Tagliani LA (1992): Restriction fragment length polymorphism and allozyme linkage map of *Cuphea lanceolata. Theor Appl Genet* 83:528–532

Weller JI (1986): Maximum likelihood techniques for the mapping and analysis of quantitative trait loci with the aid of genetic markers. *Biometrics* 42: 627–640

Weller JI (1987): Mapping and analysis of quantitative trait loci in *Lycopersicon* (tomato) with the aid of genetic markers using appropriate maximum likelihood methods. *Heredity* 59:413–421

Wright S (1921): Correlation and causation. *J Agric Res* 20:557–585

Wright S (1931): Evolution in Mendelian populations. *Genetics* 16:97–159

Wright S (1932): The roles of mutation, inbreeding, crossbreeding and selection in evolution. *Proc VI Internat Genet Cong* 1:356–366

Wright S (1980): Genic and organismic selection. *Evolution* 34:825–843

Wright S (1982): Character change, speciation, and the higher taxa. *Evolution* 36:427–443

Yokoyama S (1995): Amino acid replacements and wavelength absorption of visual pigments in vertebrates. *Mol Biol Evol* 12:53–61

Yokoyama R, Yokoyama S (1990): Convergent evolution of the red- and green-like visual pigment genes in fish, *Astyanax facsiatus*, and human. *Proc Natl Acad Sci USA* 87:9315–9318

Zeng Z-B (1993): Theoretical basis of separation of multiple linked gene effects on mapping quantitative trait loci. *Proc Natl Acad Sci USA* 90:10972–10976

Zeng Z-B (1994): Precision mapping of quantitative trait loci. *Genetics* 136: 1457–1468

Zhang Q, Maroof MAS, Kleinhofs A (1993): Comparative diversity analysis of RFLPs and isozymes within and among populations of *Hordeum vulgare* ssp. *spontaneum. Genetics* 134:909–916

9

Integrating Genetics, Phylogenetics, and Developmental Biology

ELIZABETH A. KELLOGG

"Nothing in biology makes sense except in the light of evolution."
—Dobzhansky

Plant molecular biology has made great strides in understanding the machinery of the cell, and increasingly, we understand how this machinery works to direct development. Molecular biology, however, has yet to explain the genetic and developmental changes that create the panoply of phenotypes that are the hallmark of evolution. This chapter outlines steps toward the goal and provides a brief discussion of what remains to be done.

When Darwin wrote about the origin of species, he was writing about the origin of phenotypes, morphological groups that are apparently hierarchically related. In the twentieth century, the New Synthesis focused attention on genes as central to the evolutionary process by showing that evolution of the phenotype is the result of accumulated genetic changes. The study of phenotypic evolution and of genetics are therefore intimately linked. Evolutionary mechanisms must at some level be genetic mechanisms.

In multicellular organisms, genetic change necessarily acts via developmental change. The idea has been stated so often that it is almost trivially obvious (Gould, 1977; McKinney and McNamara, 1991; Raff and Wray, 1989). Hence the comparative study of development and its genetic basis is central to the study of evolution. But how has development been modified? In most (perhaps all) cases we have only vague ideas.

There has been debate in the literature about the nature of evolutionary change, whether by many small accumulated changes—multiple genes of minor effect—or by macromutations—one or a few genes of major effects. Such questions would be settled by a detailed understanding of the genetic basis of morphological changes.

The Impact of Plant Molecular Genetics
BWS Sobral, Editor

Models of selection depend not only on assumptions about the number of genes, however, but also on such aspects as the extent of epistasis and pleiotropy. If we want to describe how (or even if) adaptive change occurs, then it is important to understand the extent of variation in the selected population and, equally important, the genetic basis of the variation. A hypothesis of adaptive change will be different if the change is the result of a genomic rearrangement, excision and reinsertion of a transposable element, production of a chimeric gene with a new promoter, or a point mutation in a structural gene.

Approaches to the Problem Using Classical Genetics

The idea of linking genetics, development, and phylogenetics is hardly new. Indeed, it has been part of the biosystematic research program from early on (although the definition of phylogeny was less precise than that used commonly today). Numerous examples are cited by Stebbins (1950) and by Grant (1981), among them studies on the inheritance of floral spots in violets (Clausen, 1926, 1951), corolla length in tobacco (East, 1916), and floral shape in snapdragons (Lotsy, 1916), as well as the widely cited studies on ecotypic differentiation in *Potentilla* (Clausen and Hiesey, 1958).

There has been persistent speculation, and some data, to suggest that major evolutionary innovations in plants are the result of changes in single genes of major effect. Gottlieb (1984) and Hilu (1983) cite numerous examples of changes that appear to be due to single genes. These include the shift from simple to compound inflorescences in tomato (Crane, 1915; MacArthur, 1928), fused to free petals in mountain laurel (Jaynes, 1974) and morning glory (Miyake and Imai, 1926), three to five carpels in melons (Rosa, 1928), and sterile to fertile flowers in grasses (Borgonakar et al, 1962), among many others. Gottlieb (1984) suggests that plants might exhibit many more single gene changes differentiating species than would animals, on the grounds that plants can tolerate macromutational changes because of their open, opportunistic growth pattern.

All this work has two limitations. First, the methods employed all require interfertile taxa. It is then necessary to invoke uniformitarianism and to extrapolate from these interfertile, closely related species to more distantly related ones, i.e., if these same processes go on long enough, then they might lead to new genera or families or phyla. Second, all these studies were done before it was possible to determine the molecular basis of the genetic changes. The studies thus show that phylogenetic change can indeed be described in terms of genetic change, but they cannot define precisely what the genetic changes are.

Use of Modern Gene Mapping Techniques

The ability to identify and describe particular genes has, of course, been one of the triumphs of plant molecular biology. Thus, the second problem, defining the genetic changes precisely, has been addressed by recent advances in molecular mapping.

In a recent review, Tanksley (1993) summarizes progress in mapping quantitative traits relative to single gene markers and, more recently, to molecular markers. While the majority of these studies have been applied to characters related to yield in crop plants, some address the question of morphological variation. There are two major points of interest to the evolutionary biologist in the data presented by Tanksley. First, polygenes have very different patterns of variation. This means that any simple model for the inheritance of a trait (at one extreme, a single locus controlling the phenotype, at the other extreme, many loci with small and approximately equal effects) cannot be assumed a priori. Each trait needs to be tested individually.

For many quantitative traits, much of the variance is controlled by only a few loci of relatively large effect. For example, in a cross between two species of tomato (*Lycopersicon esculentum* and *L. pennellii*), such characters as days to first flower and plant height are, as expected, controlled by multiple QTL with consistently small effects (deVincente and Tanksley, 1993). For number of buds in the inflorescence, however, a single QTL explains 34% of the variance, and for leaflet length to width ratio (a trait commonly recorded by plant systematists) three QTL account for approximately 30%, 21%, and 12% of the variance, respectively.

The second point of interest is that transgressive variation is common. This results when some of the alleles in one parent have the opposite effect from that predicted by the parental phenotype, e.g., in a cross between a tall and a short plant, the short parent often has alleles that create an increase in height. With recombination in the F_2 and subsequent generations, plants often have a phenotype more extreme (i.e., taller or shorter) than either of the parents. Such variation obviously could lead to novel phenotypes over evolutionary time. (These results also suggest that use of quantitative characters in phylogenetic analyses might be somewhat hazardous because character states of ancestors cannot be reliably inferred from those of their descendants.)

Doebley and Stec (1991, 1993) have produced what remains the single best example of using gene mapping techniques to examine an evolutionary problem. They were concerned with the origin of a cultigen (*Zea mays* ssp. *mays*), but their approach could be more widely applied. They addressed the question of whether the differences between maize and its immediate ancestors (*Z. m.* ssp. *mexicana* and *Z. m.* ssp. *parviglumis*) were created by many genes of minor effect or a few genes of major effect. They created segregating F_2s from crosses between *Z. m.* ssp. *mays* and *Z. m.* ssp *mexicana* (Doebley and Stec, 1991) and ssp. *mays* and ssp. *parviglumis* (Doebley and Stec, 1993). These they scored for nine morphological characters and 82 molecular marker loci and then used multivariate regression and interval mapping to assess correlations between the morphological characters and the molecular markers. They found that the differences between the cultigen and its putative ancestors were due primarily to five regions of the genome, confirming the data of Beadle (1939, 1980) years earlier.

One of the major morphological loci, *teosinte glume architecture*, has been investigated in detail, with the maize allele transferred to the teosinte background and vice versa (Dorweiler et al, 1993). The gene clearly affects several aspects of spikelet development, such as rachilla angle, lignification of glumes, and silica in the long cells on the glume abaxial surface.

This method easily could be applied to other grasses. The genomes throughout the grass family appear to be largely colinear, and mapping probes are available that will hybridize to many members of the family (Bennetzen and Freeling, 1993). Genome mapping work is underway in wheat, barley, rye, ryegrass, fescue, oats, rice, pearl millet, foxtail millet, maize, sorghum, and sugar cane. This means that any segregating population in the grasses can in principle be mapped and the genetic basis of the differences determined.

The approach could also be used for any interfertile taxa in *Solanum* or *Nicotiana*, for example, because the maps and probes are already available. It would be reasonable to set up a mapping population of interfertile species that differ by some intriguing set of morphological characteristics—fruit morphology and leaf compounding come to mind.

Taxa That Are Not Interfertile

Unfortunately, many evolutionary questions involve deeper branches of the evolutionary tree and therefore include taxa that are distantly related and also not interfertile. This is the more common situation. Fortunately, plant molecular biology provides methods that allow us to circumvent, at least partially, the need to work in interfertile populations.

As far as I know, the literature of plant biology does not address the direct use of molecular biological techniques to answer evolutionary questions. Sequence data are used to infer organismic phylogeny, but molecular genetics has rarely been used to understand the basis of complex adaptations. So far, evolutionary ideas are by-products of studies of pure molecular genetics, but are rarely the *raison d'être* of the investigation.

With this in mind, I outline below how one might use molecular and developmental biology to address directly some evolutionary questions. I see two basic questions concerning the evolution of development, and each suggests a slightly different research program, although I would hope they lead to the same end.

How Have Developmental Pathways Changed Through Evolutionary Time?

This approach would take any pathway that has been studied in some detail and compare its components across a range of species. The experimental outline would be as follows:

(1) For each gene in the pathway, sequence all members of the gene family from a set of increasingly distant taxa. Each of these gene trees can be seen as

a standard molecular evolution project, involving production of gene trees, assessment of within- and between-species variation, location of polymorphisms in the molecule, and inferences about patterns of selection and constraint.

(2) Examine upstream and downstream regions to compare regulatory sequences.

(3) Examine tissue specificity using a combination of Northern blots, Western blots, and in situ hybridizations.

(4) Where alterations in gene structure or promoter sequence are found, use these to transform the model system to attempt to mimic the phenotype of the wild species.

This strategy would be less risky than the approach outlined under the second question below, in that it would be almost certain to turn up something of interest. It would, however, be much more laborious, particularly in the initial stages.

There are some genes or gene pathways that could be easily analyzed by this approach, such as the anthocyanin pathway or the C_4 pathway. I will discuss work on the C_4 pathway at some length, but note that the same methods could be applied to other pathways as well.

EXAMPLE 1, C_4 PHOTOSYNTHESIS

C_4 photosynthesis is a high-efficiency photosynthetic pathway that has arisen multiple times in the evolution of flowering plants. The C_4 pathway involves changes in the tissue-specific expression of several photosynthetic enzymes, as well as histological changes in the internal structure of the leaf. The details of the pathway have been studied extensively in several model systems (Edwards and Walker, 1983; Nelson and Langdale, 1992).

Within the grass family alone there have been at least four origins of the C_4 photosynthetic pathway (Kellogg and Campbell, 1987; Barker et al, 1995). Multiple gene trees and morphological trees agree that the earliest diverging branches in the grasses are the bamboos, including rice and its relatives, and the pooid grasses, including wheat, barley, rye, and oats (Kellogg and Linder, 1995). All are C_3. The remainder of the grasses are in a large clade that includes four subfamilies, with a mix of C_3 and C_4 members. One C_4 lineage includes maize, sugar cane, and sorghum, a second includes finger millet and other grasses with the NAD-ME subtype of C_4, a third lineage includes only *Aristida* and *Stipagrostis*, and a fourth *Eriachne* and *Pheidochloa* (Figure 1). The four lineages differ in C_4 subtype, whether or not the chloroplasts have well-developed grana (affected by the expression of genes in photosystem II), and whether the veins are surrounded by one or two bundle sheaths (Hattersley and Watson, 1992).

In the C_3 pathway, carbon assimilation and carbon reduction occur in the mesophyll (the nonvascular portion of the leaf); the primary carbon acceptor is Rubisco. In the C_4 pathway, Rubisco expression and the standard C_3 pathway are relegated to the bundle sheath surrounding the vascular tissue. Carbon is still assimilated in the mesophyll, but the primary carbon acceptor is PEP carboxylase.

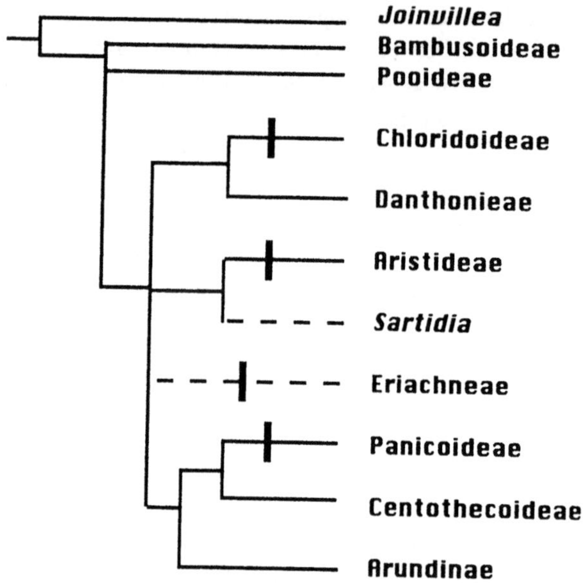

Figure 1. Simplified phylogeny of the grass family, relative to its sister group, Joinvillea. Phylogenetic relationships are based on data summarized by Kellogg and Campbell (1987), Kellogg and Linder (1995), and Barker et al (1995). Origins of C_4 photosynthesis are indicated by heavy black bars. Placement of Eriachneae and *Sartidia* (dashed lines) is approximate, and based only on morphological data.

The resulting 4-carbon compound is then shunted to the bundle sheath, where it is decarboxylated, and the carbon molecule is taken up by Rubisco and sent through the standard C_3 pathway (Nelson and Langdale, 1992).

This is accomplished by modifying tissue specific expression of PEP carboxylase, Rubisco, and multiple other photosynthetic enzymes, such as malic enzyme (for decarboxylation), PPDK (restores phosphoenol pyruvate), and LHCP (part of the light-harvesting complex). In maize, PEP carboxylase expression is increased and Rubisco decreased in the mesophyll, ME is increased in the bundle sheath, LHCP is decreased, and PPDK is (surprisingly) not regulated in a tissue-specific manner.

We have found recently (Sinha and Kellogg, submitted) that not all lineages share the same pattern of gene expression as maize. In fact, only Rubisco and PEP carboxylase are similar. This is perhaps not surprising, given that each lineage represents a separate origin of the pathway; we might expect that it would be re-invented slightly differently each time. This does suggest, though, that the commonalities lie in the regulation of carbon assimilation (Rubisco and PEP carboxylase).

All of the enzymes are part of small multigene families, except for the large subunit of Rubisco which is chloroplast-encoded. Therefore, gene trees need to be constructed for each gene. Preliminary gene trees for PEP carboxylase indicate that across angiosperms, different C_4 lineages have coopted different members of the gene family for carbon assimilation.

Similar projects could be undertaken with the small subunit of Rubisco, the two malic enzymes (NAD-dependent, and NADP-dependent), and PPDK. These molecules are all involved in various aspects of intermediary metabolism and photosynthesis, and are thus of some intrinsic evolutionary and biochemical interest. Each of the gene tree projects could therefore stand alone, even though it feeds into the larger project.

Once we know which members of a gene family have been coopted for use in the C_4 pathway, we can then examine the upstream and downstream regions of the genes to determine if there are common elements. Various studies have identified regulatory sequences both 5′ (LHCP: Bansal et al, 1992) and 3′ (*rbc*S: Viret et al, 1994) of the genes. We expect that these sequences will be shared among members of a C_4 lineage, but may be different across different lineages.

Immunolocalization and in situ investigations could be conducted in early developmental stages, and in leaf sheaths, as well as in mature leaves. It is not yet known what genes may be involved in controlling vein spacing or the regulation of chloroplast development in bundle sheaths. However, even when candidate genes are identified, it will be important to have precise descriptions of the phenotype in each of the C_4 lineages, again to determine the commonalities. Major developmental genes are most likely to be identified in maize, or possibly in rice; these results will then have to be extrapolated to the rest of the family.

Can Differences in Adult Morphologies Be Ascribed to Particular Changes in Developmental Pathways?

This is a slightly different way of framing the genetic/phylogenetic question, and may be somewhat harder to test. It begins with the observation of similarity between a mutant phenotype in a model organism and a natural phenotype in a wild species. The hypothesis is then that the same gene or pathway is involved in both the mutant and the wild species. Experiments would involve:

(1) isolating RNA from the appropriate tissue of the wild species and its nearest relatives and probing a Northern blot with the gene of interest. If the gene is involved in producing the phenotype, then differences in expression should be apparent.

(2) isolating the relevant gene from the wild species using sequence similarity (PCR or probing a library);

(3) sequencing it to see if there is a lesion in the wild plant similar to that in the mutant;

(4) examining the 5′ and 3′ regions for changes in putative regulatory sequences;

(5) examining the sequences and regulatory environment of factors both upstream and downstream of the target gene; and

(6) transforming the wild species with the non-mutant gene from the model species to try to complement the phenotype.

EXAMPLE 2, LIGULE DEVELOPMENT

Grass leaves are made up of a basal sheathing portion and a distal blade; at the junction of sheath and blade a small flap of tissue, the ligule, develops. The ligule is formed relatively late in leaf development (ca. plastochron 4 or 5 in maize) from epidermal tissue (Sharman, 1942), and develops from the margin of the leaf toward the midrib, eventually forming a complete rim across the leaf (Hake et al, 1985). The cell divisions and expansion leading to ligule formation occur relatively late in development, at approximately the same time that cells in the blade complete differentiation (Sylvester et al, 1990).

Various grasses, among them species of *Acroceras*, *Brachiaria*, *Echinochloa*, *Hubbardia*, *Jansenella*, *Neostapfia*, *Orcuttia*, and *Streptostachys*, lack ligules entirely (Watson and Dallwitz, 1988). The boundary of sheath and blade is still detectable in these grasses, but is less regular than that in grasses with ligules (Kellogg and Sinha, 1994). In adult morphology, the taxa without ligules all appear to be normal, with growth patterns and leaf numbers similar to those of their close (ligulate) relatives. None of these taxa, however, has been studied developmentally.

The recessive *liguleless-1* gene in maize creates a phenotype similar to that found in the wild liguleless taxa (Becraft et al, 1990; Sylvester et al, 1990). It is tempting to speculate that *liguleless-1* may be involved in the naturally occurring loss of ligules, but until clones of the gene are available, this cannot be tested. Additional problems may come from the fact that many of the wild taxa are polyploid, e.g., *Orcuttia*, *Neostapfia*, and *Tuctoria* are tetraploid (Reeder, 1982), and liguleless *Echinochloa* species may be tetraploid or hexaploid (Hilu, 1994).

EXAMPLE 3, FLORAL DEVELOPMENT

The description of genes controlling floral organ identity (transcription factors containing a MADS box) has led to an exciting model for describing floral evolution (Coen and Carpenter, 1993; Coen and Meyerowitz, 1991). Doyle (1994) has constructed a gene tree for the MADS box genes and has identified some orthologous groups that also share expression patterns. Anomalies remain, however, and the structure of the tree indicates that more such genes remain to be isolated from many systems. The *Arabidopsis* model is thought to be similar to that in *Antirrhinum*. Other MADS box genes have been found in tomato (Pnueli et al, 1994) and petunia (vander Krol et al, 1993), however, with rather different expression patterns, indicating that the basic model may be more complex, and/or that it is extensively modified in other systems.

An attempt to use the *Arabidopsis* model in a comparative context has been described recently for *Silene* (white campion), a dioecious species (Hardenack et al, 1994). The authors hypothesize that organ identity genes might play a role in sex expression in *Silene*; none, however, is linked to the Y (sex determining) chromosome. Expression patterns are similar to those in *Arabidopsis* and *Antirrhinum* indicating that the MADS box genes are not involved in the origin of dioecy. Despite this negative result, the authors make some potentially intriguing observations on the distribution of the *squamosa* and *leafy* orthologues, which are distributed throughout the determinate inflorescence.

The difficulty with such a research program is the likelihood of a negative result. Except in the simplest cases there will almost certainly be multiple ways to generate a particular phenotype. Given our current knowledge of genes and gene pathways, there is a fair chance that the phenotype of the wild species is conditioned by some gene other than that causing the mutant. In the study of *Silene* mentioned above, for example, the authors point out that *TASSELSEED2*, which is known to affect sex expression in maize, might be another candidate for sex determination (DeLong et al, 1993). Furthermore the difficulty of developing a new transformation system is often appreciable. This approach may still be useful if the genes or the species are sufficiently intriguing, as in the case of *Silene dioica*, but in many cases it may require much investment for potentially low return.

The Importance of a Phylogeny

None of the foregoing is possible without some fairly detailed knowledge of phylogeny. The work of Doebley and Stec is compelling in part because it rests on the firm foundation of a set of phylogenetic studies identifying *Zea mays* ssp. *parviglumis* and *Z. m. mexicana* unequivocally as the nearest relatives of cultivated maize (Figure 2). If one were to undertake the same experiments comparing *Z. m.* ssp. *mays* and *Tripsacum dactyloides*, the loci identified would include not only the loci differentiating ssp. *mays* from the wild subspecies of *Z. mays*, but also the loci differentiating *Z. mays* from the other three species of *Zea*, as well as the loci differentiating *Zea* from *Tripsacum*. This would also be a fascinating result, but harder to interpret. Furthermore, it could only be interpreted with knowledge of the phylogeny.

Similarly, the research on C_4 photosynthesis described above rests entirely on the observation that the pathway has arisen multiple times in the history of the grasses. This assertion is based on the phylogeny.

The Main Obstacles

For full integration of genetics, phylogenetics, and development, we need not only good plant molecular biology, but also a clear picture of relationships (phylogeny) and a very precise description of phenotypic change (good developmental morphology). Work on the basic mechanisms—receptors, signal transduction, regula-

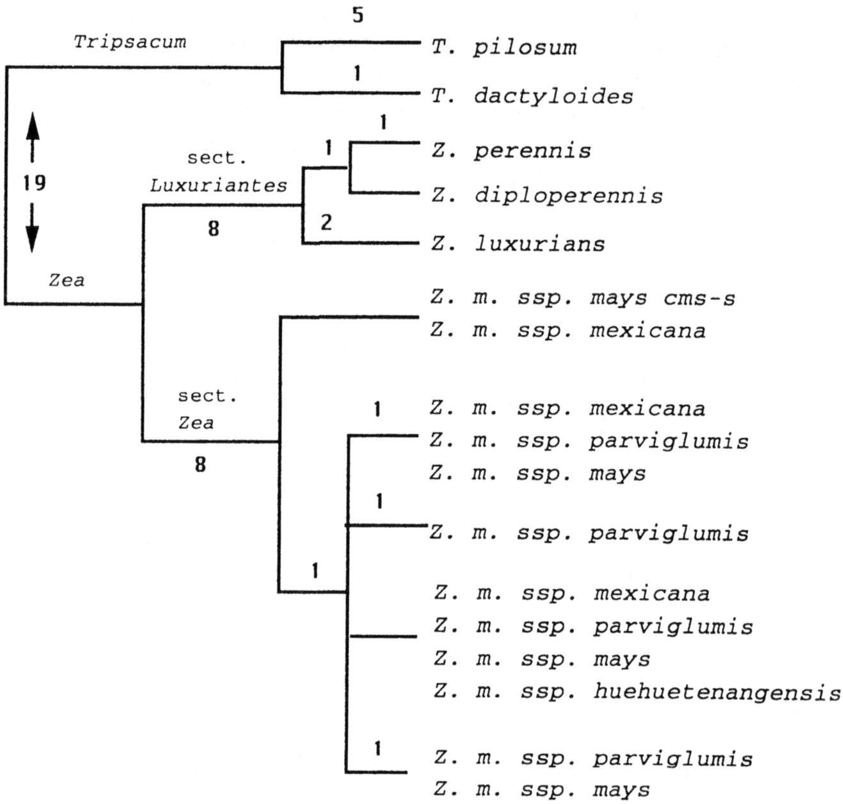

Figure 2. Phylogeny of the chloroplast genome for members of the genus *Zea* and two species of its sister genus, *Tripsacum*. Characters are restriction site changes. Numbers of branches indicate branch lengths. Redrawn by permission from *Econ Bot* 44 (Suppl 3):12, J Doebley, Copyright 1990, The New York Botonical Garden.

tory cascades, biosynthetic pathways—is proceeding rapidly. The problem is in the gene trees, and developmental biology of systems that are outside the model systems. There are two problems that I see in implementing this research program. The first is the lack of critical comparative data, and the second is a fundamental difference in approach between evolutionary and molecular biologists.

There are far too many good comparative developmental studies in the literature to be listed in this short paper. In general, however, studies of plant development have proceeded independently of either plant molecular biology or phylogenetics. The exceptions are, of course, developmental work on the model systems and a comparative handful of studies interpreting plant development in a phylogenetic context (Tucker et al, 1993; Hufford, 1988; Endress, 1992). This means that developmental data are often not available for the same groups for which we have phylogenetic and molecular data, and conversely, we lack good phylogenetic and/or molecular genetic data for taxa that have been studied in detail developmentally. The classic study by Kaplan (1980) on phyllodes in *Aca-*

cia, or Rutishauser's (1984) work on stipules in Rubiaceae are good examples, in which the development is described but no phylogeny or molecular genetic data are available.

We currently lack extensive comparative gene sequence data for many developmentally interesting genes. The systematic community has, quite reasonably, focused on extracting phylogenetic information from tractable chloroplast and ribosomal genes and has yet to pursue nuclear genes in any detail. There is a practical reason for this. To extract meaningful comparative information from a gene, it must be easily sequenced, because we will need a detailed understanding of genes from many disparate organisms. This will often mean that multiple alleles of multiple genes will have to be sequenced from many species. Although this effort may begin by screening genomic or cDNA libraries, ultimately researchers will have to switch to a PCR-based approach. In many cases the sequence of the entire gene will be needed, not just a handy bit of 500 bp from some place in the middle.

The various genome projects should help in identification, cloning, and preliminary sequencing of candidate genes. However, much effort remains in order to understand how a gene or gene family has evolved over time. There are currently few such projects ongoing in plants.

The other major obstacle for such studies stems from the differences in scientific culture between molecular and evolutionary biology. One notable one is the nature of proof. In molecular biology, linking a particular gene with a particular phenotype requires at the very least cosegregation of the mutant allele with the phenotype in crosses of mutant and wild type and, preferably, ability of the wild type gene to rescue mutants. The former test requires a segregating population, the latter a transformation system. The evolutionary biologist, on the other hand, is interested in generality, which requires making the same observations multiple times in multiple taxa, especially in a phylogenetically critical set of organisms. Time does not permit creating a new transformation system or even a segregating F_2 population for each species. This means that the evolutionary biologist cannot perform what a molecular biologist would consider a definitive experiment, and conversely the molecular biologist cannot undertake what the evolutionary biologist would consider definitive comparisons.

This leads almost inevitably to difficulties in funding. This sort of research program is generally too genetic for evolutionary biologists and too comparative for geneticists. It would require lots of system development which is chronically hard to fund. For specific questions, both approaches outlined above are problematical, the first because it is a fishing expedition without a clear hypothesis to test, and the second because it is risky.

One should not be too pessimistic, however. Molecular biologists are increasingly looking beyond their model systems, phylogeneticists are increasingly trying to understand the mechanistic bases of the characters they study, and developmental biologists are increasingly incorporating genetic and phylogenetic data. We may indeed be approaching another New Synthesis of systematics and genetics.

ACKNOWLEDGMENTS

I thank D. Baum and N. Sinha for helpful discussion.

REFERENCES

Bansal KC, Viret J-F, Haley J, Khan BM, Schantz R, Bogorad L (1992): Transient expression of *cab-m*1 and *rbc*S-*m*3 promoter sequences is different in mesophyll and bundle sheath cells in maize leaves. *Proc Nat Acad Sci USA* 89:3654–3658

Barker NP, Harley E, Linder HP (1995): Phylogeny of Poaceae based on *rbc*L sequences. *Syst Bot*: In press

Beadle G (1939): Teosinte and the origin of maize. *J Hered* 30:245–247

Beadle G (1980): The ancestry of corn. *Sci Am* 242:112–119

Becraft PW, Bongard-Pierce DK, Sylvester AW, Poethig RS, Freeling M (1990): The *liguleless-1* gene acts tissue specifically in maize leaf development. *Devel Biol* 141:220–232

Bennetzen JL, Freeling M (1993): Grasses as a single genetic system: genome composition, collinearity and compatibility. *Trends Genet* 9:259–261

Borgaonakar DS, Harlan JR, deWet JMJ (1962): A cytogenetical study of hybrids between *Dicanthium annulatum* and *D. fecundum*. II. *Proc Okla Acad Sci* 42:13–16

Clausen J (1926): Genetical and cytological investigations on *Viola tricolor* L. and *V. arvensis* Murr. *Hereditas* 8:1–156

Clausen J (1951): *Stages in the evolution of plant species*. Ithaca, NY: Cornell University Press

Clausen J, Hiesey WM (1958): Experimental studies on the nature of species. IV. Genetic structure of ecological races. In: *Carnegie Institute Washington Publication* 615

Coen ES, Carpenter R (1993): The metamorphosis of flowers. *Plant Cell* 5:1175–1181

Coen ES, Meyerowitz EM (1991): The war of the whorls: Genetic interactions controlling flower development. *Nature* 353:31–37

Crane MB (1915): Heredity of type of inflorescences and fruits in the tomato. *J Genet* 5:1–11

DeLong A, Calderon-Urrea A, Dellaporta SL (1993): Sex determination gene *TASSELSEED2* of maize encodes a short-chain alcohol dehydrogenase required for stage-specific floral organ abortion. *Cell* 74:757–768

deVincente MD, Tanksley SD (1993): QTL analysis of transgressive segregation in an interspecific tomato cross. *Genetics* 134:585–596

Doebley J (1990): Molecular evidence and the evolution of maize. *Econ Bot* 44(suppl 3):6–27

Doebley J, Stec A (1991): Genetic analysis of the morphological differences between maize and teosinte. *Genetics* 129:285–295

Doebley J, Stec A (1993): Inheritance of the morphological differences between maize and teosinte: comparison of results for two F_2 populations. *Genetics* 134:559–570

Dorweiler J, Stec A, Kermicle J, Doebley J (1993): *Teosinte glume architecture 1*: a genetic locus controlling a key step in maize evolution. *Science* 262: 233–235

Doyle JJ (1994): Evolution of a plant homeotic multigene family: toward connecting molecular systematics and molecular developmental genetics. *Syst Biol* 43:307–328

East EM (1916): Studies on size inheritance in *Nicotiana*. *Genetics* 1:164–176

Edwards GE, Walker DA (1983): *C_3, C_4: mechanisms and cellular and environmental regulation of photosynthesis*. Oxford: Blackwell Scientific

Endress PK (1992): Evolution and floral diversity: The phylogenetic surroundings of *Arabidopsis* and *Antirrhinum*. *Int Jour Plant Sci* 153:106–122

Gottlieb LD (1984): Genetics and morphological evolution in plants. *Amer Nat* 123:681–709

Gould SJ (1977): *Ontogeny and phylogeny*. Cambridge, MA: Harvard University Press

Grant V (1981): *Plant Speciation, 2nd edition*. New York: Columbia University Press

Hake S, Bird RM, Neuffer MG, Freeling M (1985): The maize ligule and mutants that affect it. In: *Plant Genetics*, Freeling M, ed. New York: Alan R. Liss

Hardenack S, Ye D, Saedler H, Grant S (1994): Comparison of MADS box gene expression in developing male and female flowers of the dioecious plant white campion. *Plant Cell* 6:1775–1787

Hattersley PW, Watson L (1992): Diversification of photosynthesis. In: *Grass evolution and domestication*, Chapman GP, ed. Cambridge: Cambridge University Press

Hilu KW (1983): The role of single-gene mutations in the evolution of flowering plants. In: *Evolutionary Biology*, Hecht MK, Wallace B, Prance GT, eds. New York: Plenum Publishing Corporation

Hilu KW (1994): Evidence from RAPD markers in the evolution of *Echinochloa* millets (*Poaceae*). *Pl Syst Evol* 189:247–257

Hufford LD (1988) The evolution of floral morphological diversity in *Eucnide* (Loasaceae): The implications of modes and timing of ontogenetic changes on phylogenetic diversification. In: *Aspects of Floral Development*, Leins P, Tucker SC, Endress PK, eds. Berlin: Gebrüder Borntraeger

Jaynes R (1981): Inheritance of flower and foliage characteristics in mountain laurel (*Kalmia latifolia* L.). *J Hered* 72:245–248

Kaplan DR (1980): Heteroblastic leaf development in Acacia: morphological and morphogenetic implications. *La Cellule* 73:137–203

Kellogg EA, Campbell CS (1987): Phylogenetic analysis of the Gramineae. In: *Grass systematics and evolution*, Soderstrom TR, Hilu KW, Campbell CS, Barkworth ME, eds. Washington, D.C.: Smithsonian Institution Press

Kellogg EA, Linder HP (1995): Phylogeny of the Poales. In: *Monocotyledons:*

systematics and evolution, Rudall PJ, Cribb PJ, Cutler DF, Humphries CJ, eds. London: Royal Botanic Gardens, Kew

Kellogg EA, Sinha NR (1994): unpublished observations

Lotsy JP (1916): *Evolution by means of hybridization.* The Hague: M. Nijhoff

MacArthur JW (1928): Linkage studies with the tomato. II. Three linkage groups. *Genetics* 13:410–420

McKinney ML, McNamara KJ (1991): *Heterochrony: the evolution of ontogeny.* New York: Plenum Press

Miyake K, Imai Y (1926): On a monstrous flower and its linkage in the Japanese morning glory. *J Genet* 16:63–76

Nelson T, Langdale JA (1992): Developmental genetics of C_4 photosynthesis. *Ann Rev Plant Physiol Plant Mol Biol* 43:25–47

Pnueli L, Hareven D, Broday L, Hurwitz C, Lifschitz E (1994): The TM5 MADS box gene mediates organ differentiation in the three inner whorls of tomato flowers. *Plant Cell* 6:175–186

Raff RA, Wray GA (1989): Heterochrony: Developmental mechanisms and evolutionary results. *Jour Evol Biol* 2:409–434

Reeder JR (1982): Systematics of the tribe Orcuttieae (Gramineae) and the description of a new segregate genus, Tuctoria. *Amer Jour Bot* 69:1082–1095

Rosa TJ (1928): The inheritance of flower types in *Cucumis* and *Citrullus*. *Hilgardia* 3:233

Rutishauser R (1984): Leaf whorls, stipules and colleters in Rubieae (Rubiaceae) in comparison with other angiosperms. *Beitr Biol Pflanz* 59:375–424

Sharman BC (1942): Developmental anatomy of the shoot of *Zea mays*. *Ann Bot* 6:245–283

Sinha NR, Kellogg EA: Parallelism and diversity in multiple origins of C_4 photosynthesis in the grass family. (submitted)

Stebbins GL (1950): *Variation and evolution in plants.* New York: Columbia University Press

Sylvester AW, Cande WZ, Freeling M (1990): Division and differentiation during normal and *liguleless-1* maize leaf development. *Development* 110:985–1000

Tanksley SD (1993): Mapping polygenes. *Ann Rev Genet* 27:205–233

Tucker SC, Douglas AW, Liang H-X (1993): Utility of ontogenetic and conventional characters in determining phylogenetic relationships of Saururaceae and Piperaceae (Piperales). *Syst Bot* 18:614–641

van der Krol AR, Brunelle A, Tsuchimoto S, Chua N-H (1993): Functional analysis of petunia floral homeotic MADS-box gene *pMADS1*. *Genes Dev* 7:1214–1228

Viret J-F, Mabrouk Y, Bogorad L (1994): Transcriptional, photoregulation of cell-type-preferred expression of maize *rbc5*–m3:3' and 5' sequences are involved. *Proc Nat Acad Sci USA* 91:8577–8581

Watson L, Dallwitz MJ (1988): *Grass genera of the world: illustrations and characters, descriptions, classification, interactive identification, information retrieval.* Canberra: Australian National University

10

Molecular Variation and the Delimitation of Species

JERROLD I DAVIS

Species are widely regarded as fundamental elements of modern biology, but practitioners of species biology—those who delimit and study the attributes of species, and seek to understand the manner in which species originate—have yet to agree on what a species is. To state the case more accurately, there is less agreement now than there has been at times in the past. The controversy over species concepts takes many forms, from quiet certitude among members of some communities that theirs is the proper view, to open, vigorous, and sometimes rancorous debate. One modern viewpoint, often unstated however, is that the very word species is an outmoded vestige of another era, perhaps more fashionable than phlogiston, but only a word to be used loosely and vaguely, as a placeholder, with little risk of argument.

This is a deplorable state of affairs, for scientific progress is inhibited when words have no meaning or have so many meanings that they are simply moving targets. At the very least it seems reasonable to suggest that those who mean little or nothing by the word species should refrain from using it.

Two recent developments in evolutionary biology have influenced dramatically the theory and practice of species biology. One of these, the rise of molecular genetics, has enabled investigators to detect DNA sequences and their patterns of variation, which has dramatically increased the number of attributes of organisms that can be observed. The other development, the establishment of phylogenetics as a unifying theme in evolutionary biology, represents a revolution in the manner in which available variation, including DNA sequence variation, is analyzed. Application of the phylogenetic paradigm at higher levels in the hierarchy of biotic relationships (i.e., among genera, families, kingdoms) has become conventional. A standing problem is the manner in which this revolution should or can be extended to the problems of species delimitation and the speciation process. This is precisely the level of biotic organization at which molecular genetics and the phylogenetic paradigm interact in a profoundly exciting

The Impact of Plant Molecular Genetics
BWS Sobral, Editor

but often perplexing manner. There is an element of hierarchy, the proper domain of phylogenetic analysis, in relationships among species, but just as clearly, populations and species are the theaters in which reticulate patterns of gene exchange and descent are found. A profound element of this problem is that hierarchic relationships may exist among nonrecombining alleles (i.e., among genes of a gene tree), in situations in which relationships among the organisms that carry these alleles are distinctly reticulate (Tateno et al, 1982; Neigel and Avise, 1986; Nei, 1987; Pamilo and Nei, 1988; Harrison, 1991; Doyle, 1992).

Contemporary Species Concepts

Virtually all species concepts in current use share a common focus upon genetically diverged communities of organisms, though they differ profoundly in the aspects of genetic divergence that are recognized as significant.

The Biological Species Concept (BSC) of Dobzhansky (1937) and Mayr (1942, 1982) is perhaps the most familiar of species concepts (though many are uncomfortable with its name, which suggests that other species concepts are lacking in biological relevance). The BSC regards as fundamental the existence of genetically determined barriers to gene flow between species, i.e., reproductive isolating mechanisms or barriers. Related species concepts, such as those of Paterson (1985) and Templeton (1989), focus on such features as the reproductive and ecological integration that exists within species, rather than on the barriers that may exist between species, which will often be accidental byproducts of evolutionary diversification.

Another set of species concepts focuses upon genetic differentiation of population systems, without reference to barriers that might prevent interspecific gene flow. Thus, genetically determined isolating mechanisms may exist, but need not, for species status to be recognized. Inherent in these species concepts is a model of genetic communities that fragment to yield two or more separate communities, which may or may not be differently adapted or genetically isolated by any barrier except geography. Patterns of genetic differentiation that develop are the direct subjects of analysis, and reasons for such differentiation are secondary questions. The potential outcome of matings between individuals of different species—establishment, survival, eventual mating success of offspring, and the fate of their offspring—are viewed as just that, potential future events, not current indicators of species status. Consider two populations, for example, that survive on opposite sides of the earth. They may be differentiated by many characteristics unrelated to survival, and they may retain the potential to interbreed. Species concepts in this category have originated within the phylogenetics community, and are consistent with the overall goals of this community of specifying biotic diversity patterns and interpreting them in terms of past events, prior to the development of hypotheses concerning the processes that drove these events (Platnick, 1977; Brady, 1985).

I resist categorizing this second family of species concepts as phylogenetic, for one of them is known as the Phylogenetic Species Concept (PSC). The PSC recognizes species boundaries where divergence has yielded genetically distinct population systems in the form of fixed genetic differences among such systems (Rosen, 1979; Nelson and Platnick, 1981; Cracraft, 1983, 1989; Nixon and Wheeler, 1990; Davis and Nixon, 1992). Thus, two population systems are recognized as separate species if for some character (e.g., a particular genetic locus) the two communities exhibit mutually exclusive sets of states (i.e., alleles of a genetic locus). In a simple manifestation one population system may be fixed for allele A, while another is fixed for allele B, or simply lacks allele A (Davis and Manos, 1991; Davis and Goldman, 1993). Species status might also be recognized if one population carries alleles A and D, while the other carries alleles B and C. Mutually exclusive sets of alleles are evidence that the population systems are not interbreeding (potential to interbreed is another question), and they serve as species-distinguishing characters. Application of the PSC obviates the occurrence of identical individuals within different species, and though a population perspective is instrumental in the delimitation and characterization (i.e., diagnosis) of species, any individual can be assigned afterward to the appropriate species.

The Genealogical Species Concept (GSC) seeks a different aspect of divergence, in the form of monophyly of individual species (Rosen, 1979; Donoghue, 1985; de Queiroz and Donoghue, 1988). This property has more recently been designated exclusivity (Baum, 1992; Baum and Shaw, 1995). Monophyly, of course, has become established in the past quarter century as a necessary attribute of higher (i.e., multispecies) taxa. A group of species is monophyletic if it includes a common ancestral species and all of its descendant species (Hennig, 1966; Farris, 1974). The goal of the GSC is to extend the standard of monophyly, or a comparable one, to the level of species. Precise standards of evidence for the recognition of genealogical species have yet to be established. However, with the goal of recognizing groups of organisms of unique ancestry, approaches have been discussed including the reconstruction of gene trees, followed by the application of methods derived from coalescence theory (Slatkin and Maddison, 1989; Hudson, 1990; Hudson et al, 1992). It is not clear whether such methods would accommodate species that did not exhibit fixed differences (i.e., whether identical individuals might exist within different species). If fixed differences are regarded as implicit in the requirement of exclusivity, the number of population systems accorded species status under the terms of the GSC would seem to be smaller than the number recognized as species under the PSC, for an attribute that is sufficient for the latter would be necessary but insufficient for the former. In any case, criterion of exclusivity proposed earlier does exist, this being presence of an autapomorphy (Rosen, 1979; Donoghue, 1985; de Queiroz and Donoghue, 1988), i.e., a derived character that occurs in just one species. Whatever else might be implied by presence of an autapomorphy, a fixed difference is implied, for the character would occur in all individuals within an autapomorphic species and would not occur in other species.

Population Differentiation—Events and Evidence

Differences between the three principal species concepts outlined above can be clarified by reference to cases in which their application to a given pattern of genetic differentiation leads to the recognition of different numbers of species, or different species boundaries. In the following paragraphs I discuss a series of known relationships both as seen by an omniscient observer and as observed and interpreted by a practitioner who samples and analyzes patterns of diversity. It should be noted that an additional factor in any analysis of this sort is the rigor with which a species concept is applied. Any available species concept can be applied strictly or not, and I hope in the following paragraphs to dissociate this practical aspect of species delimitation from the basic issue of which concept is applied.

Consider, first, a simple sequence of population divergence (Figure 1, top). Two local populations are initially present (at time 1), each a group of individuals that constitute an integrated genetic community at a particular location. The two local populations are interconnected by occasional gene flow, and thus constitute a single metapopulation that departs, as a whole, from panmixia. Also, this metapopulation does not exchange genes with any other populations and has some number of unique attributes that distinguish all of its constituent individual organisms from all organisms of other populations. This system of two loosely connected local populations might have been generated by dispersal from a single

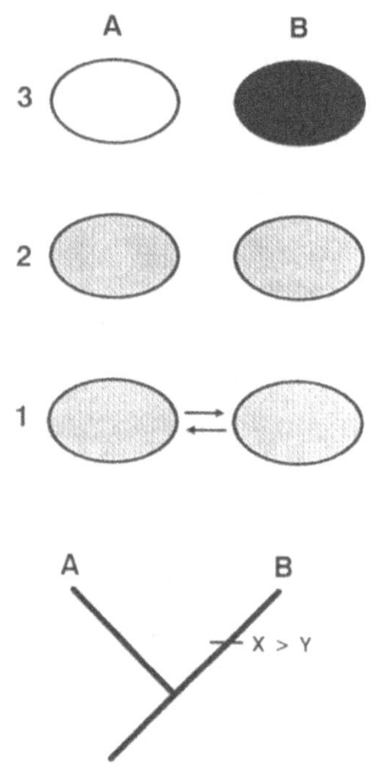

Figure 1. Two local populations (A and B) at three successive points in time (1–3), and a cladogram depicting their divergence. Shading of populations signifies polymorphism of two alleles (X and Y) at a locus, white signifies fixation of allele X, and black signifies fixation of allele Y. Arrows signify occasional gene flow between the populations. At time 1 both populations are members of a single genetic community or metapopulation. At time 2 interpopulational gene flow has ceased. At time 3 populations A and B have become fixed for alleles X and Y, respectively. Allele X occurs in related species; allele Y, a derived form of X, occurs only in population B, and is depicted as an autapomorphy of that population.

population, which could be one of the two that still exist, or some third population that no longer survives. Alternatively, an originally continuous population that encompassed the ranges of both local populations might have become fragmented as the intervening region ceased to be occupied.

The omniscient observer notes that gene flow between populations A and B, originally infrequent, eventually ceases entirely, so that by time 2 there still are two local populations, but each is now an independent genetic community. At that point the omniscient observer produces a cladogram that depicts the rather uncomplicated descent history of the two populations that exist at time 2 (Figure 1, bottom). The common ancestor at the node of the cladogram is the metapopulation from which the two independent populations descended after the cessation of interpopulational gene flow.

At the moment that interpopulational gene exchange ceases there may be quantitative differences (i.e., allele frequency differences) between the two populations, but because of the homogenizing effect of the occasional interpopulational gene flow that has existed up to this moment there are, initially, no fixed differences. One of many shared polymorphisms of populations A and B is that of alleles X and Y at some locus. Allele X is the plesiomorphic allele, and it also occurs, and is fixed, in related lineages, but allele Y occurs only in these two populations (this gene clearly is not one of the characters that distinguish these two populations from all others). When gene exchange ceases, microevolutionary forces such as mutation, selection, and drift, which influenced allele frequencies prior to this point, will continue to impinge and eventually will cause the genetic compositions of the populations to diverge further. Various alleles become fixed in one population or another, among them allele X in population A, but this does not in itself create a fixed difference between the populations, for population B still may be polymorphic for X and Y. Thus, allelic fixation in one population does not in itself make the populations distinct. Eventually, however, at time 3, perhaps after several fixation events in each population, involving different loci, allele Y becomes fixed in population B, and the two populations exhibit their first fixed difference. From this point forward, identical individuals cannot arise in the two populations. At this moment the omniscient observer adds a character state transition to the previously drawn cladogram, which is the autapomorphic occurrence of allele Y in population B. The earlier fixation of allele X is <u>not</u> recognized as an autapomorphy of population A, for as noted above this allele also occurs and is fixed in related populations.

We now consider this situation from the standpoint of an investigator who relies upon evidence in order to discern pattern and to infer history. At times 1 and 2, frequency differences exist for alleles at any number of loci. If the frequency differences are substantial in number, and if populations are inadequately sampled (i.e., rare alleles are not observed), it is possible, and perhaps likely, that one or more of the frequency differences will be interpreted as a fixed difference (Davis and Nixon, 1992). Thus, an observer who applies the PSC may erroneously recognize two phylogenetic species before a fixed difference has arisen. Ironically, if this false reading occurs at time 2, the observer might infer correctly that gene flow has ceased between the two populations, though no explicit character evidence yet exists in the form of a fixed difference. At time 3 there is a fixed difference, and if it is discovered, the observer will infer correctly that the system comprises two phylogenetic species. However, if this character is not observed, the investigator will conclude that no fixed difference exists, and that the system consists of just one phylogenetic species.

Rigor of application, as noted earlier, might allow different investigators to draw different conclusions from the same observations. Nearly fixed differences in one or more characters, especially if observed in many characters, might cause some investigators to argue that gene exchange between the populations, though perhaps nonzero, is trivial, or that with so many extreme frequency differences already observed, there is likely to be one or more fixed differences among the genes that have not been examined. Therefore, the investigator might conclude that a species boundary should be recognized. Alternatively, some investigators might argue that the observation of just one fixed difference is insufficient grounds for the recognition of a species boundary, and that additional characters must be identified before species status is warranted. Implicit in such interpretations, at least for some, is a statistical argument concerning the status of the enormous background genome from which such a small number of observations actually are made. Thus, for some investigators the inferred existence of many extreme frequency differences may count for more than the observed existence of one fixed difference.

The same observations, if interpreted on the basis of the BSC, and if not involving a character related to reproductive isolation, would not signify the presence of a species boundary, unless perhaps the presence of a fixed difference, coupled with physical proximity of the two populations, argued for the existence of an as yet unobserved isolating mechanism. However, if the observed difference were in a character directly interpretable as an isolating mechanism, that character would support the recognition of a species boundary, again subject to the possible inaccuracies inherent in undersampling of populations and characters, and dependent on the rigor with which the concept was applied. Regarding rigor of application, the presence in each population of a few individuals with the genetic makeup typical of the other population might be deemed insignificant (i.e., the investigator might still recognize two species) if occasional interspecific hybridization was deemed a trivial point.

The GSC, in a significant way, poses a different criterion than the other two species concepts discussed here, i.e., exclusivity of descent as a source of historical unity, versus distinctness of genetic communities as a condition that exists at present. As noted above, it remains uncertain whether distinctness is an implicit antecedent of exclusivity. If it is, only a subset of population systems that are recognized as species under the terms of the PSC will be recognized as species under the terms of the GSC. Exclusivity then would require a fixed difference *and* a derived character (or a related sort of evidence). In the present instance (Figure 1), population B would exhibit this necessary combination of attributes, but population A, lacking an autapomorphy, would not. Proponents of genealogical species (often, however, in the older literature, using the name "phylogenetic species") usually have consigned nonautapomorphic systems to a different status, as metaspecies (Donoghue, 1985; de Queiroz and Donoghue, 1988). In the present case, however, the observed autapomorphy in population B cannot be taken as evidence for a different ontological status (i.e., exclusivity of descent) from that of population A. This is because alleles X and Y co-occurred as a polymorphism in the ancestral population that existed at time 1, and continued to exist as a polymorphism of both populations at time 2 after interpopulational gene exchange had ceased. It is purely a matter of chance that X became fixed in population A, and Y in population B! Consider how interpretations might change if, by chance, it had been allele Y that became fixed in population A and allele X in population B. Under the terms of the PSC, one would recognize two species, just as before, while under the terms of the GSC, where presence of an autapomorphy is taken to be crucial, the perceived ontological status of the two populations would be reversed: population A would now be a species and population B a metaspecies. The problem here is that the topology of the gene tree (allele Y being a descendant of allele X) has been misinterpreted as evidence of an ancestor-descendant relationship between the populations in which the alleles occur. Thus, the logical relationship between observed autapomorphies and inferred exclusivity depends upon the restrictive assumption that an autapomorphic allele that occurs in one of two closely related population systems originated after the two systems ceased to exchange genes.

A final variant of the events depicted in Figure 1 involves a three-allele situation, where allele X again is the plesiomorph that occurs both within this population system and in close relatives, while alleles Y and Z originate before time 2 (i.e., when there is one metapopulation that includes populations A and B). The gene tree for these three alleles, (Figure 2) indicates that allele Z is a transformed state of Y, just as Y is of X. Consider now the fixation of Y in population A, coupled with the loss of this allele and the continued occurrence of alleles X and Z in population B. In this case, two phylogenetic species again are recognized, for identical individuals cannot be drawn from the two populations. As noted above, a phylogenetic species need not be fixed for any single character; it is only necessary that it exhibit a combination of alleles that does not occur elsewhere, and in a two species system, that the two carry mutually exclusive sets of alleles. Once again, the topology of the gene tree is not indicative of relationships between the species. All that has occurred at the population level is divergence of

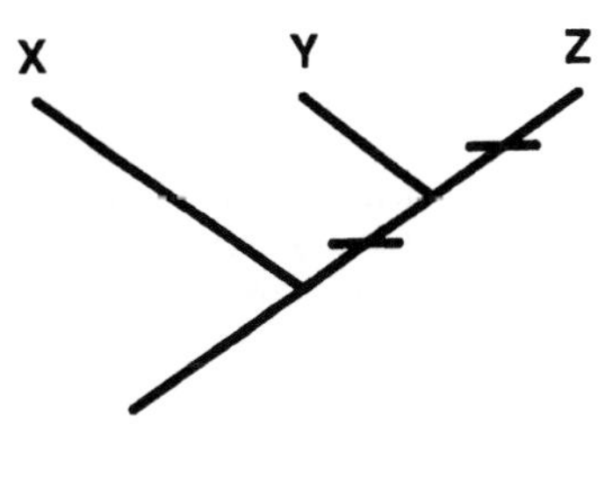

Figure 2. Gene tree showing history of origination of alleles Y and Z from allele X. Horizontal bars represent nucleotide substitutions.

two gene pools, followed by the loss of alleles. Because all three alleles evolved before the divergence of the populations, ancestor-descendant relationships among alleles are unrelated to ancestor-descendant relationships among the populations in which they occur.

Consider now a slightly more complex sequence of events (Figure 3, top), starting at time 1 with three local populations (A, B, C) interconnected as a metapopulation by occasional gene flow. Again, assume that this set of populations is distinct from all other related populations. The omniscient observer notes that gene flow first ceases between populations A and B, so that by time 2 there are still three local populations, but the metapopulation includes only local populations B and C. Later, gene flow ceases between populations B and C, and three separate populations remain. At this point the omniscient observer produces a cladogram that depicts the descent history of the three independent populations that exist at time 3 (Figure 3, bottom).

Note that no characteristics of the organisms are depicted in this figure. It is comforting that so much can be known of the history of these populations without any reference to evidence. Although the described sequence of events is more complicated than in the previous example with just two populations, the present case still is a simple instance of sequential separation of gene pools, uncompli-

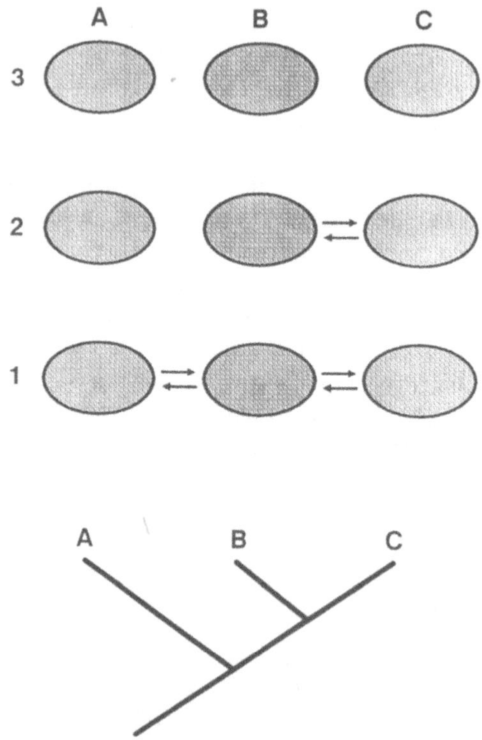

Figure 3. Three local populations (A–C) at three successive points in time (1–3), and a cladogram depicting the sequence of cessation of interpopulational gene flow. Arrows signify occasional gene flow among the populations. At time 1 all three populations are members of a single genetic community or metapopulation. At time 2 gene flow continues to occur between populations B and C, but has ceased between them and population A. At time 3 gene flow between populations B and C has ceased.

cated by such events as later hybridization (e.g., an isolated instance of gene flow between populations A and C, after time 3).

Let us now consider some of the observations that might be made by an investigator. As in the previous example, frequency differences may exist between the populations, in alleles at any number of loci. Leaving aside the problem of undersampling, an observer at any time during this sequence might infer that interpopulational gene flow is rare, but does rarity of gene flow imply a species boundary, and if so, how rare is rare enough?

Assume now that after time 3, a unique character (e.g., an allele) originates and is fixed in population B. From this moment forward, a comprehensive survey of genetic variation at the proper locus would detect the presence of two phylogenetic species, one (species B) consisting of population B, the other (species AC) comprising populations A and C. If the character that distinguishes species B from species AC is one that confers reproductive isolation, two species will be recognized under the terms of the BSC as well. Once again, the GSC yields a different result. In this case, unlike the previous example, we have declared that the species-distinguishing character both originated and became fixed within population B. The investigator might suspect, as the omniscient observer knows, that all individuals in species AC do not have an exclusive ancestry (the omniscient observer knows that population B shares a more recent common ancestry with one of its constituent populations [C] than do populations A and C with each other). Hence, on the basis of information provided by the omniscient observer, species AC is deemed a metaspecies.

Consider, then, what happens if an autapomorphy arises in population A, rather than B. In this instance, application of the PSC leads to the recognition of two species, A and BC, and again, the BSC yields a similar result if the character represents an isolation mechanism. Application of the GSC leads to the recognition of species A on the basis of its autapomorphy, but species BC, which *does* have an exclusive ancestry (according to our omniscient observer), is regarded as a metaspecies, on account of absence of an autapomorphy.

Conclusions

The PSC has been criticized for having been formulated with regard to evidence rather than with regard to actual historical relationships (Baum and Donoghue, 1993). However, the actuality that underlies the PSC, fully differentiated population systems, is clear. There surely are many noninterbreeding population systems in existence, some undoubtedly exclusive in origin, but in the absence of evidence individual cases fall outside the realm of observation, and thus of science.

Though the GSC and PSC are often deemed to be closely related concepts, apparently by virtue of their having originated within the phylogenetics community, they reflect quite different goals, and in some ways the PSC is more akin to the BSC. Under the terms of both the PSC and BSC, investigators evaluate existing diversity patterns and discern lines of demarcation between population systems. When such a boundary is discovered, the groupings on both sides of the boundary are recognized as species. In contrast, adherents of the GSC seek patterns of hierarchical descent within population systems. If a boundary line is discovered, just one species may be delimited, and what remains is of unspecified status. There will always be historical events that cannot be traced, and under the terms of the GSC, there may always be populations that cannot be assigned to a species.

As noted above, general approaches to the discovery of genealogical species have been suggested as promising (i.e., the use of coalescence theory), but to date, the only specific criterion to be proposed is autapomorphy. I have argued in the preceding pages that autapomorphy is an unreliable criterion of exclusivity, for its interpretation as evidence of exclusivity assumes origin of the autapomorphic character after cessation of gene exchange between populations. Where this is not what has occurred, autapomorphy is, in fact, an arbitrary discriminator because the ontological status of two sister population systems may be identical, and the presence of an autapomorphy in one and of a plesiomorphy in the other may be simply a matter of chance.

The topologies of gene trees can be reconstructed, and like autapomorphies, patterns of occurrence of the alleles of a gene tree can be observed within and among individuals, populations, and population groups (whether or not they are called species). Though evolutionists have acknowledged the distinction between gene trees and species trees, continued efforts to specify their precise implications, when alleles co-occur within populations, are founded on ancillary assumptions concerning rates of origin and migration of alleles, and effective population sizes. These efforts promise much, and surely should continue, but when such estimates are unavailable, it is unclear how such methods can lead to the discrimination of individual species in diverse natural settings.

The PSC, when applied strictly, specifies that identical individuals cannot be drawn from different species. If evidence for species status under the GSC continues to be expressed in terms of autapomorphies, a false ontological distinction is made between species and metaspecies, but species (as well as metaspecies?) will at least be distinct. If, however, a statistical approach is adopted, in which population averages overrule the evidence provided by each individual, a version of the GSC might emerge that does allow two species (or a species and a related metaspecies) to include identical individuals. One might then imagine a challenge, in the form of a creature related to Maxwell's demon, who would interchange identical individuals between two species to determine whether this affects an investigator's perceptions of species boundaries. If a GSC emerges, however, that implicitly accepts fixed differences as prior conditions of species status, then its goal would seem to be to evaluate the attributes of groupings that have been identified by this criterion beforehand as phylogenetic species.

One of the most difficult of practical problems in the area of species analysis is to identify the less inclusive groupings known as local populations and metapopulations. A species consists of one or more populations, and although these terms are convenient hooks for our general understanding of interbreeding individuals arrayed across the landscape, precise specification of the boundaries and genetic structures of natural populations would seem to remain one of the greatest challenges to the application of any species concept.

REFERENCES

Baum DA (1992): Phylogenetic species concepts. *Tr Ecol Evol* 7:1–2

Baum DA, Donoghue MJ (1993): Choosing among alternative phylogenetic species concepts. *Am J Bot* 80:115 (abstract).

Baum DA, Shaw KL (1995): Genealogical perspectives on the species problem. In: *Experimental and Molecular Approaches to Plant Biosystematics*, Hoch PC, Stevenson AG, Schaal BA, eds. St. Louis, MO: Missouri Botanical Garden

Brady RH (1985): On the independence of systematics. *Cladistics* 1:113–126

Cracraft J (1983): Species concepts and speciation analysis. *Curr Ornithol* 1:159–187

Cracraft J (1989): Speciation and its ontology: the empirical consequences of alternative species concepts for understanding patterns and processes of differentiation. In: *Speciation and Its Consequences*, Otte D, Endler JA, eds. Sunderland, MA: Sinauer Associates

Davis JI, Goldman DH (1993): Isozyme variation and species delimitation among diploid populations of the *Puccinellia nuttalliana* complex (Poaceae): character fixation and the discovery of phylogenetic species. *Taxon* 42:585–599

Davis JI, Manos PS (1991): Isozyme variation and species delimitation in the *Puccinellia nuttalliana* complex (Poaceae): an application of the phylogenetic species concept. *Syst Bot* 16:431–445

Davis JI, Nixon KC (1992): Populations, genetic variation, and the delimitation of phylogenetic species. *Syst Biol* 41:421–435

De Queiroz K, Donoghue MJ (1988): Phylogenetic systematics and the species problem. *Cladistics* 4:317–338

Dobzhansky T (1937): *Genetics and the Origin of Species*. New York: Columbia University Press

Donoghue MJ (1985): A critique of the biological species concept and recommendations for a phylogenetic alternative. *Bryologist* 88:172–181

Doyle JJ (1992): Gene trees and species trees: molecular systematics as one-character taxonomy. *Syst Bot* 17:144–163

Farris JS (1974): Formal definitions of paraphyly and polyphyly. *Syst Zool* 23:548–554

Harrison RG (1991): Molecular changes at speciation. *Annu Rev Ecol Syst* 22:281–308

Hennig W (1966): *Phylogenetic Systematics*. Urbana, IL: University Illinois Press.

Hudson RR (1990): Gene genealogies and the coalescent process. *Oxford Sur Evol Biol* 7:1–44

Hudson RR, Slatkin M, Maddison WP (1992): Estimation of levels of gene flow from DNA sequence data. *Genetics* 132:583–589

Mayr E (1942): *Systematics and the Origin of Species.* New York: Columbia University Press

Mayr E (1982): *The Growth of Biological Thought.* Cambridge, MA: Harvard University Press

Nei M (1987): *Molecular Evolutionary Genetics.* New York: Columbia University Press

Neigel JE, Avise JC (1986): Phylogenetic relationships of mitochondrial DNA under various demographic models of speciation. In: *Evolutionary Processes and Theory,* Karlin S, Nevo E, eds. New York: Academic Press

Nelson G, Platnick N (1981): *Systematics and Biogeography: Cladistics and Vicariance.* New York: Columbia University Press

Nixon KC, Wheeler QD (1990): An amplification of the phylogenetic species concept. *Cladistics* 6:211–223

Pamilo P, Nei M (1988): Relationships between gene trees and species trees. *Mol Biol Evol* 5:568–583

Paterson HEH (1985): The recognition concept of species. In: *Species and Speciation,* Vrba ES, ed. Pretoria, South Africa: Transvaal Museum Monograph 4

Platnick NI (1977): Cladograms, phylogenetic trees, and hypothesis testing. *Syst Zool* 26:438–442

Rosen DE (1979): Fishes from the uplands and intermontane basins of Guatemala: revisionary studies and comparative geography. *Bull Amer Mus Nat Hist* 162:267–376

Slatkin M, Maddison WP (1989): A cladistic measure of gene flow inferred from the phylogenies of alleles. *Genetics* 123:603–614

Tateno Y, Nei M, Tajima F (1982): Accuracy of estimated phylogenetic trees from molecular data. I. Distantly related species. *J Mol Evol* 18:387–404

Templeton AR (1989): The meaning of species and speciation: a genetic perspective. In: *Speciation and Its Consequences,* Otte D, Endler, JA, eds. Sunderland, MA: Sinauer Associates

Part III

Microorganisms in Agriculture: Two Examples

11

Application of the Polymerase Chain Reaction to the Detection of Plant Pathogens

RHONDA HONEYCUTT AND MICHAEL MCCLELLAND

Introduction

This chapter will focus on polymerase chain reaction (PCR)-based techniques applied to plant pathogens. Readers are also referred to the chapter by Kate J. Wilson on the subject of molecular approaches to the field ecology of microorganisms.

Although there is overlap in the terms detection and diagnosis, for the purpose of this chapter we distinguish detection from diagnosis based on: whether the identity of the pathogen is known *a priori*; whether one is interested in the presence of a particular pathogen, as may be the case for quarantine testing (detection); or whether one must determine the agent of disease from among a number of possibilities (diagnosis). In the former, detection tests that allow the specific pathogen to be identified are performed, and in the latter several candidate pathogens must be considered, usually by performing several tests. This distinction of goals is important in directing the investigator to which of the available detection and diagnostic tools are appropriate to use.

Techniques currently available to plant pathologists and diagnosticians have been reviewed recently (Hansen and Wick, 1993; Maclean et al, 1993; Swaminathan and Matar, 1993) and include immunological assays, isozyme profiles, fatty acid profiles, morphological keys, selective culturing, and differential host inoculations. These techniques will not be discussed further, except to note that all have some advantages as well as limitations. In general the advantages are relatively lower costs and technical input, and the limitations are sensitivity, selectivity and, in some cases, speed.

In recent years, many studies have used nucleic acid methods for identifying and detecting pathogens at various taxonomic levels: genus, species, sub-

The Impact of Plant Molecular Genetics
BWS Sobral, Editor

species, or pathovar. The level of sensitivity and selectivity required influences the choice of methodologies. RNA cataloging (Palleroni et al, 1973) and DNA:DNA hybridzations studies (Hildebrand et al, 1990), for example, have contributed much to taxonomic classifications of major bacterial plant pathogens. However, these techniques have not become routine for pathogen detection, largely because they are time consuming and technically challenging. The use of DNA probes in dot-blot assays is very simple but lacks sensitivity and, depending on the probe used, selectivity (Rasmussen and Reeves, 1992). Today, restriction fragment length polymorphisms (RFLPs)(Botstein et al, 1980) combined with the polymerase chain reaction (PCR)(Saiki et al, 1985) offer the greatest level of sensitivity and selectivity for detection and identification in all groups of plant pathogens. In this chapter we will focus on these technologies, their advantages and limitations, and give examples of their applications to different groups of pathogens. This chapter is not intended to be an exhaustive literature review of this vast subject, rather we point to specific references as examples of the various approaches that have been used. A forthcoming book entitled *Methods and Practice of Plant Disease Diagnosis* (Semer, 1995) provides the most up-to-date list of diagnostics for plant pathogens.

Considerations for Pathogen Detection Strategies

The ideal test always detects a pathogen if present (100% sensitivity) and never gives a false positive signal when the pathogen is not present (100% selectivity). No real-world test is likely to be both completely sensitive and completely specific. For example, without an internal positive control it is possible that the detection effort is negative due to failures in the reagents.

One problem confounding sensitivity and selectivity is that the spectrum of the detection system is dependent on the taxonomy of the organism. Unexpected results occur when a variant of the pathogen is sufficiently different that it is not detected or when a nonpathogenic variant is detected. Indeed, a feature of diagnostics that is often overlooked by those not directly involved in product development is that diagnosis is only as good as the current state of taxonomy. A diagnostic tool that has a poorly understood spectrum of detection is not very good. Thus, the importance of continuing phylogenetic studies cannot be underestimated. Unfortunately, this aspect of diagnostics does not receive the same attention as the development of a detection kit, even if the kit is phylogenetically inadequate.

The level of sensitivity and selectivity required for detection of a given pathogen is influenced by the biology of the organism. On the one hand, for organisms that are closely related to common nonpathogenic strains, selectivity can be most important. On the other hand, maximum sensitivity for the detection assay is required for pathogens that can exhibit a latency phase and remain, for a time, without inducing visible symptoms in the host. This is crucial in the quarantine setting where the first line of defense is to prevent import or transport to the field of infected plant materials. However, we must not lose sight of the fact that agriculture differs from medicine in that the most important component of a detection strategy has to be cost, so even relatively inefficient detection strategies can sometimes be deemed cost effective.

PCR-Based Methods

PCR-based detection potentially circumvents many of the problems currently associated with other detection methods used for plant pathogens. The main advantage of PCR over other methods is that PCR can be highly sensitive, amplifying a target over 10^9 times. This means that the organism does not need to be cultured as required for methods such as Southern blot analysis. However, the sensitivity of the PCR method can also be a problem because of the potential for false positive reactions due to carry-over contamination from previous positive experiments. Various ways to minimize this potential problem have been used. For example, all materials are UV irradiated except the target. Readers are referred to a careful review of the important factors in PCR (Persing, 1993).

The standard PCR strategy requires the design of primers that anneal perfectly to specific targets at high stringency. Primer design requires prior knowledge of DNA sequences of the targeted region. These targets, which permit discrimination of pathogens, include protein coding genes, ribosomal genes and their flanking spacer sequences, tRNAs and their spacer sequences, or plasmid sequences. In addition, specific primers can be derived from sequences of arbitrarily selected genomic regions originally produced using PCR and arbitrary primers.

Important criteria for the selection of a target include genetic stability of the targeted sequence and conservation of the sequence. The target is not ideal if the target is lost in some pathogenic strains or if the sequence evolves quickly so that primers may not work in some pathogens.

If the goal is detection of a specific pathogen, then the PCR targets can be ranked based on specificity. Genes involved in disease process or pathogenicity are expected to be superior to sequences not involved in pathogenicity. However, if the goal is a differential diagnosis between a number of related potential pathogens, then there are few DNA-based tests available as yet. We will argue that ribosomal and tRNA targets may be useful in this regard.

Genes Involved in Disease Process or Pathogenicity

The simplest approach to detection by PCR is to select primers that will only work in the strains or species of interest. When a gene associated with the disease process or pathogenicity is known, this makes an ideal target for diagnostic PCR because presence of the gene implies the disease, and absence or loss of the gene probably results in loss of pathogenicity. A few examples are presented below.

Genes involved in hypersensitivity and pathogenicity (*hrp* Leite et al, 1994) and toxin production (Prosen et al, 1993) have been targeted for the development of diagnostic primers. These genes are characteristic of pathogenic strains of bacterial species and are highly conserved. Leite et al (1994) found primers based on *hrp* genes from *Xanthomonas campestris* pv. vesicatoria identified pathogenic isolates of 28 *X. campestris* pathovars as well as *X. fragariae*, but no amplification products were observed for two *X. campestris* pathovars tested or for *X. albilineans* and the opportunistic pathogen *stenotrophomonas maltophilia*, which presumably does not contain homologous *hrp* sequences. They suggest that restriction enzyme analysis of the *hrp*-specific amplified products may facilitate further classification of *X. campestris* pathovars.

Similarly, the production of phaseolotoxin is characteristic of pathogenic isolates of *Pseudomonas syringae* pv. phaseolicola, and has been targeted by Prosen et al (1993) for the development of pathovar-specific primers for this bean pathogen. Amplification using these primers yielded product only in *P. syringae* pv. phaseolicola and not in *P. syringae* pv. syringae or *Xanthomonas campestris* pv. phaseoli, which are closely related pathogens of the same plant host. The sensitivity of their assay, 10–20 colony forming units per milliliter from water extract of bean seeds, was sufficient to detect the presence of the pathogen in commercial seed lots in which the pathogen was undetected by selective culturing or serological methods. RFLPs from the amplified polygalacturonase gene product allowed Gillings et al (1993) to rapidly distinguish strains of *Pseudomonas solanacearum.*

Specific genes have also been targeted for identification of plant viruses. Degenerate primers derived from sequence alignments have been used to amplify the variable region between the nuclear inclusion-b protein and the 5′ terminus of the coat protein (Langeveld et al, 1991; Colinet et al, 1993). These primers yield products whose lengths are specific for distinct potyviruses, which can cooccur in the same plant host. Similar strategies have been used to design primers that distinguish strains within a potyvirus, as demonstrated for sugarcane mosaic virus (Claude Fauquet, ILTAB, personal communication).

The only disadvantages of this approach are the disadvantages of PCR itself, namely false positives due to potential carry-over contamination and false negatives due to poor template preparation or defective reagents. The latter at least can be known if an internal positive control is incorporated such as a longer artificial template that uses the same primers but which is added at so low a concentration that it will amplify well only if the reaction does not contain the pathogen target of interest.

Sequences Not Involved in Pathogenicity

Often the genes associated with pathogenicity are not yet known for a particular pathogen. In such cases sequences that are associated with a particular phylogenetic grouping can be selected to design a PCR-based detection system. In some instances, identification to the level of species suffices, whereas other situations require identification at the subspecies or pathovar level. The investigator hopes that a sequence can be found that identifies all pathogens in a group and no nonpathogens.

GENES NOT ASSOCIATED WITH PATHOGENICITY

If a PCR target sequence for a pathogen is not associated with a disease-related gene, then caution is needed. There is always the possibility that a nonpathogenic strain will give a positive result or that a pathogenic strain will be negative. Nonetheless, there are instances in which sequences unrelated to pathogenicity have been successfully targeted for detection. For example, Schneider et al (1993) cloned a plasmid sequence with promoter activity from *Clavibacter michiganesis* subsp. sepedonicus and developed subspecies-specific primers for this potato pathogen. Although the amplified sequence does not encode any known pathogenicity or virulence factor, products were detected only in the targeted subspecies and not in a plasmid-free isolate of *C. michiganesis* subsp. sepedonicus, in other *Clavibacter* species tested, nor in *Erwinia carotova* subsp. carotovora which causes a similar disease in potato. A series of PCR primers based on plasmid sequences were designed by Hartung et al (1993) for use in detecting specific pathotypes of *Xanthomonas campestris* pv. citri. One primer pair was successful in distinguishing pathovar citri from other *X. campestris* pathovars responsible for disease in citrus, while another primer pair distinguished pathotype A from the other pathotypes causing the same disease. The plasmid sequence on which these primers are based was found in only 85% of the pathotype A strains, so detection based on these primers could be used in conjunction with other methods for the detection of pv. citri for international quarantine (Hartung et al, 1993).

REPETITIVE SEQUENCES

In the genomes of prokaryotes, as well as eukaryotes, some sequences are reiterated many times. In addition to the ribosomal and tRNA genes discussed later, other repeated sequences include insertion elements (IS), the BOX (a highly conserved bacterial repeat element), enterobacterial repetitive intergenic consensus (ERIC) sequences, and repetitive extragenic palindromic (REP) elements of bacterial genomes, and so-called satellite regions. Some have been used as probes for RFLP analyses of plant pathogens. So far very few have been targeted for the development of PCR primers.

There are many examples of the use of repetitive elements as probes on Southern blots, but in very few cases have these been converted to PCR-based diagnostics. Repeat elements cloned from *Xanthomonas campestris* pv. oryzae (Leach et al, 1990), present in that genome in about 80 copies, were used in RFLP studies to distinguish populations of *X. campestris* pv. oryzae from *X. campestris* pv. oryzicola. These studies suggested a clonal origin for *X. campestris* pv. oryzae strains from different geographical locations. These elements were also found in fewer copies in other pathovars of *X. campestris*. For *Phytophthora infestans*, genetic diversity and migration of populations was monitored using RFLPs based on cloned species-specific chromosomal (moderately repetitive sequence) probes (Sujkowski et al, 1993).

Tandem repeat loci within satellite DNA regions were the basis for probes used by DeScenzo and Harrington (1994) to detect polymorphisms in the fungus *Heterobasidion annosum*. Fingerprints using the $(CAT)_5$ probe revealed more variability than isozyme markers on a collection of isolates and permitted the distinction of sexually versus asexually produced isolates. The same microsatellite probe was used to identified haplotypes in *Colletotrichum orbiculare* (Correll et al, 1993). Similarly, satellite DNA was used to develop a species-specific probe, which is sensitive enough to detect a single nematode in direct hybridization assays, for *Bursaphelenchus xlophilus* (Tares et al, 1994), a pathogen of pine trees.

Once a repeat is identified that has the appropriate taxonomic distribution, it is desirable to convert such probes for PCR, if possible. Primers designed from two cloned species-specific chromosomal repetitive sequences (Ersek et al, 1994) allow detection and distinction of *Phytophthora* species which may cocolonize the same host.

ANONYMOUS REGIONS

One way to identify anonymous regions that have the desired taxonomic distribution is arbitrarily primed PCR (Welsh and McClelland, 1990; Welsh and McClelland, 1991b) or RAPDs (Williams et al, 1990). This strategy produces a characteristic fingerprint of PCR products using an arbitrary primer. The method has been widely used to identify isolates, and for phenetic analysis (Welsh et al, 1992; Ralph et al., 1993a). For *Heterobasidion annosum*, fingerprint profiles were correlated with intersterility groups for population and epidemiology studies (Garbelotto et al, 1993), and distinguished pathogenic from nonpathogenic isolates of *Fusarium oxysporum* f. sp. dianthi (Manulis et al, 1994). Fingerprint profiles were useful in subdividing mtDNA RFLP groups in another fungal pathogen *Colletotrichum orbiculare* (Correll et al, 1993).

These arbitrarily primed PCR strategies are probably of most interest in an epidemiological rather than detection setting. However, by converting PCR products found only in the strains of interest to Sequence Characterized Amplified Regions (SCARs) (Paran and Michelmore, 1993) one may sometimes be able to provide a specific diagnostic if the product happens to be more than just a primer binding site polymorphism. Wiglesworth et al (1994) used this strategy to develop species-specific primers for *Peronospora tabacina*, the causal agent of tobacco blue mold, from a cloned arbitrarily selected polymorphism. Their assay was sufficiently sensitive to detect the presence of single spores, and an outbreak was predicted by attaching double-sided tape to a car and driving the car around a potential outbreak area to collect the air-borne spores.

Another possible approach for detecting anonymous polymorphisms is Amplification Fragment Length Polymorphisms (AFLPs) in which genomic DNA is cleaved with a restriction enzyme, linkers are ligated to the ends, and primers that are homologous to the linker and have an additional one or more arbitrary bases at the 3′ end are used to select a subset of fragments for amplification (Zabeau, 1992).

Additional studies using PCR fingerprinting of repetitive elements (Versalovic et al, 1991) to distinguish *Xanthomonas* and *Pseudomonas* pathovars were reported by Louws et al (1994). Primers have been based on the REP (Higgins et al, 1982), ERIC (Hulton et al, 1991), and BOX elements (Martin et al, 1992). Using three different primer sets, they were able to distinguish pathovars and assess the polymorphisms among strains within a pathovar.

The disadvantage of trying to derive taxonomically appropriate PCR diagnostics from arbitrarily amplified products is that many polymorphisms are due to changes in the binding site of the arbitrary primer and not in the length or presence of the DNA in between the primer binding sites. Thus, many potential diagnostics developed using these internal sequences will not distinguish the taxonomic group of interest. Nevertheless, this approach can work and is appropriate if alternatives have failed or are otherwise undesirable.

rRNA Genes and the 16*S*–23*S* rRNA Spacer Regions

Ribosomal genes have long been used as target regions for the identification and taxonomy of eukaryotic and prokaryotic organisms (Fox et al, 1980; Appels and Honeycutt, 1986). The very large and still growing resource of sequences for these genes means that primers can be designed that target sequences conserved in an entire kingdom, or that target less conserved regions found in a particular genus or species. The spacer region is much more variable in sequence and length than the rRNA genes themselves and can provide variability within species.

PCR-based strategies targeted to rRNA genes are a logical extension of ribotyping which uses Southern blots and rRNA gene probes. Ribotyping needs relatively large amounts of genomic DNA and is quite laborious. Nevertheless, ribotyping is still used for epidemiology. For example, genetic diversity among 300 strains of *Xanthomonas campestris* pv. manihotis was studied using RFLPs revealed by a rRNA probe (Verdier et al, 1993). Further subdivisions of ribotypes, obtained using probes based on chromosomal sequences and a plasmid-borne sequence involved in pathogenicity, suggested a common center of origin for strains from different geographic locations. Based on RFLPs using cloned rRNA probe, investigators were able to distinguish among *Verticillium* species. (Typas et al, 1992; Griffen et al, 1994)

The sensitivity of the RFLP method can be improved by performing analysis on PCR products. Several studies have combined the use of PCR followed by restriction enzyme analysis for the identification of plant pathogens based on ribosomal sequences. Generally, the ribosomal units are amplified with either universal or specific primers, then the products are digested with restriction enzymes to reveal site polymorphisms. This approach is referred to as Amplified rDNA Restriction Analysis (ARDRA) (Vaneechoutte et al, 1993). When the RE sites are mapped then these Mapped Restriction Site Polymorphisms (MRSPs) can also be useful phylogenetic markers (Ralph et al, 1993b; Ralph and McClelland, 1994; Al-Janabi et al, 1994).

Some examples of rRNA gene amplification followed by RFLP analysis include applications to Mollicutes, (Rawadi and Dussurgiet, 1994) and *Mycoplasma* like organisms (MLOs) (Griffiths et al, 1994; Lee et al, 1993). Group-specific primers from 16S rRNA sequences have been used to detect mixed infections of

MLOs in plants (Lee et al, 1994; Namba et al, 1993). PCR based on the more variable rRNA intervening transcribed spacers were used for quantitation of the pathogen *Verticillium tricorpus* in plants (Moukhamedov et al, 1994) and to detect *Colletotricum gloesporioides* (Mills et al, 1992). Inter- and intraspecific variation of the symbiotically-associated fungi *Laccaria laccata* and *L. bicolor* was shown by analyzing restriction site polymorphisms in amplified ITS and intergenic spacer (IGS) regions (Martin et al, 1993). Further, this study, performed using capillary electrophoresis, demonstrated the feasibility of automating the identification and detection process.

tRNA GENE CLUSTERS

tRNA genes have features that can be exploited for a diagnostic test: (1) each amino acid acceptor type is highly conserved in sequence; (2) there can be more than 100 genes that usually appear in a number of tandem clusters; (3) the order of tRNAs is conserved between species in a genus but varies between more distantly related organisms; and (4) adjacent tRNA genes are separated by a spacer region that varies in sequence and in length between species.

Primers directed against conserved sequences in tRNA genes generate fingerprints of products (Welsh and McClelland, 1991a). Differences in the size of some or all of these products between species or strains generally represent differences in the length of the spacer regions between genes. Those products that display the taxonomic division of interest are cloned from the gel, and primers are designed to work specifically on the taxonomic group of interest (McClelland et al, 1992; Welsh and McClelland, 1992; Honeycutt et al, 1995). Figure 1

Figure 1. tRNA-intergenic length polymorphisms.

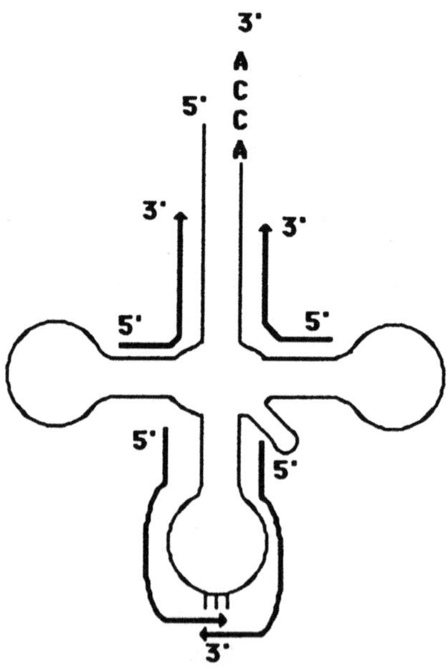

Figure 1A. tRNA genes are organized in tandem repeats. Consensus tRNA gene primers to amplify between tRNA genes can be located either near the 5′ and 3′ ends of the tRNA consensus gene or located to overlap the anticodon loop in particular tRNA amino acid acceptor types. Four positions for primers are shown aligned relative to a tRNA transcript.

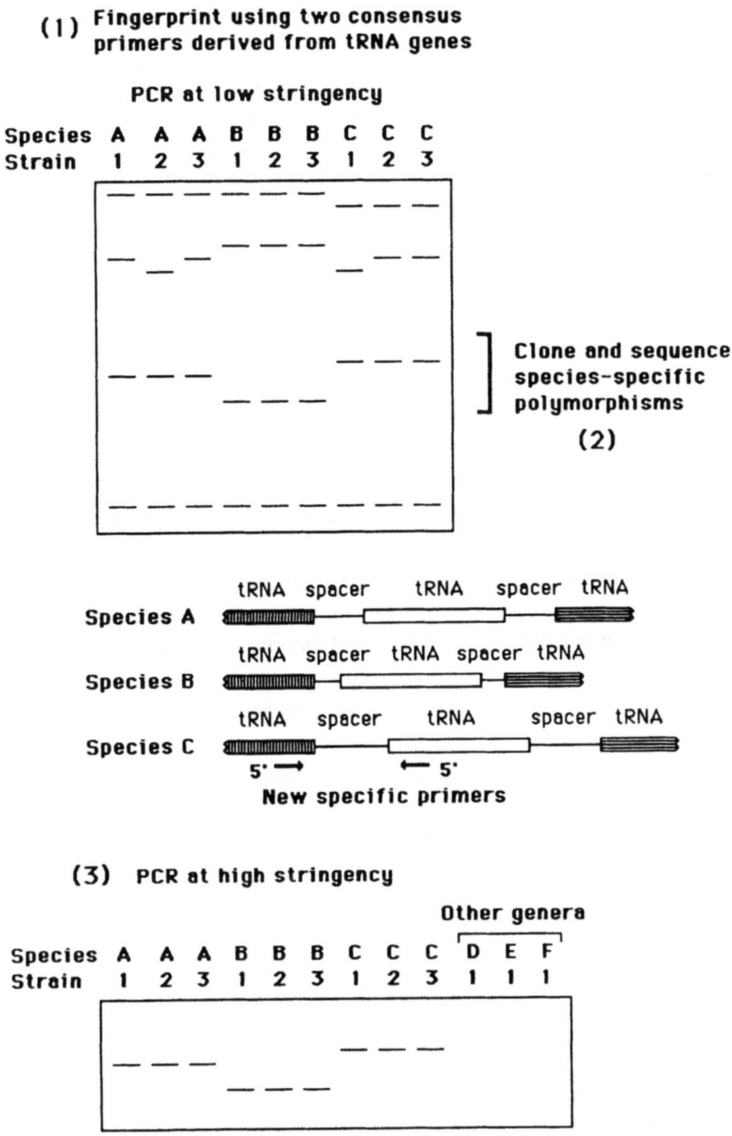

Figure 1B. The steps involved in developing a PCR diagnostic test are outlined. In step (1) a pair of primers from ***Figure 1A*** are used for PCR at moderate stringency. In step (2) PCR products that show the desired taxonomic variation in length are cloned. In step (3) primers are designed that amplify between particular pairs of tRNAs in the whole genus or family to give a characteristic product size in each taxonomic grouping. Alternatively, primers can be devised that lie in the spacer region and which only amplify in one taxonomic group.

outlines the strategy. The advantage of this strategy is the wide choice of potential PCR products that can be surveyed for the taxonomic distinction desired from all the spacer regions in the genome. Furthermore, the final high stringency PCR-based detection system can be designed to give a PCR product in a set of species or strains and distinguish among them based on product length. The system is a differential diagnostic within a genus or species because it generates a product in many species that can be distinguished by product length.

A tRNA-intergenic length polymorphism diagnostic system has been developed for *Staphylococci* (Welsh and McClelland, 1992), *Streptococci* (McClelland et al, 1992), *Acinetobacter* (Wiedmann-Al-Ahmad M et al, 1994), *Xanthomonas albilineans* (Honeycutt et al, 1995) and *Mollicutes* (McClelland and Welsh, unpublished data).

Differential Diagnostics

PCR methods that generate a product in more than one pathogen that can then be distinguished, for example, by length or by a restriction digest have the potential to detect not only a pathogen but to give a differential diagnosis between pathogens in a single experiment. While AP-PCR (RAPDs) and methods based on Southern blotting can produce diagnostic patterns in a wide variety of strains and species, these methods usually require culturing of organisms and thus are useful in epidemiology but are not generally suitable for rapid detection. On the other hand, methods such as PCR of rRNAs, tRNAs, and their spacer regions allow for primers that generate PCR products from many species. In some cases these can be distinguished simply by length. In other cases a restriction digest will give a diagnostic pattern.

It is perhaps surprising, given the bigger bang for the buck that differential diagnostics can provide, that very little in the way of differential diagnostic tests have been applied except in epidemiology. Almost all PCR tests currently available are what we would call detection tests.

Future Prospects

One of the most important future developments may be to make PCR-based detection systems more user-friendly and much cheaper so they can be adopted more widely as a routine tool for pathogen detection. The National Institute of Standards and Technology has begun an Advanced Technology Program focus area on diagnostics. We hope this means there will be more funding for improvements in pathogen detection.

One possibility is micromachining to make volumes smaller and so make PCR cheaper and faster. Detection methods that avoid gels can greatly improve speed and throughput. For example, advances in PCR-based detection include "TaqMan" which uses a probe with attached fluorescent dyes. When attached to the probe, the fluorescein is quenched by rhodamine. However, if the probe binds to a template within a target for PCR, then Taq polymerase 5′ to 3′ exonuclease activity releases the fluorescein, and luminesecence can be measured. The ligation chain reaction can also be adapted to avoid gels by modifying oligos with capture and detection haptens.

While we have concentrated on PCR-based technologies there are other nucleic acid detection technologies that have been successfully developed or are very exciting in their early stages of development. Other amplification methods include nucleic acid sequence based amplification (NASBA) (Kievits et al, 1991), an in vitro isothermal nucleic acid amplification technology in which amplification of target RNA or DNA sequences is accomplished by the simultaneous enzymatic activity of AMV reverse transcriptase, T7 RNA polymerase and RNase H, and SDA (strand displacement amplification) (Walker, 1993).

One area we have not discussed is direct hybridization using DNA oligonucleotides or PNA (with peptide instead of phosphate linkages) that carry fluorescent or other reporter groups and are directed to pathogen-specific DNA. Hybridization is detected under a microscope. DNA from the pathogen might also be enriched by affinity to such oligos.

Solid-state chips that carry a set of oligonucleotides to detect multiple organisms are also being developed (Pease et al, 1994). Material amplified by PCR, using conserved sequences such as rRNAs or tRNAs, could then be classified into species by hybridizing to particular sequences on the chip. Sensitivity of detection of such hybridization may rise to the point where unamplified material can be used directly.

ACKNOWLEDGMENTS:

This work was supported in part by grants from the National Institutes of Health: AI34829, NS33377 and CA68822, and a grant from the International Consortium for Sugarcane Biotechnology to Bruno WS Sobral.

REFERENCES

Al-Janabi SM, McClelland M, Peterson C, Sobral BWS (1994): Phylogenetic analysis organellar DNA sequences in the *Andropogoneae: Saccharinae. Theor Appl Genet* 88:933–944

Appels R, Honeycutt RL (1986): rDNA: Evolution over a billion years. In: *DNA Systematics*, Vol. 2: *Plants*, Dutra SK, ed. Boca Raton: CRC Press

Botstein D, White RL, Skolnick M, Davis RW (1980): Construction of a genetic linkage map in man using restriction fragment length polymorphisms. *Am J Hum Genet* 32:314–331

Colinet D, Kummert J, Lepoivre P, Semal J (1993): Identification of distinct potyviruses in mixedly-infected sweetpotato by the polymerase chain reaction with degenerate primers. *Phytopath* 84:65–69

Correll JC, Rhoads DD, Guerber JC (1993): Examination of mitochondrial DNA restriction fragment length polymorphisms, DNA fingerprints, and randomly amplified polymorphic DNA of *Colletotrichum orbiculare*. *Phytopath* 83: 1199–1204

DeScenzo RA, Harrington TC (1994): Use of (CAT)5 as a DNA fingerprinting probe for fungi. *Phytopath* 84:534–540

Ersek T, Schoelz JE, English JT (1994): PCR amplification of species-specific

DNA sequences can distinguish among *Phytophthora* species. *Appl Environ Microbiol* 60:2616–2621

Fox GE, Stackebrandt E, Hepsell RB, Gibson J, Maniloff J, Dyer TA, Wolfe RS, Balch WE, Tanner RS, Magrum LJ, Zablen LB, Blakemore R, Gupta R, Bonen L, Lewis BJ, Stahl DA, Luehrsen KR, Chen KN, Woese CR (1980): The phylogeny of prokaryotes. *Science* 209:457–463

Garbelotto M, Bruns TD, Cobb FW, Otrosina WJ (1993): Differentiation of intersterility groups and geographic provenances among isolates of *Heterobasidion annosum* detected by random amplified polymorphic DNA assays. *Can J Bot* 71:565–569

Gillings M, Fahy P, Davies C (1993): Restriction analysis of an amplified polygalacturonase gene fragment differentiates strains of the phytopathogenic bacterium *Pseudomonas solanacearum. Let Appl Microbiol* 17:44–48

Griffen AM, Heale JB, Bainbridge BW (1994): Cloning and characterization of the ribosomal RNA gene complex from the plant pathogen *Verticillium alboatrum. FEMS Microbiol Let* 118:291–296

Griffiths HM, Sinclair WA, Davis RE, Lee I-M, Dally EL, Guo Y-H, Chen TA, Hibben CR (1994): Characterization of mycoplasmalike organisms from *Fraxinus, Syringa,* and associated plants from geographically diverse sites. *Phytopath* 84:119–126

Hansen MA, Wick RL (1993): Plant disease diagnosis: Present status and future prospects. *Adv Plant Path* 10:65–126

Hartung JS, Daniel JF, Pruvost OP (1993): Detection of *Xanthomonas campestris* pv. citri by the polymerase chain reaction method. *Appl Environ Microbiol* 59:1143–1148

Higgins CF, Ames GF, Barnes F, Clement JM, Hofnung M (1982): A novel intercistronic regulatory element of prokaryotic operons. *Nature* 298:760–762

Hildebrand DC, Palleroni NJ, Schroth MN (1990): Deoxyribonucleic acid relatedness of 24 xanthomonad strains representing 23 *Xanthomonas campestris* pathovars and *Xanthomonas fragariae. J Appl Bacteriol* 68:263–269

Honeycutt RJ, Sobral BWS, McClelland M (1995): tRNAs reveal species-specific markers in *Xanthomonas. Microbiol*: in press

Hulton CSJ, Higgins CF, Sharp PM (1991): ERIC sequences: a novel family of repetitive elements in the genomes of *Escherichia coli, Salmonella typhimurium* and other enterobacteria. *Mol Microbiol* 5:825–834

Kievits T, Van Gemen B, Van Strijp D, Schukkink R, Dircks M, Adriaanse H, Malek L, Sooknanan R, Lens P (1991): NASBA Isothermal enzymatic in-vitro nucleic acid amplification optimized for the diagnosis of HIV-1 infection. *J Virolog Meth* 35:273–286

Langeveld SA, Dore JM, Memelink J, Derks AFLM, VanDer Vhagt CIM, Asjes CJ, Bol JF (1991): Identification of potyviruses using the polymerase chain reaction with degenerate primers. *J Gen Virol* 72:1531–1541

Leach JE, White FF, Rhoads ML, Leung H (1990): A repetitive DNA sequence differentiates *Xanthomonas campestris* pv. oryzae from other pathovars of *X. campestris. Mol Plant-Microbe Interact* 3:238–246

Lee I-M, Gundersen DE, Hammond RW, Davis RE (1994): Use of mycoplasmalike organism (MLO) group-specific oligonucleotide primers for nested-PCR assays to detect mixed-MLO infections in a single host plant. *Phytopath* 84:559–566

Lee I-M, Hammond RW, Davis RE, Gundersen DE (1993): Universal amplification and analysis of pathogen 16S rDNA for classification and identification of mycoplasmalike organisms. *Phytopath* 83:834–842

Leite RP, Minsavage GV, Bonas U, Stall RE (1994): Detection and identification of phytopathogenic *Xanthomonas* strains by amplification of DNA sequences related to the *hrp* genes of *Xanthomonas campestris* pv. vesicatoria. *Appl Environ Microbiol* 60:1068–1077

Louws FJ, Fulbright DW, Taylor Stephens C, deBruijn FJ (1994): Specific genomic fingerprints of phytopathogenic *Xanthomonas* and *Pseudomonas* pathovars and strains generated with repetitive sequences and PCR. *Appl Environ Microbiol* 60:2286–2295

Maclean DJ, Braithwaite KS, Manners JM, and Irwin JAG (1993): How do we identify and classify fungal plant pathogens in the era of DNA analysis? *Adv Plant Path* 10:207–244

Manulis S, Kogan N, Reuven M, Ben-Yephet Y (1994): Use of the RAPD technique for identification of *Fusarium oxysporum* f. sp. dianthi from carnation. *Phytopath* 84:98–101

Martin B, Humbert O, Camara M, Guenzi E, Walker J, Mitchell T, Andrew P, Prudhomme M, Alloing G, Hackenbeck R, Morrison DA, Boulnois GT, Claverys JP (1992): A highly conserved repeated DNA element located in the chromosome of *Streptococcus pneumonie. Nucleic Acids Res* 20:3479–3483

Martin F, Vairelles D, Henrion B (1993): Automated ribosomal DNA fingerprinting by capillary electrophoresis of PCR products. *Anal Biochem* 214: 182–189

McClelland M, Peterson C, Welsh J (1992): Length polymorphisms in tRNA intergenic spacers detected using the polymerase chain reaction can distinguish streptococcal strains and species. *J Clin Microbiol* 30:1499–1504

Mills PR, Sreenivasaprasas S, Brown AE (1992): Detection and differentiation of *Colletotrichum gloeosporioides* isolates using PCR. *FEMS Microbiol Let* 98:137–144

Moukhamedov R, Hu X, Nazar RN, Robb J (1994): Use of polymerase chain reaction-amplified ribosomal intergenic sequences for the diagnosis of *Verticillium tricorpus. Phytopath* 84:256–259

Namba S, Kato S, Iwanami S, Oyaizu H, Shiozawa H, Tsuchizaki T (1993): Detection and differentiation of plant-pathogenic mycoplasmalike organisms using polymerase chain reaction. *Phytopath* 83:786–791

Palleroni NJ, Kunisawa R, Contopooulou R, Dooudoroff M (1973): Nucleic acid homologies in the genus *Pseudomonas. Int J Syst Bacteriol* 23:333–339

Paran I, Michelmore RW (1993) Development of reliable PCR-based markers linked to downy mildew resistance gene in lettuce. *Theor Appl Biol* 85:985–993

Pease AC, Solas D, Sullivan EJ, Cronin MT, Holmes CP, Fodor SPA (1994):

Light-generated oligonucleotide arrays for rapid DNA sequence analysis. *Proc Nat Acad Sci USA* 91:5022–5026

Persing DH (1993): Target selection and optimization of amplification reactions. In: *Diagnostic Molecular Microbiology*, Persing DH, Smith TF, Tenover FC, White TJ, eds. Washington DC: ASM Press

Prosen D, Hatziloukas E, Schaad NW, Panopoulos NJ (1993): Specific detection of *Pseudomonas syringae* pv. phaseolicola DNA in bean seed by polymerase chain reaction-based amplification of a phaseolotoxin gene region. *Phytopath* 83:965–970

Ralph D, McClelland M (1994): Phylogenetic evidence for horizontal transfer of an intervening sequence between species in a spirochete genus. *J Bacteriol* 176:5982–5987

Ralph D, McClelland M, Welsh J, Baranton G, Perolat P (1993a): *Leptospira* categorized by arbitrarily primed PCR and by mapped restriction polymorphisms in PCR-amplified rDNA. *J Bacteriol* 175:973–981

Ralph D, Postic D, Baranton G, Pretzman C, McClelland M (1993b): Species of *Borrelia* distinguished by restriction site polymorphisms in 16S rRNA genes. *FEMS Micro Let* 111:239–244

Rasmussen OF, Reeves JC (1992): DNA probes for the detection of plant pathogenic bacteria. *J Biotechnol* 25:203–220

Rawadi G, Dussurget O (1995): Advances in PCR-based detection of mycoplasmas contaminating cell cultures. *PCR Methods and Applications* 4:199–208

Saiki RK, Scharf S, Faloona F, Mullis KB, Horn GT, Erlich HA. Arnheim N (1985): Enzymatic amplification of beta-globin genomic sequences and restriction site analysis for diagnosis of sickle cell anemia. *Science* 230:1350–4

Schneider BJ, Zhao J, Orser CS (1993): Detection of *Clavibacter michiganensis* subsp. sepedonicus by DNA amplification. *FEMS Microbiol Let* 109:207–212

Semer CR (1995): *Methods and Practice of Plant Disease Diagnosis*. Washington DC: American Phytopathological Society Press

Sujkowski LS, Goodwin SB, Dyer AT, Fry WE (1994): Increased genotypic diversity via migration and possible occurrence of sexual reproduction of *Phytophthora infestans* in Poland. *Phytopath* 84:201–207

Swaminathan B, Matar GM (1993) Molecular Typing Methods. In: *Diagnostic Molecular Microbiology*, Persing DH, Smith TF, Tenover FC, White TJ, eds. Washington DC: ASM Press

Tares S, Lemontey J-M, deGuiran G, Abad P (1994): Use of species-specific satellite DNA from *Bursaphelenchus xylophilus* as a diagnostic probe. *Phytopath* 84:294–298

Typas MA, Griffen AM, Bainbridge BW, Heale JB (1992): Restriction fragment length polymorphisms in mitochondrial DNA and ribosomal RNA gene complexes as an aid to the characterization of species and sub-species populations in the genus *Verticillium*. *FEMS Microbiol Let* 95:157–162

Vaneechoutte M, deBeenhouwer H, Claeys G, Verschraegen G, deRouck A, Paepe N, Elaichouni A, Portaels F (1993): Identification of *Mycobacterium* species by using amplified ribosomal DNA restriction analysis. *J Clin Microbiol* 31: 2061–2065

Verdier V, Dongo P, Boher B (1993): Assessment of genetic diversity among strains of *Xanthomonas campestris* pv. manihotis. *J Gen Microbiol* 139: 2591–2601

Versalovic J, Thearith K, Lupski JR (1991) Distribution of repetitive DNA sequences in eubacteria and application to fingerprinting of bacterial genomes. *Nucleic Acids Res* 19:6823–6831

Walker GT (1993): Empirical aspects of strand displacement amplification. *SO PCR Meth Appl* 3:1–6

Welsh J, McClelland M (1990): Fingerprinting genomes using PCR with arbitrary primers. *Nucleic Acids Res* 18:7213–7218

Welsh J, McClelland M (1991a):Genomic fingerprints produced by PCR with consensus tRNA gene primers. *Nucleic Acids Res* 19:861–866

Welsh J, McClelland M (1991b): Genomic fingerprinting with AP-PCR using pairwise combinations of primers: Application to genetic mapping of the mouse. *Nucleic Acids Res* 19:5275–5279

Welsh J, McClelland M (1992): PCR-amplified length polymorphisms in tRNA intergenic spacers for categorizing staphylococci. *Mol Microbiol* 6:1673–1680

Welsh J, Pretzman C, Postic D, Saint Girons I, Baranton G, McClelland M (1992): Genomic fingerprinting by arbitrarily primed PCR resolves *Borrelia burgdorferi* into three distinct groups. *Int J Sys Bacteriol* 42:370–377

Wiedmann-Al-Ahmad M, Tichy H-V, Schoen G (1994) Characterization of *Acinetobacter* type strains and isolates obtained from wastewater treatment plants by PCR fingerprinting. *Appl Environ Microbiol* 60:4066–4071

Wiglesworth MD, Nesmith WC, Schardl CL, Li D, Siegel MR (1994): Use of specific repetitive sequences in *Peronospora tabacina* for the early detection of the tobacco blue mold pathogen. *Phytopath* 84:425–430

Williams JGK, Kubelik AR, Livak KJ, Rafalski JA, Tingey SV (1990): DNA polymorphisms amplified by arbitrary primers are useful as genetic markers. *Nucleic Acids Res* 18:6531–6535

Zabeau M, Vos P, inventors; KEYGENE N.V., applicant (1992): Selective restriction fragment amplification: a general method for DNA fingerprinting. European Patent Application number 92402629.7

12

Molecular Approaches to Understanding and Manipulating Field Ecology of Microorganisms in Agriculture

KATE J. WILSON

Introduction: How Diverse is a Mouthful of Soil?

In Hindu mythology, Yashoda, the mother of the god Krishna, berates him as a child for eating mud. As she moves to catch him, he opens his mouth and she falls down in praise of her son, for she can see the whole universe in his mouth. If we take the story literally, it can also have a great deal of truth, for the diversity of organisms in one gram of soil is staggering.

Counts of bacteria in soil can be as high as 1.5×10^{10} per gram dry soil using a measurement based on acridine orange staining (Torsvik et al, 1990). It is impossible to obtain reliable figures on the diversity of species that this encompasses because estimates of the proportion of bacteria that we have been able to culture and identify ranges from $<1\%$ to a maximum of 10% (Torsvik et al, 1990; Hawksworth and Mound, 1991; Amann et al, 1994). However, one set of data has indicated that the complexity of DNA isolated directly from the bacterial community in soil is 200-fold greater than the complexity of DNA isolated from a population of bacteria cultivated from the same soil, leading to an estimate of about 4,000 different species, each represented by an average of 3.8×10^6 individual bacteria per gram dry soil (Torsvik et al, 1990). This extraordinary diversity and complexity is borne out by molecular analyses of the eubacterial content of soil samples, as discussed later in this chapter.

These figures only consider bacteria. Bacteria, including the mycelial actinomycetes, are included among the soil microflora, along with fungi, and algae. All of these have critical roles in global nutrient cycling (Leung et al, 1994). Actinomycetes, algae, and fungi additionally play an important role in the for-

The Impact of Plant Molecular Genetics
BWS Sobral, Editor

mation and maintenance of soil structure. The proportion of total soil biomass contributed by the different groups may vary considerably. In a number of natural ecosystems, 90%–95% of microfloral biomass in the upper soil horizons comprised fungal mycelium, spores and yeast cells, with 25% of the remaining bacterial biomass comprising filamentous actinomycetes (Zvyagintsev, 1994). By contrast, in an agricultural ecosystem, the biomass of bacteria exceeded that of fungi 100-fold (Bloem et al, 1994). Soil microfauna include protozoa, nematodes, and small mites whose impact on agriculture is primarily by the effect of their predation on the microflora.

The title of this chapter thus encompasses an extraordinarily broad area. The discussion referred to above encompasses only soil microorganisms. Additionally, phylloplane microbes, or microorganisms that affect animal growth and health, all have an impact on agriculture. How then can we ever understand, let alone develop the ability to manipulate, specific microorganisms that play a particularly significant role in agricultural production, no matter how many novel molecular techniques we may have? The literature is increasingly filled with technical developments, but their application to real-world ecological questions is still in its infancy. It is critical that molecular biologists truly capture the imagination of the field scientists, and vice versa, so that the two groups of scientists can work closely together to ensure that new tools and approaches really work when applied to a field-level question and are then actually put to use.

In this chapter I will not exhaustively cover all novel molecular methodologies, as these are excellently reviewed elsewhere (Drahos, 1991; Pickup, 1991; Atlas et al, 1992; Kloepper and Beauchamp, 1992; Sayler et al, 1992; Ward et al, 1992; Amann et al, 1994; Wilson, 1995). Instead I will discuss a few key areas, and try to highlight actual or potential examples in which molecular biology will help us to understand and manipulate the field ecology of microorganisms, primarily bacteria, in agriculture.

Identification of Novel Microorganisms

One of the most novel areas in which molecular biology has begun to contribute to microbial ecology is in the identification and detection of previously unknown microorganisms through analysis of specific DNA sequences. In most cases these are microorganisms that have not yet been cultured, and therefore have not been identified by any other means. A number of sequences have been found to provide taxonomically valid and useful information across very distant species boundaries (Schlegel, 1994), the most widely used being the sequence of 16*S* rRNA genes. These sequences provide powerful tools for taxonomic purposes because they comprise a mixture of highly variable and highly conserved regions. The conserved regions allow the use of universal primers to amplify regions of the gene for cloning and/or sequencing (Weisburg et al, 1991), and the variable regions provide taxonomic information. The information gleaned from 16*S* rRNA sequence is not on its own sufficient to define species (Stackebrandt and Goebel, 1994), but does at least give a very reliable indication of species groups.

The greatest novelty of this approach is the ability to detect microbial species solely on the basis of DNA isolated from the environment. The first examples of the use of this technique looked at microbial populations in two environments, an algal mat in Yellowstone National park in the United States, which had been characterized extensively using conventional techniques of microbial ecology (Ward et al, 1990), and a microbial population in the Sargasso Sea (Giovannoni et al, 1990). In both cases all the sequences obtained were novel 16*S* rRNA sequences. In the case of the algal mat from Yellowstone, it was possible to compare the sequences derived from DNA obtained directly from the environment with those obtained from 15 different species that had been cultured from that site. None of the environmental sequences corresponded to those of the cultured organisms, and in most cases the differences were substantial. This indicates that the variety of prokaryotes within that algal mat is far greater than previously imagined, and hence that our understanding of the ecology of even that single ecosystem is limited.

If DNA sequence is obtained from uncultured organisms, tools for studying the ecology of these organisms can be developed. The most powerful tools are the use of labelled oligonucleotides that hybridize to taxonomically specific regions of the genome and can be detected by virtue of the attached label, generally a radionucleotide or a fluorescent tag. It is possible to develop probes that specifically detect kingdoms, phyla, major bacterial groupings such as the different subdivisions of the proteobacteria, or closely related species groups (Giovannoni et al, 1988; DeLong et al, 1989; Amann et al, 1990b). The use of fluorescently labelled oligonucleotides is particularly powerful as it can enable detection of the microbes in situ in the environment (Amann et al, 1991). It is also applicable to high-throughput methods of enumerating bacteria such as flow-cytometry (Amann et al, 1990a). An example of this approach would be the study of magnetotactic bacteria; fluorescently labelled oligonucleotide probes to 16*S* rRNA have been used to study the ecology of an uncultured, magnetotactic bacterial species in a freshwater lake in Upper Bavaria in Germany (Spring et al, 1993; Amann et al, 1994).

The implications of these techniques for our understanding of the diversity of soil microorganisms are clear, and in theory we now have a tool to examine the true diversity of soil microorganisms. This information could be of considerable agronomic importance, as it is likely that the effects of a number of agricultural management strategies are mediated, at least in part, by shifts in soil microbial content (Pankhurst et al, 1994; Bloem et al, 1994).

Attempts have been made to analyze the diversity of members of the domain Bacteria in soil by sequencing cloned 16S rRNA gene sequences obtained by PCR amplification of DNA extracted directly from soil. These studies again emphasized the extent of the as yet unknown diversity of microbes in the environment (Liesack and Stackebrandt, 1992; Stackebrandt et al, 1993). In the first of these publications, none of 30 16*S* rRNA sequences obtained were identical to known sequences, and seven sequences appeared to belong to members of an entirely novel main line of descent with the domain Bacteria (Stackebrandt and Liesack, 1992). Analysis of further clones only revealed greater diversity, with little of it corresponding to species expected to predominate in the acid soil conditions from which the samples were obtained (Stackebrandt et al, 1993).

The substantial discrepancy between earlier information on microbial community diversity obtained from culturing techniques and those obtained from analysis of environmental DNA are discussed in Stackebrandt et al (1993) who point out the many remaining limitations of applying this technology to analysis of soil microbial community diversity. A key problem is the reliability of techniques that extract DNA from soil. Two general approaches are available. The cell extraction approach first isolates cells from the soil and subsequently lyses them, thus yielding relatively large amounts of reasonably clean DNA (Ogram et al, 1987; Holben et al, 1988). The major problem is the perceived bias in the types of cells that are readily extracted from the soil. By contrast, in direct lysis methods cells are lysed in situ, and DNA is isolated directly from the environment (Ogram et al, 1987; Steffan et al, 1988). This technique is thought to reduce bias, but it requires greater treatment to obtain DNA of sufficient quality for PCR reactions. In either case, it is extremely hard to be certain just how representative of the soil community a particular DNA preparation is.

Other potential biases are introduced with the selection of primers for PCR amplification. Generally primers that are thought to amplify 16*S* rRNA sequences from all members of the domain Bacteria are used. However, small nucleotide differences could markedly influence the efficiency of amplification of different template sequences, giving rise to over- or underrepresentation in clone libraries. A separate problem is that these methods will invariably lead to repeated analysis of the most physically abundant PCR products, whether these represent the most abundant organisms or simply the most accessible and amplifiable target sequences, and so some form of subtractive procedures may have to be developed to reduce redundant analyses and allow detection of rare species.

Techniques for in situ detection are also more problematic in soil. Detection through hybridization with fluorescently labelled oligonucleotides generally relies on hybridization to the abundant rRNA molecules present in metabolically active cells. If cells are quiescent, as many in soil are likely to be, the sensitivity of detection is dramatically reduced (Hahn et al, 1992). This limitation can be overcome to a degree by the use of protocols to amplify the detection signal. For example, probes labelled with biotin, which were detected by binding to fluorescein-labelled avidin, enhanced the sensitivity of detection of marine nanoplankton protists (Lim et al, 1993), and a similar technique employing digoxygenin-labelled probes detected with a digoxygenin-specific antibody/alkaline phosphatase conjugate, enhanced the sensitivity of detection of specific *Frankia* strains in symbiosis with plants (Hahn et al, 1993). However, in situ hybridization and detection techniques in soil are more complicated because the autofluorescence of organic material present in soil obscures reliable detection of cells (Hahn et al, 1992). Moreover, methods that rely on detection of a color change, as in the detection of alkaline phosphatase by formation of a blue product from the substrate 5-bromo-4-chloro-3-indolyl phosphate, are obviously not reliable in soil. An interesting alternative might be to use a system in which the detected signal is light, as experience with the luciferase system indicates that there is little background emission of light in soil, and that light emission can be reliably detected, at least in dilute soil suspensions (Rattray et al, 1990). This approach would be possible using, for example, the digoxygenin system and detecting the alkaline phosphatase using standard commercially available chemiluminescent substrates for alkaline phosphatase.

Shifts in microbial community structure have been analyzed using rRNA-specific probes in certain well-characterized ecosystems. Perhaps the most detailed is an analysis of shifts in microbial community structure in activated sewage sludge (Wagner et al, 1994). In these experiments, fluorescent probes were used that were targeted against: *Acenitobacter* sp.; bacteria of the *Cytophaga-Flavobacterium* cluster; Gram-positive bacteria with high G + C content; bacteria of the alpha, beta, or gamma subdivision, respectively, of the proteobacteria; and a generic probe to detect all eubacteria. Experiments based on culturing of microorganisms had indicated that *Acenitobacter* species might play a predominant role in sewage plants with enhanced biological phosphate removal. Community structure was analyzed by counting the proportion of microscopically visible cells that hybridized with each fluorescently-labelled, phylogenetically-targeted probe. This analysis indicated that the proportion of the total bacterial community comprising *Acenitobacter* species did not alter significantly with different treatments shown to enhance biological phosphate removal. By contrast, the numbers of cells hybridizing with the probes specific for Gram-positive bacteria with high G + C content appeared much more strongly correlated with treatment strategies that enhanced biological removal of phosphates. In this example molecular tools were used to provide valuable information about an environmentally important microbial community.

An interesting alternative to taxonomic surveys is the use of nucleic acid sequences to survey particular physiological or metabolic traits. An example is the use of *nifH* sequence analysis to survey nitrogen-fixing species associated with rice roots. This gene encodes one of the subunits of the enzyme nitrogenase, responsible for nitrogen-fixation, and is strongly conserved among species allowing design of primers to conserved regions for amplification of environmental *nifH* genes. Moreover, there appears to be a good correlation between taxonomic data derived from 16S rRNA sequence analysis and *nifH* gene sequence analysis, implying little, if any, horizontal gene transfer, which might complicate matters (Young, 1992). Culturing techniques have implicated *Azospirillum* species as the dominant nitrogen-fixing species associated with rice roots. However, analysis of partial *nifH* sequences obtained from DNA extracted from rice roots only produced one out of 23 *nifH* sequences that appeared to derive from a member of the alpha subdivision of proteobacteria, which is the group that contains *Azospirillum* (Ueda et al, 1995). While this work is far from definitive, it does suggest that the data implicating *Azospirillum* as the major contributor of biologically fixed nitrogen to rice could be reexamined. Interestingly, the authors also searched for *nifH* sequences in DNA extracted from soil surrounding the roots, but they were unsuccessful in obtaining PCR amplification products using the *nifH* primers. By contrast, the same DNA preparation from soil did yield amplification products with ribosomal RNA gene primers, indicating that the DNA preparation procedure was successful. The implication is that diazotrophs are far more abundant in the rhizosphere than in the surrounding soil. Clearly these results are preliminary and need further validation, but they suggest one means by which DNA amplification and sequence analysis from environmental samples could be used to examine questions of agricultural relevance.

One of the limitations in applying these methodologies to ecological analysis is the throughput of sequencing. One possibility for increasing this is to use the new chip-based sequencing technologies in a resequencing strategy. Microchip sequencing uses oligo hybridization to deduce DNA sequence, termed sequencing by hybridization (SBH) (Fodor et al, 1993). In this example, short oligonucleotides (e.g. 8–10 nucleotides in length) constituting the sequence of the 16*S* rRNA would be put onto a chip. The chip would include oligonucleotides comprising all known permutations of the 16*S* rRNA sequence. 16*S* rRNA genes cloned from environmental DNA would then be labelled, sheared, and hybridized to the chip. Hybridization to specific oligos on the chip could then be read from the pattern of fluorescence emissions on the chip using laser confocal detection systems. The chip would thus serve as a "biodiversity chip", giving an immediate read-out of the sequence of the 16*S* rRNA gene being analyzed. This would increase the throughput of sequencing, and hence of gaining information that could be used for ecological studies, by orders of magnitude.

These techniques based on detection and analysis of specific sequences clearly have a great deal to contribute to our knowledge of the microbial content and ecology of diverse environments. However, they will only become relevant to agriculture once a number of methodological constraints have been addressed, and once the methods become sufficiently high-throughput to enable routine screening of soil samples for biodiversity content, coupled with indications of the relative abundance of different species. One of the key constraints is the difficulty of extracting representative and usable DNA from soil; it is no accident that the majority of papers using these tools in microbial ecology refer to aquatic systems. It is a tall order, but, in the long run if these problems can be solved, it is possible that these techniques will be used in a very practical sense to assess the impact of application of different chemicals, or of different tilling regimes, on soil microbial content, and the possible correlation of beneficial or detrimental changes in soil properties with the emergence or disappearance of particular soil species.

Discrimination of Genotypes Within Species

Molecular biology has probably had its greatest impact to date on microorganisms and agriculture through its ability to discriminate closely related microorganisms with a specificity not previously possible. This may be achieved through a variety of techniques, including: total protein profiles (Sen, 1994); restriction enzyme fingerprints (Sadowsky, 1994); restriction fragment length polymorphisms (RFLPs) (Lazo et al, 1987); fingerprinting by hybridization with repeated sequences (Levy et al, 1991; Leach et al, 1992); random amplification of polymorphic DNA (RAPDs)/arbitrarily primed polymerase chain reaction (AP-PCR) (Williams et al, 1990; Welsh and McClelland, 1990); repetitive element PCR (rep-PCR) (Versalovic, 1991; de Bruijn, 1992); and variations on or combinations of the above.

These techniques have made possible discrimination of different fungal pathovars (Michelmore and Hulbert, 1987; Levy et al, 1991), rhizobial strains (de Bruijn, 1992), and bacterial pathogens (Leach et al, 1992; Louws et al, 1994), to cite only a few selected examples. The methods vary in their ease of use and ability to discriminate different strains of bacteria or fungi. However, all provide a level of sensitivity that greatly exceeds the phenotype classification that was used previously. For example, pathogens have been previously classified into pathovars or pathotypes according to their reaction on a standard set of host cultivars, and symbiotic bacteria have been classified according to the groups of host legumes that they could nodulate. Fingerprinting techniques allow much more reliable discrimination between strains.

Of these techniques, the rep-PCR techniques, which fingerprint bacterial genomes by PCR amplification of short conserved regions of the bacterial genome, are among the most robust and straightforward to use. The use of PCR is far simpler than any technique that relies on hybridization as there is no need for probe preparation. Additionally, there is often no need for DNA preparation because adequate lysis of bacterial cells occurs on heating a PCR reaction to 95°C to release sufficient genomic DNA to act as a template for amplification. The use of rep-PCR is more reproducible than the use of RAPDs/AP-PCR because longer primers are used (18–22 mers, compared to 8–12 mers), and the primers are targeting known sequences. This enables the use of much more stringent PCR conditions, thus reducing the likelihood of any background nonspecific PCR products and reducing the variability that occurs when short arbitrary primers hybridize to imperfectly matched sequences in the genome (Louws et al, 1994).

These tools, when applied to groups of organisms with well-characterized background information, can provide considerable information about the ecology and evolution of symbionts and pathogens in the field. One example would be a study in the Philippines in which strains of the agent of bacterial blight on rice, *Xanthomonas oryzae* pv. Oryzae, had been collected from different locations over a 20-year period, during which the cultivars of rice preferentially grown in the Philippines had changed. These strains had been previously classified in pathovar groupings according to their reaction on five different host cultivars. Analysis using hybridization with a repeated sequence indicated that the genetic diversity greatly exceeded the diversity characterized through conventional pathogenicity typing procedures. These techniques allowed the influence of factors such as host cultivar, climate, and cropping system in the evolution of new pathogen genotypes to be examined.

Similarly, fingerprinting of *Magnaporthe grisea*, the rice blast fungus, was able to resolve a mass of contradictory data about the diversity and rapidity of origin of pathotypes as defined on seven or more differential rice cultivars (Levy et al, 1991). This analysis showed that, in the United States, genomic fingerprints obtained by hybridization with a genomic repeat sequence of *M. grisea* accurately delimited the pathotypes defined by conventional analysis and indicated that the pathogen genotypes were quite stable, contrary to the claims of other workers. This stability means that the genetic fingerprints can be used to identify isolates from distinct geographic locations and hence provide reliable markers for studying the ecology and evolution of this important pathogen. This tool is now being used as part of a strategy to breed for durable resistance to rice blast fungus (Tohme et al, 1992).

In the future, substantial increase in throughput of these fingerprinting approaches may be achieved using the microchips described above that have been designed for sequencing by hybridization. Hybridization of amplified regions of genomic DNA to chips arrayed with oligonucleotides of defined sequence could generate characteristic fluorescence (i.e. hybridization) patterns which will be diagnostic of specific genomes.

By and large these fingerprinting tools do not have taxonomic value. However, in a few cases they do. For example, in bacteria, it has been possible to use PCR of ribosomal 16*S* RNA genes, followed by digestion of amplified fragments, to generate diagnostic patterns of RFLPs that can differentiate among different species of rhizobia (Laguerre et al, 1994). Likewise, RFLP analysis of PCR products of ribosomal DNA spacer regions can also provide species identifications (Vilgalys and Hester, 1990; Jensen et al, 1993). However, these techniques can only be used for species identifications if a database exists for a group of closely related species.

How Can We Improve the Performance of Known Beneficial Microorganisms

Known beneficial microorganisms include: symbiotic nitrogen-fixing bacteria (*Rhizobium, Bradyrhizobium, Azorhizobium* species); free-living and associative nitrogen-fixing bacteria; fungi and bacteria that mobilize phosphate in soil; and fungi and bacteria that are antagonistic to other, disease-causing or otherwise deleterious bacteria. In some cases the strains of bacteria are well defined; in others they remain to be closely identified. In each case, however, there is the potential to enhance the contribution of these beneficial microorganisms to agriculture through microbial inoculation.

The concept of microbial inoculation to improve crop growth dates back at least 100 years (Campbell and Macdonald, 1989), but remains a highly troublesome technology. Commercial methods of application of *Rhizobium* application, for example, may lead to 99% of applied cells dying within 24 hours of inoculation (Herridge et al, 1987), and hence research is still continuing into the most appropriate means of inoculation (Brockwell et al, 1988). Problems to be addressed in developing any inoculation technology include: devising a means to culture the bacteria or fungi; devising a suitable inoculum carrier and means to produce the inoculum; deciding the optimal means of applying the inoculum; and monitoring the success of inoculation in regard to its effect on plant growth and establishment of the strain in the field. Application of biocontrol agents is often even less successful because in most cases the introduced organisms have to compete against an already established population in the phyllosphere (Andrews, 1992).

How can molecular biology help? The problem in testing all these variables is knowing how they affect the end result, i.e., establishment of a specific strain in the field or on the crop. Methods of identifying the inoculant strain have included antigenic methods or use of strains marked with specific patterns of antibiotic resistance, but these methods are all extremely labor intensive. For example, in the case of *Rhizobium* the contents of individual nodules must be painstakingly analyzed. Hence, logistics have dictated very small sample sizes and have mitigated against extensive analysis of the success of inoculation technology as a routine measurement in inoculation trials (Wilson, 1995).

Recently, a number of marker gene systems have been developed for tracking specific strains of bacteria in the field (Drahos, 1991; Pickup, 1991; Kloepper and Beauchamp, 1992; Wilson, 1995). A gene encoding a protein that is easy to assay, generally an enzyme, is introduced into the specific strain to be studied and acts as a molecular tag, distinguishing the marked strain from other, closely-related strains of bacteria or fungi. The advantages of using a reporter gene depend largely on the properties of the reporter gene product. To date, reporter genes used as markers for Gram-negative bacteria in microbial ecology have included *lacZ* encoding β-galactosidase, the *xylE* gene, encoding catechol 2,3-dioxygenase and the different sets of luciferase genes, the bacterial *luxAB* genes or the *luc* gene from fireflies. Each has different advantages and limitations (Drahos, 1991; Wilson, 1995). A particular advantage of the *lux* system when studying free-living bacteria in soil is the ability to unambiguously identify colonies of marked bacteria in a high background of unmarked bacteria (Cebolla et al, 1993).

In the case of *Rhizobium* strains, and many other bacteria that interact with plants, the most powerful marker gene available at present for studies of microbial ecology is the *gusA* gene encoding β-glucuronidase (GUS). This is because there is no background activity in plants or in most bacteria or fungi that interact with plants, and additionally, there are a number of straightforward assays available for GUS that allow spatial localization of marked bacteria. For example, if *gusA* is introduced into a specific *Rhizobium* strain, which is then inoculated onto an appropriate host plant, nodules formed by the marked strain turn blue when the nodulated roots are incubated in buffer containing the substrate X-Gluc, whereas those induced by other strains show no colour change (Wilson et al, 1991; Streit et al, 1992; Wilson et al, 1995; Streit et al, 1995).

Using this method, we have processed more than 100 intact roots of *Phaseolus vulgaris* (common bean), each with approximately 200 nodules, and within twenty-four hours of harvest have been able to gain a rapid qualitative assessment of the extent of nodulation of each root by the GUS-marked inoculum strain. Previously, typical sample sizes for determination of percentage nodule occupancy by an inoculated strain were 20–30 nodules per replication, and a maximum of 600 nodules could be analyzed for nodule occupancy in a full day's work. The throughput of nodule typing can therefore be increased at least 30-fold using an appropriate marker gene, with a consequent increase in statistical accuracy.

The technique works equally well in cases in which only a very few nodules are formed by the inoculant strain, in some cases only three or four nodules out of a total of approximately 200 nodules per plant, and in which 100% of the nodules are formed by the inoculant strain. Importantly, it also shows the physical location of nodules on the root, which provides information about the movement of the inoculant strain in the soil. Inoculum is almost always applied at the level of the seed, and there is little information about how rapidly the inoculant strain travels through the soil as the roots germinate and grow downwards.

This method can be used to detect colonization of the rhizosphere by individual marked bacteria (Katupitiya et al, 1994). Appropriate marker genes open the way for extremely detailed study of the early stages of infection and for colonization of root and leaf surfaces by rhizosphere and phyllosphere bacteria. Such information is needed to develop either theoretical or empirical understanding of optimal methods of inoculum application (Kloepper and Beauchamp, 1992; Andrews, 1992).

Genetic Manipulation to Achieve Better Inoculation Success

As well as providing new tools for studying inoculation success, genetic manipulation can be used to enhance the establishment of microbes in the environment. There are a number of examples from *Rhizobium* inoculation technology. For example, strains can be engineered to produce toxins to which they are resistant but which kill other, competing strains of bacteria: a *Rhizobium leguminosarum* bv. *trifolii* strain with an introduced trifolitoxin gene was found to have enhanced competitiveness on clover (Triplett, 1990). Another strategy is to manipulate strains with particular advantageous properties, to try to render them better able to survive in specific conditions. An example would be the transfer of a symbiotic plasmid from an acid-tolerant *R. leguminosarum* bv. *trifolii* strain into a more effective, but acid-sensitive strain, to confer acid resistance on the preferred strain (Chen et al, 1991).

A very interesting concept is that of engineering specificity between the host plant and the inoculant strain. This has been achieved using genetic recognition factors for the symbiosis between *R. leguminosarum* bv. *viciae* and the host plant pea (*Pisum sativum*). The work is based on the observation that wild pea cultivars from Afghanistan and the Middle East fail to form nodules from rhizobial strains present in western European and in North American soils. This has been shown to be due to a plant gene, *sym-2*, which is present in the Afghanistan cultivars and confers resistance to nodulation by European rhizobial strains. However, this nodulation exclusion can be overcome by the product of a nodulation gene, *nodX*, present in local strains of rhizobia. It has been possible to introduce the *sym-2* gene into commercial pea lines through crosses with the Afghanistan peas, and the *nodX* gene has been cloned and introduced into commercial *R. leguminosarum* bv. *viciae* inoculant strains. The result is a pea cultivar that totally failed to nodulate with indigenous strains of *R. leguminosarum* bv. *viciae* present in North American soils, and would only nodulate with strains that had been engineered to contain the *nodX* gene to overcome this nodulation exclusion (Fobert et al, 1991). This could be a very efficient means of ensuring the competitive success of a highly effective inoculum strain.

Summary and Future Prospects

In this chapter I have tried to give an overview of some of the ways in which molecular biology may be able to contribute to our understanding and control of the field ecology of beneficial and detrimental microorganisms in agriculture. In each case I have not exhaustively reviewed all of the techniques nor reviewed all the examples of the applications, but have tried to illustrate how they may be used to further our ability to understand the ecology of microbes in the environment. Of the areas discussed, one set of techniques, those that enable us to identify and study organisms through their DNA sequences, allows us to gain entirely new information about microbial ecology. To date, these techniques have had little or no impact on understanding the microbial ecology of agricultural systems, but in the long run they may provide entirely new ways of analyzing the impacts of different agricultural practises on microbial biodiversity.

Molecular fingerprints and introduced marker genes do not generate entirely new information, but both provide a far more efficient and far more accurate means of identifying and distinguishing microbial strains in the environment. These techniques are perhaps the most likely to contribute to our understanding of the diversity, evolution and ecology of agriculturally important microorganisms in the short term, if applied to agriculturally relevant questions for which there are already considerable background agronomic data.

Finally, I have discussed a couple of examples whereby molecular biology can be used to improve the ability of a microbial strain to establish and compete in the environment. This is probably the most contentious area because the impact on the environment and on microbial diversity of releasing genetically engineered organisms is unknown (Wilson and Lindow, 1993; Leung et al, 1994). However, these examples do illustrate interesting applications of the tricks of molecular biology, not to engineer a better product, but to enhance the ability of a natural product to survive and compete in the field.

REFERENCES

Amann RI, Binder BJ, Olson RJ, Chisholm SW, Devereux R, Stahl DA (1990a): Combination of 16S rRNA-targeted oligonucleotide probes with flow cytometry for analyzing mixed microbial populations. *Appl Environ Microbiol* 56:1919–1925

Amann RI, Krumholz L, Stahl DA (1990b): Fluorescent oligonucleotide probing of whole cells for determinative, phylogenetic and environmental studies in microbiology. *J Bacteriol* 172:762–770

Amann R, Ludwig W, Schliefer K-H (1994): Identification of uncultured bacteria: a challenging task for molecular taxonomists. *Am Soc Microbiol News* 60:360–365

Amann R, Springer N, Ludwig W, Görtz H-D, Schliefer K-H (1991): Identification *in situ* and phylogeny of uncultured bacterial endosymbionts. *Nature* 351:161–164

Andrews JH (1992): Biological control in the phyllosphere. *Ann Rev Phytopathol* 30:603–635

Atlas RM, Sayler G, Burlage RS, Bej AK (1992): Molecular approaches for environmental monitoring of microorganisms. *BioTechniques* 12:706–717

Bloem J, Lebbink G, Zwart KB, Bouwman LA, Burgers SLGE, Devos JA, Deruiter PC (1994): Dynamics of microorganisms, microbivores and nitrogen mineralization in winter wheat fields under conventional and integrated management. *Agric Ecosyst Environ* 51:129–143

Brockwell J, Gault RR, Herridge DF, Morthorpe LJ, Roughley RJ (1988): Studies on alternative means of legume inoculation: microbiological and agronomic appraisals of commercial procedures for inoculating soybeans with *Bradyrhizobium japonicum. Aust J Agric Res* 39:965–972

Campbell R, Macdonald RM, eds. (1989): *Microbial Inoculation of Crop Plants*, Special publications of the Society for General Microbiology, Vol. 25. Oxford: IRL Press

Cebolla A, Ruiz-Berraquero F, Palomares AJ (1993): Stable tagging of *Rhizobium meliloti* with the firefly luciferase gene for environmental monitoring. *Appl Environ Microbiol* 59:2511–2519

Chen H, Richardson AE, Gartner E, Djordevic MA, Roughley RJ, Rolfe BG (1991): Construction of an acid-tolerant *Rhizobium leguminosarum* bv *trifolii* strain with enhanced capacity for nitrogen fixation. *Appl Environ Microbiol* 57:2005–2011

de Bruijn FJ (1992): Use of repetitive (repetitive extragenic palindromic and enterobacterial repetitive intergeneric consensus) sequences and the polymerase chain reaction to fingerprint the genomes of *Rhizobium meliloti* isolates and other soil bacteria. *Appl Environ Microbiol* 58:2180–2187

DeLong EF, Wickham GS, Pace NR (1989): Phylogenetic stains: ribosomal RNA-based probes for the identification of single microbial cells. *Science* 243:1360–1363

Drahos DJ (1991): Current practices for monitoring genetically engineered microbes in the environment. *AgBiotech News Info* 3:39–48

Fobert PR, Roy N, Nash JHE, Iyer VN (1991): Procedure for obtaining efficient root nodulation of a pea cultivar by a desired *Rhizobium* strain and preempting nodulation by other strains. *Appl Environ Microbiol* 57:1590–1594

Fodor SPA, Rava RP, Huang XC, Pease AC, Holmes CP, Adams CL (1993): Multiplexed biochemical assays with biological chips. *Nature* 364:555–556

Giovannoni SJ, Britschgi TB, Moyer CL, Field KG (1990): Genetic diversity in Sargasso Sea Bacterioplankton. *Nature* 345:60–63

Giovannoni SJ, DeLong EF, Olsen GJ, Pace NR (1988): Phylogenetic group-specific oligonucleotide probes for identification of single microbial cells. *J Bacteriol* 170:720–726

Hahn D, Amann RI, Ludwig W, Akkermans A, Schliefer K-H (1992): Detection of micro-organisms in soil after *in situ* hybridization with rRNA-targeted, fluorescently labelled oligonucleotides. *J Gen Microbiol* 138:879–887

Hahn D, Amann RI, Zeyer J (1993): Whole-cell hybridization of *Frankia* strains with fluorescence or digoxygenin-labeled 16S rRNA-targeted oligonucleotide probes. *Appl Environ Microbiol* 59:1709–1716

Hawksworth DL, Mound LA (1991): Biodiversity databases; the crucial significance of collections. In: *The Biodiversity of Microorganisms and Invertebrates: Its Role in Sustainable Agriculture*, Hawksworth DL, ed. Wallingford, UK: CAB International

Herridge DF, Roughley RJ, Brockwell J (1987): Low survival of *Rhizobium japonicum* inoculant leads to reduced nodulation, nitrogen fixation and yield of soybean in the current crop but not in the subsequent crop. *Aust J Agric Res* 38:75–82

Holben WE, Jansson JK, Chelm BK, Tiedje JM (1988): DNA probe method for the detection of specific microorganisms in the soil bacterial community. *Appl Environ Microbiol* 54:703–711

Jensen MA, Webster JA, Straus N (1993): Rapid identification of bacteria on the basis of polymerase chain reaction-amplified ribosomal DNA spacer polymorphisms. *Appl Environ Microbiol* 59:945–952

Katupitiya S, New PB, Elmerich C, Kennedy IR (1994): Improved nitrogen fixation in 2,4-D treated wheat roots associated with *Azospirillum lipoferum*: studies of colonization using reporter genes. *Soil Biol Biochem* 27:477–452

Kloepper JW, Beauchamp CJ (1992): A review of issues related to measuring colonization of plant roots by bacteria. *Can J Microbiol* 38:1219–1232

Laguerre G, Allard M-R, Revoy F, Amarger N (1994): Rapid identification of rhizobia by restriction fragment length polymorphism analysis of PCR-amplified 16S rRNA genes *Appl Environ Microbiol* 60:56–63

Lazo GR, Roffey R, Gabriel DW (1987): Pathovars of *Xanthomonas campestris* are distinguishable by restriction fragment length polymorphism. *Int J Syst Bacteriol* 37:214–221

Leach JE, Rhoads ML, Vera Cruz CM, White FF, Mew TW, Leung H (1992): Assessment of genetic diversity and population structure of *Xanthomonas oryzae* pv oryzae with a repetitive DNA element. *Appl Environ Microbiol* 58:2188–2195

Leung K, England LS, Cassidy MB, Trevors JT, Weir N (1994): Microbial diversity in soil: effect of releasing genetically engineered micro-organisms. *Mol Ecol* 3:413–422

Levy M, Romao J, Marchetti MA, Hamer JE (1991): DNA fingerprinting resolves pathotype diversity in the rice blast fungus. *Plant Cell* 3:95–102
Liesack W, Stackebrandt E (1992): Occurence of novel groups of the domain Bacteria as revealed by analysis of genetic material isolated from an Australian terrestrial environment. *J Bacteriol* 174:5072–5078
Lim EL, Amaral LA, Caron DA, DeLong EF (1993): Application of rRNA-based probes for observing marine nanoplanktonic protists. *Appl Environ Microbiol* 59:1647–1655
Louws FJ, Fulbright DW, Taylor-Sptephens C, de Bruijn FJ (1994): Specific genomic fingerprints of phytopathogenic *Xanthomonas* and *Pseudomonas* pathovars and strains generated with repetitive sequences and PCR. *Appl Environ Microbiol* 60:2286–2295
Michelmore RW, Hulbert S (1987): Molecular markers for genetic analysis of phytopathogenic fungi. *Ann Rev Phytopathol* 25:383–404
Ogram A, Sayler GS, Barkay T (1987): The extraction and purification of microbial DNA from sediments. *J Microbiol Meth* 7:57–66
Pankhurst CE, Doube BM, Gupta VVSR, Grace PR, eds. (1994): *Soil Biota: Management in Sustainable Farming Systems*. Melbourne: CSIRO Information Systems
Pickup RW (1991): Development of molecular methods for the detection of specific bacteria in the environment. *J Gen Microbiol* 137:1009–1019
Rattray EAS, Prosser JI, Kilham K, Glover LA (1990): Luminescence-based nonextractive technique for *in situ* detection of *Escherichia coli* in soil. *Appl Environ Microbiol* 58:2444–2488
Sadowsky MJ (1994): DNA fingerprinting and restriction fragment length polymorphism analysis. In: *Methods of soil analysis, Part 2. Microbiological and biochemical properties*. Weaver RW, Angle S, Bottomley P, Bezdicek D, Smith S, Tabatabai A, Wollum A, eds. Madison, WI: Soil Science Society of America
Sayler GS, Nikbakht K, Fleming JT, Packard J (1992): Applications of molecular techniques to soil biochemistry. In: *Soil Biochemistry, Vol. 7*. Stotzky G, Bollag J-M, eds. New York: Marcel Dekker
Schlegel M (1994): Molecular phylogeny of eukaryotes. *Tr Evol Ecol* 9:330–335
Sen D (1994): Whole-cell protein profiles of soil bacteria by gel electrophoresis. In: *Methods of soil analysis, Part 2. Microbiological and biochemical properties*. Weaver RW, Angle S, Bottomley P, Bezdicek D, Smith S, Tabatabai A, Wollum A, eds. Madison, WI: Soil Science Society of America
Spring S, Amann R, Ludwig W, Schliefer K-H, van Gemerden H, Peterson N (1993): Dominating role of an unusual magnetotactic bacterium in the microaerobic zone of a freshwater sediment. *Appl Environ Microbiol* 59:2397–2403
Stackebrandt E, Goebel BM (1994): Taxonomic note: a place for DNA-DNA reassociation and 16*S* rRNA sequence analysis in the present species definition in bacteriology. *Int J Syst Bacteriol* 4:846-849
Stackebrandt E, Liesack W, Goebel BM (1993): Bacterial diversity in a soil sam-

ple from a subtropical Australian environment as determined by 16S rDNA analysis. *FASEB J* 7:232–236

Steffan RJ, Goksory J, Boj AK, Atlas RM (1988): Recovery of DNA from soils and sediments. *Appl Environ Microbiol* 54:2908–2915

Streit W, Botero L, Werner D, Beck D (1995): Competition for nodule occupancy on *Phaseolus vulgaris* by *Rhizobium etli* and *Rhizobium tropici* can be efficiently monitored in an ultisol during the early stages of growth using a constitutive GUS gene fusion. *Soil Biol Biochem* 27:1075–1081

Streit W, Kosch K, Werner D (1992): Nodulation competitiveness of *Rhizobium leguminosarum* bv *phaseoli* and *Rhizobium tropici* strains measured by glucuronidase (GUS) gene fusions. *Biol Fertil Soils* 14:140–144

Tohme J, Correa-Victoria F, Levy M (1992): *Know your enemy: a novel strategy to develop durable resistance to rice blast fungus through understanding the genetic structure of the pathogen population*, CIAT working Document no 140. Cali, Colombia

Torsvik V, Goksøyr J, Daae FL (1990): High diversity in DNA of soil bacteria. *Appl Environ Microbiol* 56:782–787

Triplett EW (1990): Construction of a symbiotically effective strain of *Rhizobium leguminosarum* bv *trifolii* with increased nodulation competitiveness. *Appl Environ Microbiol* 56:98–103

Ueda T, Suga Y, Yahiro N, Matsuguchi T (1995): Remarkable N_2-fixing bacterial diversity detected in rice roots by molecular evolutionary analysis of *nifH* gene sequences. *J Bacteriol* 177:1414–1417

Versalovic J, Koeuth T, Lupski JR (1991): Distribution of repetitive DNA sequences in eubacteria and application to fingerprinting of bacterial genomes. *Nucl Acids Res* 19:6823–6831

Vilgalys R, Hester M (1990): Rapid genetic identification and mapping of enzymatically amplified ribosomal DNA from several *Cryptococcus* species. *J Bacteriol* 172:4238–4246

Wagner M, Erhart R, Manz W, Amann R, Lemmer H, Wedi D, Schliefer K-H (1994): Development of an rRNA-targeted oligonucleotide probe specific for the genus *Acinetobacter* and its application for in situ monitoring in activated sludge. *Appl Environ Microbiol* 60:792–800

Ward DM, Bateson MM, Weller R, Ruff-Roberts AL (1992): Ribosomal RNA analysis of microorganisms as they occur in nature. *Adv Microb Ecol* 12:219–286

Ward DM, Weller R, Bateson MM (1990): 16S rRNA sequences reveal numerous uncultured microorganisms in a natural community. *Nature* 345:63–65

Weisburg WG, Barns SM, Pelletier DA, Lane DJ (1991): 16S ribosomal DNA amplification for phylogenetic study. *J Bacteriol* 173:697–703

Welsh J, McClelland M (1990): Fingerprinting genomes using PCR with arbitrary primers. *Nucl Acids Res* 18:7213–7218

Williams JGK, Kubelik AR, Livak KJ, Rafalski JA, Tingey SV (1990): DNA polymorphisms amplified by arbitrary primers are useful as genetic markers. *Nucl Acids Res* 18:6531–6535

Wilson KJ (1995): Molecular techniques for the study of rhizobial ecology in the field. *Soil Biol Biochem* 27:501–514

Wilson KJ, Giller KE, Jefferson RA (1991): β-glucuronidase (GUS) operon fusions as a tool for studying plant-microbe interactions. In: *Advances in Molecular Genetics of Plant-Microbe Interactions, vol 1*, Hennecke H, Verma DPS, eds. Dordrecht, Neth: Kluwer Academic Publishers

Wilson KJ, Sessitch A, Corbo J, Giller KE, Akkermans ADL, Jefferson RA (1995): β-glucuronidase (GUS) transposons for ecological and genetic studies of rhizobia and other Gram-negative bacteria. *Microbiology* 141:1691-1705

Wilson M, Lindow SE (1993): Release of recombinant microorganisms. *Ann Rev Microbiol* 47:913–944

Young JPW (1992): Phylogenetic classification of nitrogen-fixing organisms. In: *Biological Nitrogen Fixation*, Stacey G, Burris RH, Evans HJ, eds. New York: Chapman and Hall

Zvyagintsev DG (1994): Vertical distribution of microbial communities in soils. In: *Beyond the Biomass: Compositional and Functional Analysis of Soil Microbial Communities*, Ritz K, Dighton J, Giller KE, eds. Chichester, UK: John Wiley

Part IV

Tools: Software and Hardware

13

Informatics and Genomic Research

CAROL BULT AND CHRIS FIELDS

Introduction

The characterization of an organism's genome is an essential and fundamental step toward understanding its biochemistry, lifestyle, and history. The genome encodes the structures of an organism's RNAs and proteins explicitly, subunit by subunit; hence the molecular memory maintained by the genome describes the organism not just as a member of a species but also as a functional individual. The genome, with its mutable structure, is the locus for evolutionary change at the individual level. The genome, however, is also a spatially-organized macromolecular complex. Together with the cytoskeleton and membranes, the genome organizes the cell and its dynamics in space and time. Genomics, the study of genomes, encompasses not just the study of individual genes and their structures, expression patterns, and functions but also encompasses the study of the structural and regulatory roles played by the genome as a whole.

The accumulation of data about genomes has precipitated a transformation of bioinformatics from an obscure theoretical discipline concerned with building mathematical models to a robust engineering discipline concerned with managing and facilitating the analysis and application of biological data (Aldhous, 1993; Cuticchia et al, 1993; Ringwald et al, 1994; Waterman et al, 1994). One of the main reasons for this is that genome data, specifically genome map and DNA sequence data, are amenable to representation and analysis by computer in a way that much traditional biological data were not with the technology of the early 1980s and in many cases still are not today. Genomics is driven to a large extent by technology, and the availability of increasingly sophisticated information systems continues to raise new possibilities for representing and manipulating genome data. Advances in information technology, and in molecular and other experimental technologies, are also creating opportunities for building bridges between the study of genomes and other areas in which information technology

The Impact of Plant Molecular Genetics
BWS Sobral, Editor

has played a significant role in biology: population studies; epidemiology; structural biology; and structural and functional imaging of cells and tissues.

This chapter reviews current genome informatics, focusing on tools for managing the data produced by large-scale genome analysis. It covers the management of data collection at the individual laboratory or consortium level as well as the development of public data resources. Tools for analyzing map and sequence data are briefly described, with references to additional sources of information(Appendix, Parts D and E). The evolution of genome data systems from limited-domain, stand-alone resources, such as the GenBank DNA sequence data bank (Benson et al, 1994), toward interoperable, network-based, cross-disciplinary systems serving diverse research and application communities is outlined and strongly advocated.

Genome Data and Their Applications

The primary types of data being generated by the international animal, plant, and microbial genome projects are genetic and physical maps and DNA sequences. Numerous journals, newsletters, computer network resources, and scientific meetings have sprung up in the last decade to present and organize these data (see Appendix, Part A); reports of such data are also commonplace in the general scientific literature and even the popular press. Within the United States, primary responsibility for organizing and supporting the development of genome data resources for plants rests with the Genome Informatics Group, Information Systems Division, a component of the National Agricultural Library (Agriculture Research Service, U.S. Department of Agriculture). The annual Plant Genome conferences highlight progress in accumulating these data by groups worldwide.

From the point of view of applications, a variety of derived data, including information on gene and genome structure, gene expression, population diversity, and evolutionary or phylogenetic information, are at least as important as the basic genome map and sequence data. These types of data are generally treated as optional annotation by the primary genome databases. They are often obtained by groups not engaged directly in genome work and may be published well after the basic data to which they refer. A major challenge for genome information resources is to develop procedures by which these data may be entered directly into databases by their authors and linked to the appropriate map, sequence, or other data already in the various databases. While methods such as the hypertext linking protocol (http) employed by the World Wide Web (WWW) provide some of the functions needed for linking data (see below), additional work to develop interoperable resources for these data is required.

The development of data structures and database organizations for DNA sequences and genetic and physical map data, while challenging in some respects, is relatively straightforward. The data sets, at a few gigabytes, are moderate by computing standards. The development of information resources and analysis tools for more complex biological data, for example, on genome structure, gene expression, biochemical processes and pathways, cellular architecture and functional organization, physiology, and organism-level biology will require considerably more sophisticated techniques and possibly new software technology. The representation of gene expression patterns, for example, in a database capable of handling images showing in situ mRNA or antibody labeling will require tools for comparing and normalizing images obtained from different specimens or with different magnification, three-dimensional reconstruction, error estimation and background subtraction, and possibly animation in time. At tens of megabytes per high-resolution image, the size of such databases can grow quite rapidly. Representations of gene expression, like functional neuroanatomy (Fox and Lancaster, 1994), is a Grand Challenge level computational problem (Ciment et al, 1993).

While the scale of the databases may be smaller, the challenges associated with laboratory data management systems may be at least as complex as those associated with community databases. In a high-throughput DNA sequencing or mapping laboratory, the data management system may be required to organize all aspects of the experimental and analytical work, from inventory control through integrating the results of various automatically-initiated analysis procedures (Kerlavage et al, 1993; Clark et al, 1994; Adams et al, 1994; Lewis, 1994; Kerlavage et al, 1995).

Genome data provide the basis for methodical and, for some systems, eventually complete characterization of gene expression, metabolic pathways, and cellular physiology. The development of manipulatable data resources for maintaining the results as these studies progress will be essential; the volume of data generated will be so large that placing it in the traditional literature would render it inaccessible and hence useless in practice. Databases for biological data at levels of organization higher than molecular structures are in their infancy. Prototype pathway databases for prokaryotic metabolism have been developed (see Appendix, part D), but these do not provide the robust simulation capabilities that will be required for understanding biochemical interactions in detail or for planning engineering modifications of such pathways. Image-based databases for gene expression (Ringwald et al, 1994) and brain structure and function (Fox and Lancaster, 1994) are also under development. These systems provide a preview of the types of information systems that are likely to dominate the exchange and analysis of biological data within a few years.

As biology moves from characterization of genomes toward complete characterization of organisms and comparative analysis of shared structures and pathways across organisms at both the molecular and supramolecular levels, the ability of databases to exchange information transparently will go from a convenience to an absolute essential. Automated data exchange is already a significant bottleneck in the analysis and management of genome data; without significant advances in both technology and data-exchange sociology, inefficient data exchange will become a crippling barrier to research in biology generally. Biologists have traditionally integrated information from different sources in their heads. As the amount and complexity of data increases, this will become impossible even within the confines of relatively narrow fields, as impossible as sequence similarity searching by eye is today. We will need to devote substantial financial and personnel resources to data management and exchange to support this data and computation-intensive type of scientific work.

Information Resources for Biology: Requirements and Technologies

The refinement of the information infrastructure to organize and provide access to the data has developed in parallel with methodologies for the generation of genomic data. Specialized biological databases are available that represent such genome-related text and graphical data as molecular sequences, chromosome mapping assignments, and coordinates for three-dimensional protein structures. A variety of data analysis programs are also available, either for local use or on remote servers. The computerization of data and the development of the Internet have resulted in the creation of a vast community resource accessable from one's desktop computer.

The manner in which data are organized and managed is key to how easily a database can scale up as the amount and diversity of data increases and how easily a researcher can integrate the most current, relevant data for a particular research question. A widely-used single-curator information management system among biologists is ACeDB (A C. elegans Data Base), developed by Richard Durbin and Jean Thierry-Mieg. As the name implies, ACeDB originally was developed for the *Caenorhabditis elegans* research community but has been adapted for many other organisms and research communities (Cherry and Cartinhour, 1994). ACeDB is a popular database model because it requires little or no computer programming skills to implement, data structures within ACeDB can easily be tailored for different research communities, and the interface provides powerful graphical browsing capabilities (Lewis, 1994; Cherry and Cartinhour, 1994).

On the downside, ACeDB has limited or no client/server capabilities. Because the interface is not independent of the database itself, ACeDB typically requires a single dedicated curator to implement and distribute updates of the data via the Internet or some other medium (CD ROM, computer disk, tape, etc.) (Lewis, 1994). Thus, while ACeDB is an excellent information management tool for a single lab or small research community, it is not the most effective data model for situations in which there are frequent updates to the data or for supporting complex queries across different databases.

The integration, updating, and release of data from diverse sources is made far easier by using structured databases which employ standardized data access protocols such as Structured Query Language (SQL) and hypertext transfer protocol (http) and a client/server architecture. Relational database management systems such as SYBASE offer the advantages of supporting multiple, ad hoc, user-defined views of the data that ACeDB data models cannot currently support. Relational databases also support more sophisticated transaction tracking and data integrity checks than does the ACeDB model (Lewis, 1994).

Information Resources for Biology: Accessing the Data

Access to databases and data resources can be achieved in a number of ways, including: (1) browsing data sets via keywords or hypertext links; (2) submitting information or data electronically for analysis at a remote site; or (3) launching a specific query to one or more databases to obtain of a well-defined set of information.

Text-based searching of structured databases, or even unstructured collections of data files, is an inefficient but easy method to determine whether a data resource exits, and to retrieve information if the resource does exist. Successfully identifying relevant information resources depends on the correct choice of a search text string. Wide Area Information Servers (WAIS) support text-based searches of data resources available on the Internet. A WAIS search returns a list of information sources that match the text issued by the user. Some WAIS servers support so-called fuzzy searches to retrieve approximate matches to the keywords chosen by the user in addition to exact matches of the search text. WAIS is a useful method for searching the Internet in cases in which the user does not know about the existence of a specific resource. The process is similar to searching a library's card catalog system with a few keywords to help narrow down the field of information that must be sorted through by hand. WAIS searches of all available Internet resources already is impractical; thus many servers offer indexed information categories to narrow the search area. Many electronically-available newsletters are WAIS indexed so that users can search for articles on specific topics or by specific authors. WAIS searching facilitates the identification of potentially useful information; however, it does not have sophisticated data integration or browsing capabilities.

Another common way to access and use databases via the Internet is by the submission of a request to an analysis server as formatted text. The server then performs an analysis and returns a report to the user. Because such services employ a client/server architecture, the user does not have to log on to and monopolize the remote site. For example, a user can send a nucleotide or protein sequence to the National Center for Biotechnology Information (NCBI) server to have it searched against multiple sequence databases using the BLAST sequence similarity algorithm (Altschul et al, 1990; see Appendix, Parts D and E). The GenQuest server at Oak Ridge National Laboratory (Appendix, Parts D and E) offers similar capabilities for a variety of sequence-similarity algorithms.

Remote logging in via telnet offers another way to access various data analysis services. With telnet, a user uses the local machine like a terminal, logs on to a remote computer, and uses data analysis tools interactively. Telnet usually requires that the user have an authorized account on the remote site whereas e-mail-servers can be accessed anonymously. Both telnet and e-mail servers provide users access to multiple data analysis tools and the power of high-performance computing platforms.

Internet conventions such as gopher and the World Wide Web (WWW, or simply, the Web) support powerful Internet browsing tools (Schatz and Hardin,

1994). WWW is particularly useful because it allows access to FTP sites, gopher sites, WAIS servers, and other Internet resources using a single interface such as Mosaic or Netscape (Appendix, Part C). Both Mosaic and Netscape support the display of text, high-resolution images, graphics, and video. In effect, the Web allows users to view the entire Internet as a single, integrated resource resident on their desktop computer.

WWW is based on the use of hypertext technology via the hyptertext transfer protocol (http) and hypertext markup language (html). Html is very flexible and allows links between information at different sites to be created on the fly by incorporating http addressing to other data resources directly into the language structure of a document. Users navigate through information on the Web by pointing and clicking on color-coded text or icons (hotlinks). Each hotlink is a pointer to another layer of information within the site or at a remote site. By providing pointers to another resource, each site retains control of the presentation and structure of its specific data resource. The use of html and the ease and ubiquity of the WWW has resulted in the creation of a loose network of links between many biological information resources. Creation of html links does not require formal understandings between databases; one database can unilaterally point to another. This loose linkage can result in disappointment if the html linked database changes its structure or address (called a Universal Resource Locator or URL) without providing prior notice to the community. However, the ease with which html links can be made has resulted in the rapid development of a biological database network courtesy of the Web (see Appendix, Part D).

To retrieve specific information from a database or databases, precise queries can be formulated using such formal query languages as SQL (Structured Query Language). The use of SQL permits a complex, structured question to be submitted to relational databases. An example of a query supported by information represented in a relational data model is, "Return all sequences available with a gene name like 'receptor' which are at least 1000 bp in length and have been obtained from nonhuman sources." Forms-based interfaces that appear to work like browsers have been developed for structured databases. The user requests information by entering text in specific fields on the form. The interface software then incorporates the entered text as part of structured language queries to the database, freeing the average database user from the often formidable task of formulating a complex query in SQL. Currently many databases offer query access via stored procedures (i.e., canned queries). To move beyond queries of individual databases to connections across multiple databases is one of the first steps toward an information infrastructure that will support synthetic biological analyses. Some examples of such queries are currently available for the Genome Sequence DataBase (See Appendix, Part D).

Database Interoperability

There are two fundamental approaches to developing information resources that span multiple domains: the development of centralized, integrated databases, and the development of decentralized federated databases. The centralized approach offers the advantages of coherent design, centralized curation, and authoritative decision making. From the point of view of the user, a centralized databases offers one-stop shopping, one source for updates, and a single site for maintenance. The fundamental problem with centralization is, however, that centralized resources cannot scale with the growth of an open domain. If data must be entered, edited, curated, or otherwise processed in any way that involves active intervention at a single site, the staff and budget of that site must increase constantly to keep up with demand. As an example, the increase in staff demands of the data acquisition and editing process at the GenBank database, then at Los Alamos National Laboratory, forced the development of the Electronic Data Publishing model for GenBank in the late 1980s (Cinkosky et al, 1991). In the Electronic Data Publishing model, responsibility for accumulating and assuring the quality of sequence data was transferred from the database staff to the submitting community, effectively decentralized. The database staff continued to perform quality assurance functions but the goal was that even these would become largely automated.

In a multidisciplinary environment, a centralized strategy will only succeed if both the disciplinary boundaries and the volume of data flow can be strictly limited. Otherwise competing data resources will inevitably arise, driven simply by the demands of the community to publish and exchange data. The recent history of the development of new World Wide Web sites illustrates this phenomenon; once the Web and Mosaic became available as a method of publishing data, small databases blossomed at many sites, offering different types of information.

The decentralized strategy takes the development of multiple, independent data resources under separate control as a given, and takes as its goal the development of conventions for these resources to exchange data and access. Decentralization obviates the resource scaling problem by spreading the work of data curation across many sites, which typically have different expertise and resource bases. The key issue for the decentralized strategy is ensuring a maximum of meaningful interoperability with a minimum of centralized control over contents or curation procedures.

An idealized, integrated approach to interoperability for distributed databases is the development of agreed-to shared data structures and semantics between two databases. For example, if molecular sequence databases could all agree on a structure and semantics for the concept of a gene, then software interfaces could be written to access any gene sequence database. All the information would be stored in the same manner and with the same semantic content, independent of the organism from which the data were derived. While such a shared database structure is appealing, the reality is that databases are built to serve multiple communities, each of which has its own view of the most effective and useful way to organize the information and often with different semantic structures for the same word or concept.

One solution to the problem of mismatched semantics and multiple views of data among different data resources is to adopt a convention under which each resource has full control of the semantics of a relatively small subset of the shared domain (primary data), and cedes control of the semantics of all other parts of the shared domain (secondary data) to other databases. Each database serves as the primary acquisition and maintenance site for data in its corner of the shared domain, and points to other databases for data about their corners of the domain, much as Web sites do currently. Unique identifiers are assigned to primary data; identifiers assigned by other databases are used for secondary data. The Genome Sequence Data Base (GSDB) is structured according to this primary-secondary design (Cinkosky et al, 1995). Nucleic acid sequences and associated annotation are primary data types; all other data are secondary. Human clone identifiers, for example, are obtained from the Genome Data Base (GDB) at The Johns Hopkins University (Fasman, 1994) as primary source; taxonomic identifiers are obtained from the Sequences, Sources, Taxa database at The Institute for Genomic Research as primary source; and protein product names are obtained from Swiss-Prot at Geneva as primary source (see Appendix, part D for information on how to access these data resources).

Splitting responsibility for semantics among multiple databases allows for the same kind of anonymous interoperability among queryable systems that shared syntax allows on the WWW. Any database can effectively point to data in any other by using the external database's unique identifiers. As long as external identifiers remain unique and are allowed to carry their full semantics, this system provides semantic coherence.

The benchmark of true interoperability among distributed databases in the decentralized model is the ability for a user to launch a request or query that retrieves information from multiple databases and returns an integrated report to the user. Such cross-database relational join capability requires software that can split a query into component parts, send the components to the appropriate database, and then rejoin the different responses before sending the results back to the user. Currently, relational joins between different databases require, in most cases, maintaining copies of each database at multiple sites. This is practical using relatively straightforward satellite-distribution methods; both GSDB and GDB, for example, distribute updates daily to fully-relational satellites around the world. Software for providing full distributed join capabilities across multiple sites is currently under development by multiple vendors, with some products already on the market. As these products mature, the need for satellite copies will be dictated only by needs for query performance in the face of network loading and downtime, and occasional access to a greater variety of relational databases will be practical.

Future Directions

The demands of information-intensive industries, from banking to entertainment, are driving technology for data communications and database interoperability. Methods and technology for supporting data exchange between databases with different contents, structures, and underlying technologies are evolving rapidly.

Interfaces that allow distributed relational joins between multiple databases, without requiring the user to know the details of the schemata of the databases involved or even their locations, are likely to become available within the reasonably near future. Such interfaces will enable a further migration of biological data from searchable and browsable archives to fully-queryable resources. A more advanced generation of interfaces allowing the construction, on the fly, of complex manipulatable objects from components spread out over multiple databases that may have different structures (hypertext, relational, object-oriented) can also be expected to emerge in the reasonably near future. Such interfaces will probably be required to support the analysis of complex systems such as metabolic, cell communication, or gene expression pathways. These tools will enable biologists to ask questions with computers that they previously could only ask using the literature, and the computer will answer with far richer sets of data than the traditional literature could produce.

REFERENCES

Adams MD, Kerlavage AR, Kelley JM, Gocayne JD, Fields C, Fraser CM, Venter JC (1994): A model for high-throughput automated DNA sequencing and analysis core facilities. *Nature* 368:474–475

Aldhous P (1993): Managing the genome data deluge. *Science* 262:502

Altschul SF, Gish W, Miller W, Myers EW, Lipman DJ (1990): Basic local alignment search tool. *J Mol Biol* 215:403–410

Benson DA, Boguski M, Lipman DJ, Ostell J (1994): GenBank. *Nucl Acids Res* 22:3441–3444

Cherry JM, Cartinhour SW (1994): ACeDB: A tool for biological information. In: *Automated DNA Sequencing and Analysis*, Adams MD, Fields C, Venter JC, eds. San Diego, CA: Academic Press

Ciment M, Scherlis W and the Committee on Physical, Mathematical, and Engineering Sciences (Walter E. Massey, Chairman) (1993): Grand challenges 1993: High performance computing and communications. Washington, DC: Office of Science and Technology Policy

Cinkosky M, Fickett J, Gilna P, Burks C (1991): Electronic data publishing and GenBank. *Science* 252:1273–1277

Cinkosky M, Fickett J, Keen G (1995): A new design for the Genome Sequence DataBase. *IEEE Eng Med Biol*: in press

Clark SP, Evans GA, Garner HR (1994): Informatics and automation used in physical mapping of the genome. In: *Biocomputing-Informatics and Genome Projects*, Smith DW, ed. San Diego, CA: Academic Press

Cuticchia J, Chipperfield M, Porter C, Kearns W, Pearson P (1993): Managing all those bytes: the Human Genome Project. *Science* 262:47

Fasman K (1994): Restructuring the Genome Data Base: A model for a federation of biological databases. *J Compu Biol* 1:165–171

Fox PT, Lancaster JL (1994): Neuroscience on the Net. *Science* 266:994–997

Hahn H, Stout R (1994): *The Internet Complete Reference.* New York: Osborn-McGraw Hill

Kerlavage AR, Adams MD, Kelley JC, Dubnick M, Powell J, Shanmugam R, Venter JC, Fields C (1993): Analysis and management of data from high-throughput expressed sequence tag projects. *Proceedings from the Twenty-Sixth Annual Hawaii International Conference on Systems Sciences*, The Institute of Electrical and Electronics Engineers. Los Alamitos, CA: Computer Society Press

Kerlavage AR, FitzHugh W, Glodek A, Kelley J, Scott J, Shirley R, Sutton G, Wai-Chiu M, White O, Adams MD (1995): Data management and analysis for high-throughput DNA sequencing projects. *IEEE Eng Med Biol:* in press

Krol E (1992): *The Whole Internet.* Sebatopol, CA: O'Reilly and Associates

Lewis S (1994): Design issues in developing laboratory information management systems. In: *Automated DNA Sequencing and Analysis*, Adams MD, Fields C, Venter JC, eds. San Diego, CA: Academic Press

Ringwald M, Baldock R, Bard J, Kaufman M, Eppig JT, Richardson JE, Nadeu JH, Davidson D (1994): A database for mouse development. *Science* 265: 2033–2034

Schatz BR, Hardin JB (1994): NCSA, Mosaic and the World Wide Web: Global hypermedia protocols for the Internet. *Science* 265:895–901

Smith U (1993): *A Biologist's Guide to Internet Resources.* Available on-line from Usenet:sci.answers. Available via gopher, anonymous FTP and e-mail from various archives. For a free copy via email, send the text "send pub/usenet/ sci.answers /biology /guide/*" to the email address: mail-servers@rtfm.mit.edu

Waterman M, Uberbacher E, Spengler S, Smith FR, Slezak T, Robbins R, Marr T, Kingsbury DT, Glina R, Fields C, Fasman K, Davison D, Cinkosky M, Cartwright P, Branscomb E, Berman H (1994): Genome Informatics I: Community databases. *J Compu Biol* 1:173–190

Appendix: Access Points for Genome Resources

A. Community Newletters

Community newsletters are good sources of information regarding conferences and workshops, meeting reports, granting agency deadlines, preliminary research reports, taxon-specific rules for gene symbols, etc. Newsletters often are geared to a specific organism (e.g. rice, *Arabidopsis*, etc.) or a particular research program area (e.g., plant genome research) and are usually available at no charge or for a nominal fee. Increasingly, newsletters (and even some refereed journals) are distributed electronically. Examples of some of the newsletter available for plant genetics and genome research are given below. Most of these newsletters are available electronically through the USDA's National Agricultural Library Web Server (see Part D).

Probe: Newsletter for the USDA Plant Genome Research Program

A quarterly publication for the plant genome research community with calendars of genome meetings and workshops, granting deadlines, and information regarding projects sponsored by the USDA plant genome program.

Probe editor,
NAL, 4th floor
1301 Baltimore Blvd
Beltsville, MD 20705
phone: 301-504-6613
FAX: 301-504-2098
e-mail: smccarth@nalusda.gov

Rice Genome: Newsletter for rice genome analysis

Information about the rice genome research program in Japan.

Editorial Office of *Rice Genome*
Rice Genome Research Program
STAFF Institute
446-1, Ippaizuka
Kamiyokoba
Tsukuba
Ibaraki 305
Japan
FAX: +81-298-38-2245 or +81-298-38-2302
email : ikka@staff.or.jp

Weeds World: The International Electronic Arabidopsis Newsletter

An electronic newsletter about research in *Arabidopsis*. This newsletter has been WAIS indexed to facilitate keyword searches at AAtDB Research Companion (gopher server at weeds.mgh.harvard.edu).

e-mail: arabidopsis@nottingham.ac.uk
WWW: http://nasc.nott.ac.uk
WWW: http://probe.nalusda.gov:8000/

The Electronic AIS

Back issues (1964–1990) of the Arabidopsis Information Service newsletter available for WAIS keyword searching and as html documents for WWW.

gopher or WWW: weeds.mgh.harvard.edu
WWW: http://probe.nalusda.gov:8000/

The Rice Genetics Newsletter

The newsletter of the Rice Genetics Cooperative. Volumes 1,2,3, and 5 are available as html documents.

surface mail: Dr. H.I. Oka
National Institute of Genetics
Mishima City
411 Japan
WWW: http://probe.nalusda.gov:8000/

Report of the TGC

A newsletter of the Tomato Genetics Cooperative (annual subscription fee of US $5). Volumes 40, 41, and 43 of the TGC newsletter are available as html documents

surface mail: Rich Zobel
1017 Bradfield Hall
Cornell University
Ithaca, NY 14853-1901
USA
WWW: http://probe.nalusda.gov:8000/

Dendrome: Forest Tree Genome Research Updates

A newsletter on forest tree genome research.
surface mail: *Dendrome* subscriptions
Institute of Forest Genetics
P.O. Box 245
Berkeley, CA 94701
USA
e-mail: dendrome@s27w007.pswfs.gov
WWW: http://s27w007.pswfs.gov:80/

B. General Information Regarding Internet Resources:

As access to the Internet has become more widespread, many books have appeared in the popular press as guides for understanding and using Internet resources (e.g. Krol, 1992; Hahn and Stout, 1994). *A Biologist's Guide to Internet Resources* (Smith, 1993) is a useful (and free) text describing the basics of telnet, gopher, FTP, WWW, and email servers.

C. User Interfaces for WWW (Mosaic and Netscape)

The World Wide Web effectively integrates Internet resources (gopher, FTP, telnet, etc.) by allowing access to all of them through a single user interface. Using hypertext technology (hypertext markup language; html) and hypertext transfer protocols (http), a user can navigate within or between sites on the Internet by simply clicking on hypertext links within an html document. Netscape and Mosaic are two http-based browsers for navigating the WWW. Both Netscape and Mosaic can display graphics, high-resolution images, text, audio and video.

Netscape

Netscape Navigator is available for UNIX, Mac, and PC platforms. The software is available free of charge for academic and not-for-profit research organizations and for a nominal fee for commercial organizations.

FTP: ftp.mcom.com/netscape/
http://home.mcom.com:80/
e-mail: info@mcom.com

Mosaic

Mosaic was developed at the National Center for SuperComputing Applications (NCSA) and is available for UNIX, Mac, and PC platforms. The software is available at no charge.

FTP:	ftp.ncsa.uiuc.edu
WWW:	http://www.ncsa.uiuc.edu
e-mail:	orders@ncsa.uiuc.edu
surface mail:	NCSA Documentation Orders 152 Computing Applications Building 605 East Springfield, IL 61820-5518 USA
phone:	217-244-4130

D. Useful Universal Resource Locators (URLs) for Biologists

The following URLs are central access points to different kinds of genome data and data analysis tools. This is not an exhaustive list and many additional resources can be accessed via hypertext links from within the Web sites listed below.

http:// probe.nalusda.gov:8000/ (also see Appendix, Part F)
National Agricultural Library, USDA

Central access point for many plant genome resources, including species specific mapping, sequence, and germplasm databases, community newsletters, Plant Genome conference announcements abstracts, and software.

http://www.gdb.org/
The Johns Hopkins University Bioinformatics Web Site

Access point for biological databases maintained at JHU, including the human gene mapping database (Genome Data Base; GDB); pointers to Prot-Web, an international collection of protein databases; general starting points for basic Internet navigation.

http://www.ncbi.nlm.nih.gov/
National Center for Biotechnology Information

Access for GenBank and other databases; support for text-based and sequence similarity searches of molecular sequence databases.

http://www.ncgr.org:80/
National Center for Genome Resources

General clearinghouse for genome-related information; access to the Genome Sequence Data Base (GSDB) and SIGMA: System for Integrated Genome Map Assembly (software).

http://golgi.harvard.edu/
Harvard Biological Laboratories—Genome Research

Information on molecular and cellular biology research at Harvard; links to many model organism databases; access to GCG software documentation; links to molecular biology servers around the world; lists of biological conferences; access to other Harvard Internet resources.

http://expasy.hcuge.ch:80/
The University of Geneva
Molecular biology server; access to Swiss Prot and other molecular sequence databases; access to on-line data analysis tools for DNA and protein sequences.

http://life.anu.edu.au
Australia National University Bioinformatics Hypermedia Service
Access point for information on biodiversity, plant viruses, and bioinformatics.

http://www.tigr.org
The Institute for Genomic Research
Molecular biological database containing DNA and protein sequences annotated with information on gene expression, cellular role, isology class, taxonomic and sample collection data for multiple species; software tools for data analysis and information management of large-scale sequencing projects; information on the annual Genome Sequencing and Analysis Conference.

http://www.mcs.anl.gov:80/
Mathematics and Computer Science at Argonne National Laboratories
Metabolic pathway database; phylogenetic tree construction algorithms; sequence analysis tools for whole genomes.

http://phylogeny.arizona.edu/tree/phylogeny.html/
The University of Arizona Tree of Life
A project to link biological information on the Internet in a phylogenetic context.

http://www.public.iastate.edu:80/ ~ pedro/./research_tools.html (also see Appendix I, Part F.)
Iowa State University: Pedro's BioMolecular Research Tools
Pointers to information and services for molecular biologists.

http://www.yahoo.com (also see Appendix, Part F)
YAHOO Web Resource Listing
Useful listing of Web Resources by broad categories.

E. Sequence and Mapping Analysis Services and Software

The following Web sites offer a variety of sequence analysis and mapping services and software for use with genome related data. Pedro's BioMolecular Research Tools and the YAHOO Web resource listings both provide a more detailed list of specialized databases and services.

Multiple Sequence Alignment

Washington University Institute for Biomedical Computing
http://ibc.wustl.edu:80/
ClustalW and MSA alignment algorithms

Southern France Human Genome Project Computing Resources Center
http://genome.eerie.fr:80/gcg/gcg-home.html
Pileup alignment algorithm (part of the Wisconsin Genetic Computer Group package)

Gene Structure Prediction

Baylor College of Medicine
http://dot.imgen.bcm.tmc.edu:9331/gene-finder/gf.html
GeneFinder algorithm for predicting gene structure
Oak Ridge National Laboratory
http://avalon.epm.ornl.gov:80/gallery.html
Grail prediction of exons and other DNA sequence features

Sequence Similarity Searches

National Center for Biotechnology Information
http://www.ncbi.nlm.nih.gov/
Text and sequence similarity searches of nucleotide and protein sequences
Baylor College of Medicine
http://kiwi.imgen.bcm.tmc.edu:8088/
BCM Search Launcher
Oak Ridge National Laboratory
http://avalon.epm.ornl.gov:80/gallery.html
GenQuest sequence comparison server for nucleotide and protein sequences.

Mapping

Whitehead Institute for Biomedical Research/MIT Center for Genome Research
http://www-genome.wi.mit.edu/
Mapmaker multipoint genetic linkage analysis software for experimental crosses
National Center for Genome Resources
http://www.ncgr.org:80/
SIGMA: (System for Integrated Genome Map Assembly) a graphical tool for building and viewing integrated genome maps.

F. Examples of Mosaic home pages from the Web

Figures 1, 2, and 3 are Mosaic screen images of three Web resources. Text which is underlined indicates a hotlink to the next layer of information or to another Web site.

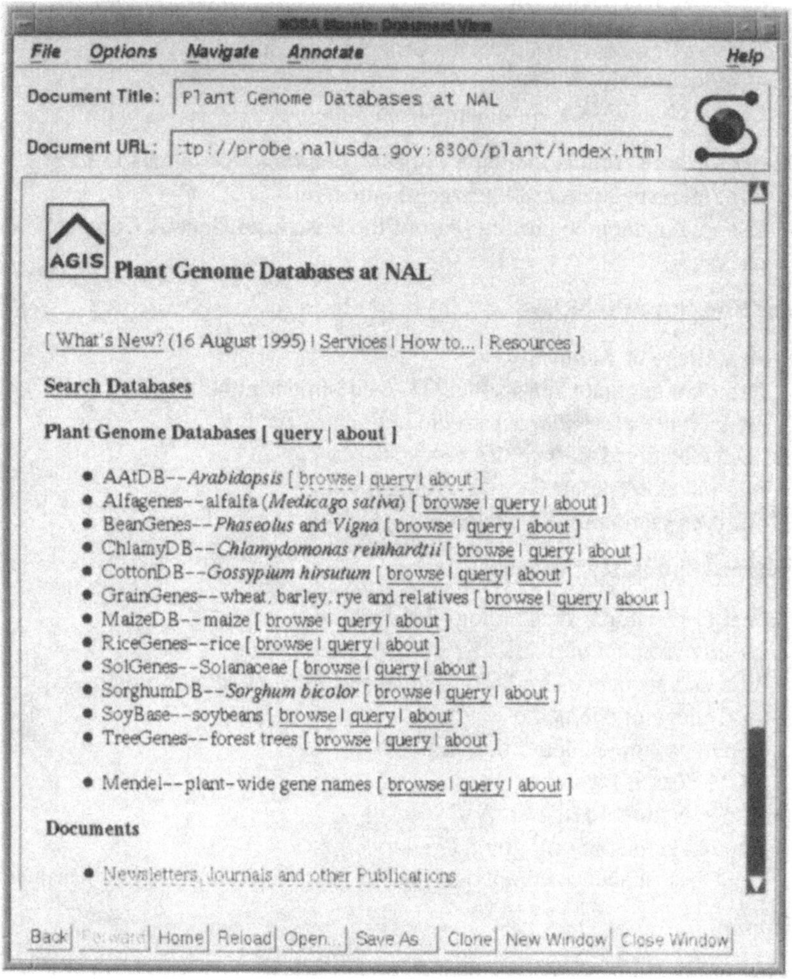

Appendix Figure 1. Agricultural Genome World Wide Web Server at the National Agricultural Library, United States Department of Agriculture. Reprinted with the permission of Dr. D. Bigwood.

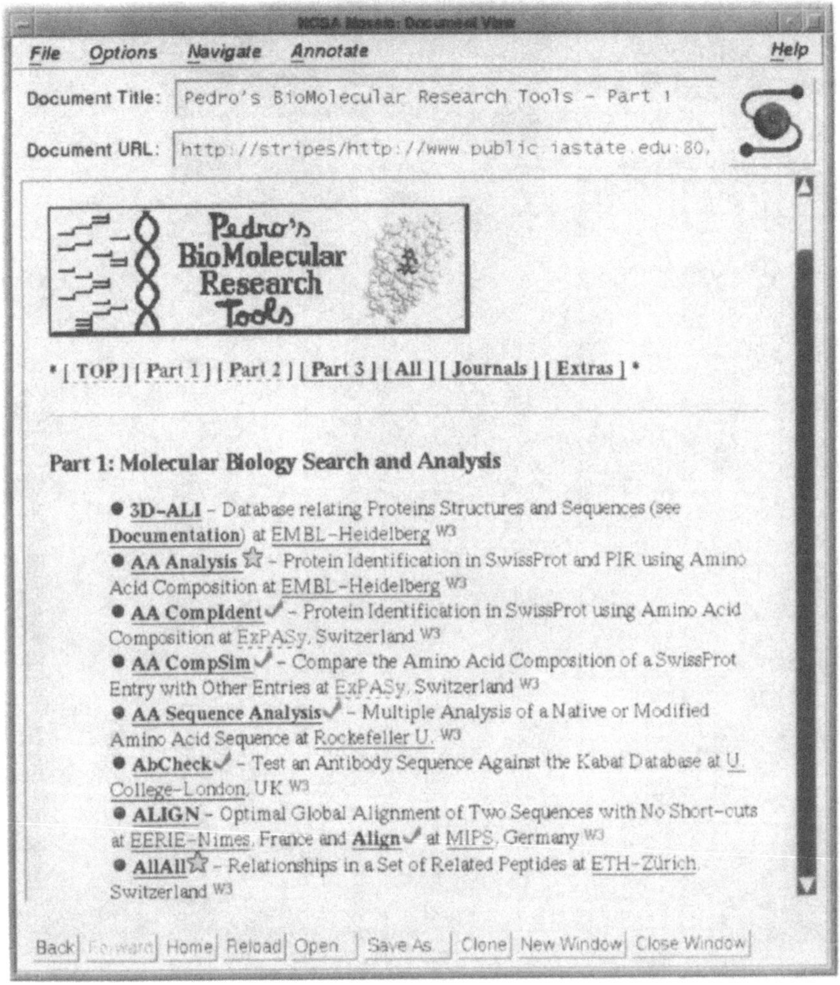

Appendix Figure 2. Pedro's BioMolecular Research Tools. Reprinted with permission of Mr. Pedro Coutinho.

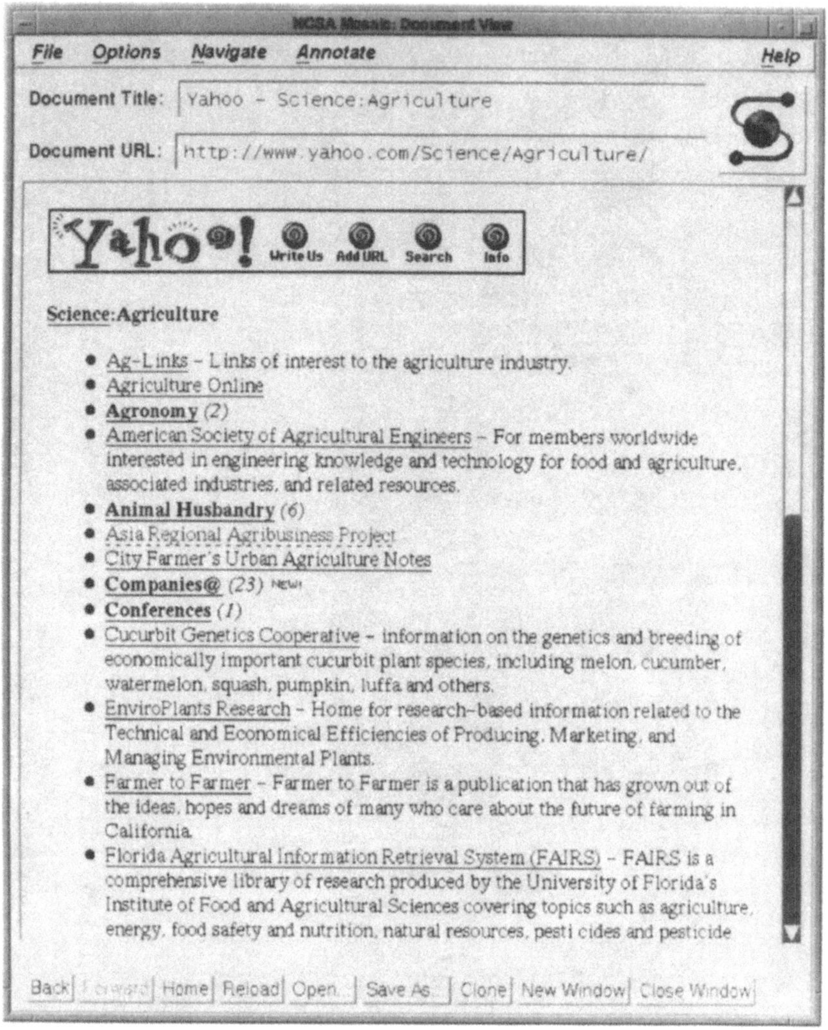

Appendix Figure 3. YAHOO resource listing for Internet resources relating to agriculture. Reprinted with the permission of Mr. Tim Brady.

14

Instrumentation for Automated Molecular Marker Acquisition and Data Analysis

STEPHEN R.E. BATES, DAVID A. KNORR, JENNIFER W. WELLER, JANET S. ZIEGLE

Introduction

Developing molecular markers has become the most rapid method for constructing detailed genetic maps. This technology can be used to streamline breeding programs, to help identify specific individuals, and to quickly discern genetic variation in large populations. Since both breeding populations and genetic diversity tend to be relatively large in plants, the development and routine use of molecular genetic markers requires large numbers of samples to be processed. Thus, automation becomes a strong consideration. Automation should be thought of as modular. A modular approach to automation imparts the greatest flexibility for incremental improvements resulting from technological advances. Issues to consider when selecting among the different types of molecular marker assays available include: cost of developing new markers; cost of routine analysis; amount of information generated per marker; amenability of the assay technology to automation; and ease of incorporating markers into data bases of general utility (Figure 1 and Table 1).

Several categories of molecular markers have emerged and include those polymorphisms detected by hybridization assays, such as restriction fragment length polymorphisms (RFLPs) and those detected using the polymerase chain reaction (PCR). PCR-based markers include those involving arbitrary or random priming such as arbitrarily-primed PCR (AP-PCR), randomly amplified polymorphic DNA (RAPD), or differentially-amplified fragments (DAFs). A more specific PCR marker technique involves detection of polymorphic regions of di-, tri-, or tetranucleotide repeats; these have been given terms such as short tandem repeats (STRs), or simple sequence repeat polymorphisms (SSRs). Amplified fragment length polymorphisms (AFLP) use restriction endonucleases to frag-

The Impact of Plant Molecular Genetics
BWS Sobral, Editor

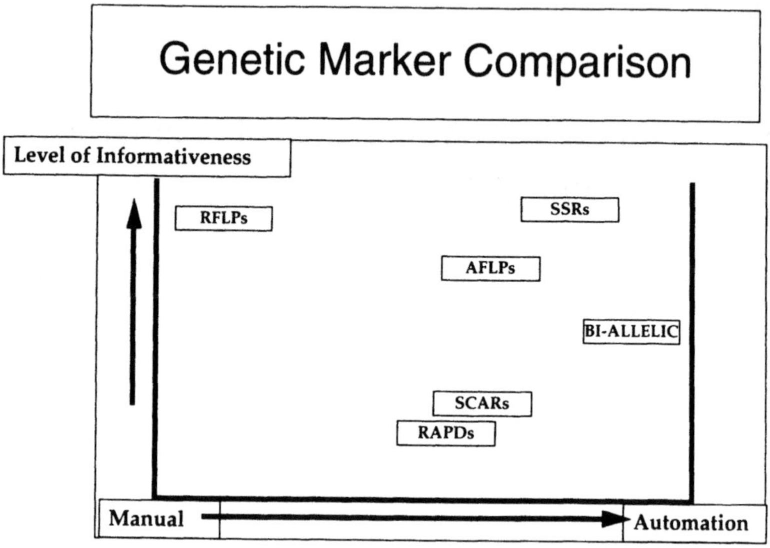

Figure 1. Genetic markers are compared for their level of information content and ease of automation.

ment the DNA but selective PCR amplification for detection. All of these methods attempt to exploit differences in DNA sequences for the purpose of defining individual differences and constructing linkage maps. While direct DNA sequencing is the most precise method, even with present automation it remains one of the most expensive and least efficient for generating useful mapping markers. A useful combination of fragment and sequence methods is illustrated in the development of sequence-confirmed amplification regions (SCARs). These markers are developed using a randomly-primed PCR to find polymorphic bands of interest.

Use of molecular genetic markers allows the entire genome to be scanned for differences linked to a gene of interest, rather than relying upon morphological or biochemical markers, which occur far less frequently and are often confounded by environmental factors or the particular developmental stage of the

Table 1. Characteristics of Molecular Marker Assays

Marker	Multiplex Ratio (the number of polymorphic products/sample reaction)	Observed Heterozygosity-information Content in Soybeans*	Marker Index Value† Multiplex Ratio × Expected Heterozygosity	Map Resolution Number of Markers Placed in a Genome#
RFLP	1–2	0.41	1.0	1000
SSR	1–2	0.56–0.68	6.0	10,000
RAPD	5–20	0.41	2.3	10,000
AFLP	20–100	0.32	6.08	>100,000

* Tingey, 1995
† Vogel, 1994
# Zabeau, 1995

organism. Each class of molecular genetic marker has unique characteristics in terms of the work needed to develop the marker before it can be applied routinely to a population, experimental requirements of using the marker, and type of data it yields. Automation can be beneficial both to speed the process of marker development and to broaden the scope of screening programs, e.g. to assist in breeding programs or to look for rare individuals in large populations

Automation can increase the efficiency and throughput of data acquisition and processing. Improvements in the sophistication and commercial availability of instrumentation for laboratory automation has facilitated rapid progress in both genetic mapping and genomic sequencing projects. Automation encompasses a range of sophistication from simple mechanical devices such as multichannel pipetters to laboratories in which everything from liquid-handling robots to high-speed data analysis is computer-controlled. When introducing automation into a project, it is important to consider the effect on overall data flow since throughput gains in one area of a project can produce bottlenecks either upstream or downstream. For instance, using robots to carry out DNA manipulations can outstrip both rates of DNA sample preparation and the capacity to analyze the resulting data.

Automation

Automation of DNA marker analysis can consist of modules comprising sample preparation, DNA manipulations, data collection and analysis, and finally, integration of the data into a genetic map or meaningful fingerprint (Figure 2).

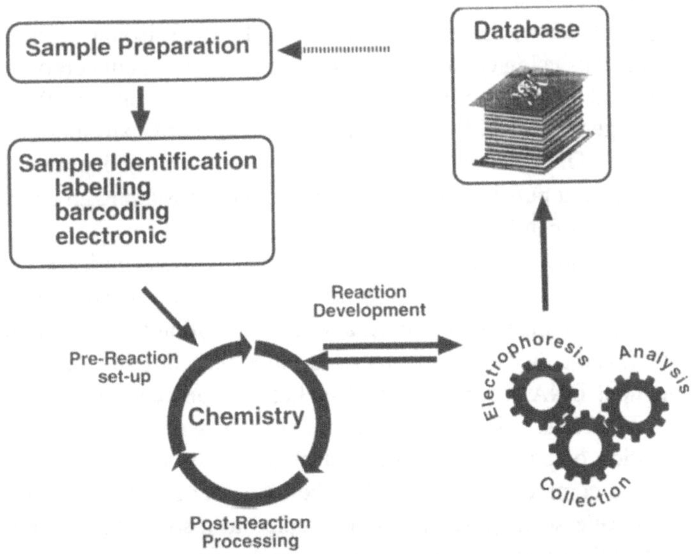

Figure 2. Automation should be viewed as a modular: Sample preparation, sample identification, chemistry, analysis and informatics.

Automation serving each of these areas is commercially available, although not all protocols are readily supported. Automating most techniques usually involves some independent development either to adapt a technique to fit a particular platform or to develop a custom platform, which takes considerable effort. Since a major component of automation is the use of batch processing, another consideration is the degree of precision required at each step. It is therefore important to develop robust, variation-tolerant protocols for automated operations.

Sample Preparation

Sample preparation is perhaps the most difficult procedure to automate. Particularly in plants, this is due to the variety of cells, tissues, and species that are studied. Inability to deal with this variation has impeded development of hardware for DNA extractions. Since PCR-based assays tend to be more forgiving because less DNA is required and higher impurities tolerated, this may dictate not only the choice of marker systems but sample preparation methods as well.

Organic extraction remains the most common method for manual purification of DNA and has been the basis for the design of several instrument configurations (e.g., ABI GenePure, Autogen 500). The throughput of these instruments generally is not sufficient to supply available types of automation downstream. High-throughput DNA extractions can be achieved on a lab-specific basis as is shown by a machine developed at DuPont De Nemours, Inc. This instrument is capable of processing a few thousand samples per day. It is unlikely that the technology will be transferred because it is custom-built.

A robotic device described by Mischiati et al, (1993) isolates genomic DNA from cultured cells using column chromatography. This system combines several commercial and custom-built components to automate one of the common processes for genomic DNA purification. Yield and purity are more than sufficient for routine DNA manipulations. It may be possible to adapt this system to the high-throughput purification of DNA from plants, assuming an adequate way to disrupt the cells can be worked out.

Perhaps the most promising format for automated DNA purification is to use solid phase methods to purify DNA by immobilizing it within an affinity support matrix. Solid phase methods are amenable to automation because the matrix itself can be manipulated. Several matrices have been developed, primarily for high-throughput DNA sequencing, and include various configurations of magnetic beads as well as silica-filled columns. Holmberg et al (1994) have demonstrated isolation of phage DNA using an oligonucleotide hybridization pullout procedure on a commercially available robot. Another recent advance is the use of peptide nucleic acids (PNAs). When coupled to a solid support, PNAs have been used to selectively purify target DNA molecules from crude plant lysates (Knorr and Otteson, 1995). Although not yet widely available, PNAs (Nielson et al, 1994) are promising because they have high affinities for specific DNA fragments and are resistant to enzymatic degradation and many of the solvents used in common DNA purification procedures.

DNA Manipulations

DNA aliquoted into microwell plates can be manipulated by robots for quantitation, dilution, and mixing of reactions. The 96-well microplate format is standard for a wide variety of assay procedures. While this format should exist for the forseeable future, the rise in throughput needs coupled with the low volume benefits of PCR, has brought about other microplate formats. Configurations with 384, 512, and 864 wells are available and are being incorporated into hardware platforms. Automation in molecular biology is most frequently used for general liquid-handling and for performing DNA manipulations such as in sequencing or PCR protocols. The Human Genome Initiative, with its emphasis on very high throughput, has fostered development of a number of robotic systems to handle both set-up and post-processing of DNA reactions (Caillat-Zucman et al, 1993; Earley et al, 1994; Garner et al, 1993; Harrison and Baldwin, 1993; Olsen et al, 1993; Zimmerman et al, 1992). Some of these robotic workstations are dedicated only for general liquid-handling, whereas others are more versatile and can be modified to include a variety of tools designed for specific applications.

The ABI Catalyst 800 and the Beckman Biomek 2000 are two of the most advanced molecular biology robots and illustrate different approaches to automation. The Catalyst 800 is completely self-contained and features a temperature controlled work surface for holding reagent tubes or microtiter plates. There is also a thermal cycler with a fixed, 96-well nondisposable, paralene-coated plate. A magnetic workstation is also available (Holmberg et al, 1994). The single nondisposable probe tip used for reagent delivery on the Catalyst is capable of surface and liquid level sensing. The Catalyst 800 is controlled by a desktop computer with software that allows design of specific parameters for each defined application, including single or multiple, back-to-back runs. Developing custom applications, particularly those involving PCR, will be available upon release of new software currently being developed. While this machine performs well, a potential drawback is inability to customize the work surface configuration. The Biomek 2000 is the most widely used instrument for a variety of high-throughput applications. The Biomek uses a semi-open architecture with a single arm, usually equipped with a multichannel pipetter. An optional configuration combines the Biomek with an additional robot (Zymate by Zymark, Inc.) in which an articulated arm travels in a circular path allowing access to a number of stations. Numerous other tools such as plate holders, thermal cyclers, and magnetic workstations have been developed by this manufacturer and others. The result is a platform that can be customized to handle almost any combination of liquid-handling and analysis manipulations, from spectrophotometry and ELISA assays to sequencing reactions and the production of membrane blots. Although updated software allows most routine types of laboratory automation to be implemented fairly quickly, customizing specific applications may involve considerable effort.

Data Acquisition

Current genetic analysis methods usually involve some form of gel electrophoresis. For RFLPs and the arbitrarily primed PCR assays, the horizontal agarose gel is the most common format. These gels are easy to pour, but fragile, so that subsequent manipulations are best done by hand. The other assays use vertical, slab polyacrylamide gels, which have more strength for subsequent manipulations but require individual attention for set-up, sample loading, and takedown.

In recent years considerable progress has been made towards automating collection and analysis of DNA fragments separated by slab gel electrophoresis. To allow the obvious advantages of computer-assisted analysis, analog data contained in electropherograms must be transformed into digital data. Photographs of gels stained with dyes or chemiluminescent probes, or standard autoradiograms can be electronically scanned to produce digital images. More sophisticated devices allow direct scanning of gels. Examples include the Stratagene Eagle Eye for stained gels, phosphorimagers from Molecular Dynamics and Fuji for radiolabelled gels, and the Fluorimager from Molecular Dynamics for fluorescently labeled DNAs. In the late 1980s, Applied Biosystems introduced a revolutionary technology for automating DNA sequencing: four-color fluorescence labeling coupled with real-time collection. In this format, DNA fragments labeled with fluorescent dyes are separated by electrophoresis. A laser positioned near the bottom of the gel excites the dyes, the emitted fluorescence is collected by a photomultiplier or charge-coupled device, and the signal is fed directly into a computer. In addition to Applied Biosystems' instruments, a similar format is also offered for single dye detection by Pharmacia and LiCor. Multicolor instruments permit multiplexing reactions, or reaction products, for running in a single gel lane. This saves time in regard to the number of PCR reactions run and the number of gels run to collect a given amount of data.

In addition to higher throughput, another advantage of using a multicolor instrument is that an internal molecular weight size standard can be included with every sample, producing an individual sizing curve for every lane. This provides a significant improvement in data quality, especially when accuracy or run-to-run reproducibility are issues, which become important when scoring large families.

An alternate form of separation is based on capillary electrophoresis (CE). The major advantage of CE is automatic filling of the capillary with the separation matrix, and sample loading by electrokinetic injection. Sample handling from start to finish can thus be handled remotely, so there is a great savings in time for the researcher. However, since currently available CE systems have a single capillary, throughput is lower than with slab gels. Approximately 50 separations per 24 hour period have been reported. Development of a four-color fluorescence CE instrument (McCollum et al, 1994), which will help increase throughput and allow simultaneous inclusion of molecular weight standards, will allow increases in reproducibility and accuracy of runs. Prototypes of instruments with multiple capillary arrays (Molecular Dynamics) show possibilities for increasing the throughput of this technology by one or two orders of magnitude. There are also hybrid slab-gel/ capillary array systems under development (Perkin Elmer, 1993).

As the field of molecular genetics has evolved, so have different approaches for detecting specific DNA sequences. Two examples of new approaches, oligonucleotide ligation assay (OLA) and the TaqMan™ PCR assay, do not require electrophoresis steps.

One form of OLA described by Nickerson et al (1992) utilizes ligation of two oligonucleotides at the site of a sequence polymorphism. Two reactions are carried out, each containing a common 5′ biotin labeled oligonucleotide and an allele-specific oligonucleotide labeled with digoxigenin. After a T4 DNA ligation, reaction products are collected in streptavidin-coated microwell plates, and the amount of digoxigenin is determined by colorimetric ELISA. Sample genotypes are determined by a computer program that calculates the mean absorbence for replicate reactions for each allele followed by a ratio of these means to determine a genotype.

A novel method for detecting PCR amplification products has been described by Holland et al (1991). A 5′ [^{32}P]-labeled oligonucleotide specific for an internal region of an amplicon was degraded during the extension phase of PCR by the 5′–3′ exonuclease activity of Taq polymerase. The method has been further improved by Lee et al (1993) and Livak et al (1995a) by constructing probes using two fluorescent dyes. In this technique, now referred to as the TaqMan™ PCR assay, fluorescence of the 5′ terminal reporter dye is quenched by a 3′ terminal quencher. Digestion occurs only during amplification of the specific target region, releasing the reporter with a concomitant rise in fluorescence. The technique is semiquantitative, and currently up to three reporter dyes can be distinguished in a single reaction. Recently Livak et al (1995b) have used the TaqMan PCR assay to determine distribution of single-base polymorphic alleles in corn. The TaqMan PCR assay is being developed into an instrument for automatic real-time detection of PCR amplifications.

Eliminating gel electrophoresis by using these types of assays can greatly increase throughput. However, the drawbacks are that only a few data points may be determined for each assay, and they require significant initial development to determine sequences for polymorphisms. On the other hand, they may be useful in situations in which sample numbers are large and relatively few markers are sufficient.

Data Analysis

Conventional autoradiographic techniques for data analysis have limitations in large-scale projects requiring thousands of genotypes, particularly since the major source of error is in the reading of the autoradiograms. Genotype assignments from autoradiograms are conventionally performed by having two persons read each gel and a third person judging any discrepancies. Interpretation of dinucleotide SSRs are further complicated by characteristic stutter bands. In the past few years several approaches have evolved for automating the SSR data analysis on sequencing instruments.

Ziegle (1992) described the development of a commercial software package to automatically size and quantify PCR fragments collected on the ABI automated sequencer. Automated sizing of SSR fragments overcomes the bottleneck of data analysis common to radioactive methods. With fluorescent technology, an internal lane size standard allows for accurate sizing in every lane of the gel and allows for comparison of individuals run on separate gels. In this study the nine loci were differentiated by both color and size range.

The Genotyper software developed by Applied Biosystems has been used by Reed et al (1994) to automate SSR allele assignment in a large scale mapping project. The Genotyper software is designed to interpret the stutter patterns characteristic of dinucleotide repeats and to automatically call alleles. The Genotyper software also produces a table of allele assignments in an ASCII format that can be exported directly to a database or downstream software analysis package, thereby eliminating transcription errors.

The problems experienced world-wide in identifying vegetatively-propagated cultivars and maintaining correctly-labeled living germplasm collections can be also addressed with automated SSR typing. Thomas et al (1993) and Thomas and Scott (1993) have isolated polymorphic SSR loci in grapevine, leading to the development of a DNA typing system using SSRs as a means of differentiating cultivars of grapevine (Thomas et al, 1994). Automation of their typing system, using fluorescent PCR primers and an automated sequencer with an internal lane size standard, has allowed absolute base pair sizes to be assigned to alleles rather than arbitrary alphabet numbers. Allele assignments for seven loci are then electronically transferred to a database. The accumulated typing of over 80 cultivars has demonstrated that cultivars that are difficult to differentiate phenotypically can be distinguished by DNA typing.

AFLP PCR products labeled with ^{32}P can be visualized by autoradiography and scored by hand, which is time consuming. Keygene has developed a semi-automated approach to analysis. ^{33}P-labeled PCR products are separated on a polyacrylamide gel, and the data are collected after electrophoresis with a Phosphorimager (Fuji). The collected data are then analyzed with custom analysis software developed at Keygene.

Weller (1995) and others have adapted AFLP technology for use with an automated sequencer by using fluorescently-labeled primers. Data are collected in real time with the ABI automated sequencer, and all fragments are sized and quantified with the Genescan software. ABI-developed Genotyper software can score the presence or absence of a polymorphic band and compare a segregating individual to one of the parents, in a semiautomated fashion, producing an ASCII formatted table that can be electronically exported to downstream software programs such as Mapmaker and Linkage (Figure 3). With this technology, the ability exists to place result files in central electronic libraries that could be accessable on-line which would result in standardization of format and increased accuracy of the information.

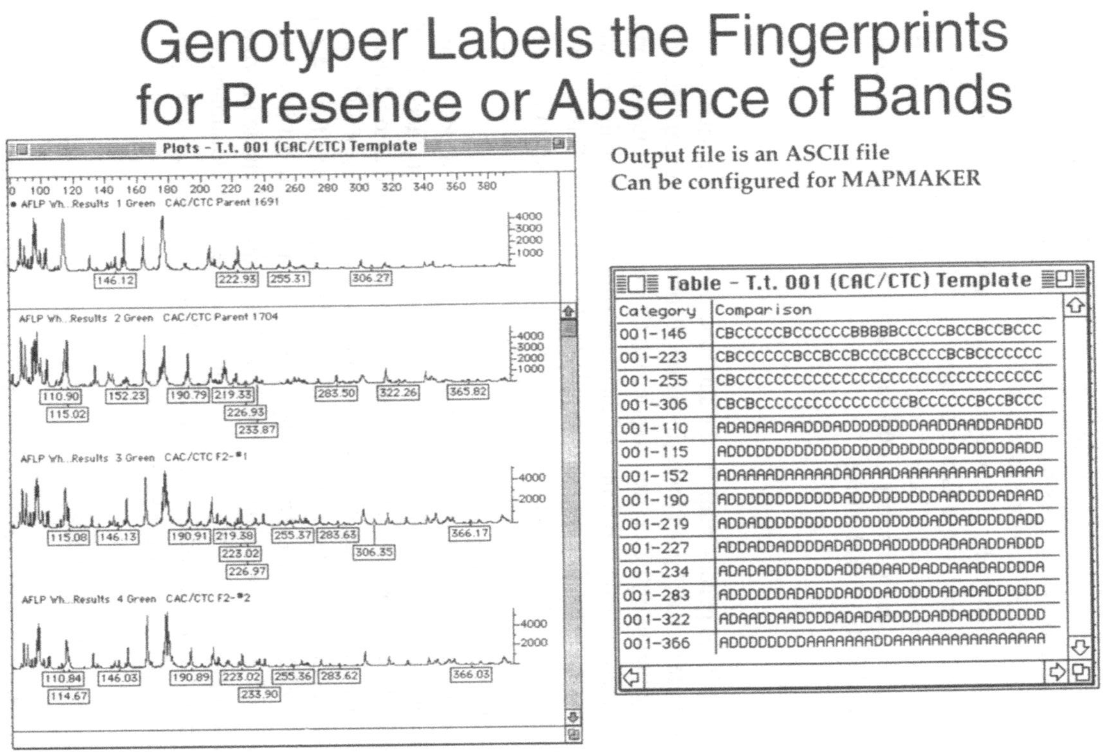

Figure 3. Polymorphic AFLP bands in an F2 population can be automatically scored using Genotyper™ software.

Automation Attributes of Molecular Markers

RFLPs

RFLP markers are perhaps the most widely used type of molecular genetic marker. They are relatively costly to develop because of the labor involved with screening, characterizing, and cloning informative probes. RFLPs provide the maximum amount of information possible because they are usually co-dominant. However, compared with PCR-based methods, RFLP assays require relatively large amounts of pure, high molecular weight genomic DNA. The process of preparing probes and performing the assays is time and labor intensive. Furthermore, probes must be maintained in libraries of bacterial cultures. Batch automation is difficult to achieve due primarily to the large amount of hands on manipulation and the individuality of each probe.

SSRs

To develop SSRs as genetic markers the surrounding sequence must be obtained. PCR primers are then designed for regions directly outside the repeat. The primers must amplify uniquely the desired region, which must also be polymorphic in the segregating population. Because of the large number of nonpolymorphic repeats, success rates for finding polymorphic SSRs is not high. DuPont has significantly increased throughput for SSR screening by using custom software to automatically analyze 95% of the data, with human interpretation required only to resolve the remaining 5%.

Schwengle et al (1994) compared fluorescent analysis of STRs to conventional autoradiography for both accuracy and efficiency, and concluded that their fluorescence-based protocol was at least as accurate as standard autoradiography. Efficiency was further increased by choosing the size and colors of STRs so that 24 STR loci could be loaded in a single lane of the gel, allowing for 864 genotypes (1728 alleles) per gel. This multiplexing gives a great boost in the information yielded per run, with many loci assessed per lane, and allelic information at each locus gathered. It does, of course, increase the initial development cost since the primers must be extensively tested, and often redesigned, to amplify all of the targets efficiently in the same reaction vessel.

Arbitrarily Primed PCR Strategies

Multiple uncharacterized annealing sites are used as PCR primer targets in the strategies variously termed random amplified polymorphic DNA (RAPDs) analysis (Williams et al, 1990), arbitrary primed PCR (AP-PCR) (Welsh and McClelland, 1990), and DNA amplification fingerprinting (DAF) (Caetano-Anolles et al, 1992). These techniques often result in several polymorphic sites per PCR reaction, and the PCR primers are available commercially (Operon Tech, Alameda,

CA) so that the initial screening of a genome with primers for those that show polymorphic products is relatively inexpensive.

Markers developed by RAPD amplification are generally dominant and therefore not as informative as co-dominant markers obtained with previously mentioned techniques. There are also context effects that can complicate analysis of these markers. In general they seem to be most useful for recurrent parent analyses (Tingey, 1995).

As RAPDs are PCR-based assays, they share the sample preparation requirements and relative ease of automation of such assays. DNA manipulations are readily automatable, and analysis can be automated on DNA sequencers running sophisticated software packages (e.g. Genotyper™ from ABI). A major drawback has been the extreme difficulty most laboratories encounter to obtain reproducible results. For this reason these types of markers are difficult to transfer between laboratories.

Sequence Confirmed Amplified Fragments (SCARs) (Michelmore et al, 1991) are a modification that allows a RAPD polymorphism to be made more robust. A RAPD DNA fragment is cloned and sequenced, to allow the investigator to develop new, longer primers that allow a much simpler and specific PCR fingerprint to be generated. This is especially valuable if there is a nonsegregating band of very similar size that makes analysis difficult. If a single product in just one of the parental lines results, a SCAR can be used with colorimetric gel-free assays.

AFLPs

Amplified Fragment Length Polymorphisms are a recently developed PCR-based fingerprinting technique (Zabeau and Voss, 1992). The DNA is cut into defined fragments using restriction endonucleases and adaptor oligonucleotides are ligated to the overhanging ends. These adaptors serve as recognition sequences for PCR primers. The complexity of the mixture of fragments is decreased using selective PCR amplification, by adding additional nucleotides to the 3′ termini of the PCR primers. The number of additional bases is adjusted such that 50–150 DNA fragments are amplified during the PCR. This can produce 10–30 polymorphisms per PCR reaction, depending upon the genomes being assayed, making AFLPs a very cost effective marker system. No prior sequence information is required so development costs reduce to screening the genome with the primers to identify those that give the most polymorphisms per reaction.

AFLP technology has an added template preparation step relative to other PCR-based assays. Slightly more genomic DNA is required, and it must be of sufficient quality to allow restriction endonuclease digestion and ligation of adaptor oligonucleotides.

AFLP reactions are readily automated using robotics and high throughput sequencers or CE based genetic analyzers (Weller et al, 1995). AFLPs transfer well between laboratories; the primers are enough longer than those of RAPDs that context effects and low annealing temperature irregularities do not appear. AFLP bands are scored as dominant markers although there is evidence that it may be possible to score heterozygotes by the appearance of peaks of half-height. Software is under develpopment to automate the scoring of these heterozygotes.

Sequence Polymorphisms

Sequence polymorphisms, including single nucleotide substitutions (Botstein et al, 1980), are fairly easy to develop by direct sequencing. They are the most frequent and widely distributed polymorphisms in a genome. However, most of these polymorphisms are biallelic, so they are not very informative individually. Once developed, they can be screened without the need for gel electrophoresis, making them highly automatable. Using the TaqMan™ PCR assay, multiple sequence polymorphisms can be multiplexed and detected on plate reading devices allowing high throughput screening. Multiple closely linked biallelic loci can be combined into haplotypes (Versluis et al, 1993) and screened using direct automated sequencing and specialized software for interpreting the haplotype assignment.

Automating Three Agronomic Applications

In order to demonstrate how the combination of marker technology and automation can be applied to agriculture three applications are illustrated.

Rapid Map Construction

To quickly build a relatively high-density genetic map for a previously uncharacterized plant, SSR or AFLP markers should be considered. Sample preparation for mapping is unlikely to be a bottleneck because the emphasis is on finding differences among relatively few individuals. Emphasis will focus on screening for markers by running large numbers of DNA reactions, using essentially different PCR primer sets. Therefore, efforts to automate the DNA manipulations, electrophoresis, and data analysis will be most rewarding. If available, SSR markers developed for a related species may be useful. However, using AFLP markers should allow rapid progress. Reaction set-up using conventional robots can quickly overwhelm subsequent steps. Modular set-up and storage of pre- and post-reaction materials should be considered. Many laboratories use microwell plates and bar-coding to facilitate sample tracking. Throughput rate for such projects is limited by availability of gel lanes to separate reaction products. Most projects using robotic sample handling also use multiple DNA sequencers to maximize throughput. Multicolor DNA sequencers further increase throughput and allow in-lane size standards. These standards allow more precise differentiation of polymorphisms and also serve as a quality check for individual gel runs. Data for different lanes in the entire project can be cross-compared, and commercial software, such as those described in the data analysis section, can be used to filter and aid in finding polymorphisms.

Varietal Identification

Varietal identification encompasses several desired ends such as determining purity of breeding stocks and typing germplasm for patent protection. Screening germplasm stocks to determine uniqueness and limit accessions, particularly of vegetatively propagated species, is another use. Any technique that shows individuality to the desired degree is acceptable. The emphasis in varietal identification is on running a relatively small set of defined markers against a large number of individuals. Sample preparation, therefore, becomes an issue for this application, and efforts to implement automation would be repaid.

It is likely that in systems with established SSR maps that these would be the marker of choice. With new systems it may be advantageous to use AFLP markers. For efficiency, the object is to determine the minimum number of reactions that place the investigator within the desired boundary for establishing uniqueness. It is also worthwhile to multiplex the reactions as much as possible, to reduce the number of reactions and the number of gel lanes required for the analysis of one sample. Once this development has been done, it is worthwhile to automate set-up and post-PCR processing of reactions. In cases where legality is a consideration, markers have to be extremely robust in order to be easily transferred; ideally such markers should continue to be useful into the future.

Marker-assisted Breeding Programs

Marker-assisted breeding programs utilize information on genetic maps to improve selection for desired phenotypes while retaining the majority of the genome as the recurrent parent. Two types of information are generally required: very detailed information in the introgressed region, usually markers that flank the gene- or genes-of-interest as closely as possible; and more widely dispersed markers on the unlinked regions to ensure that the genetic material from the recurrent parent is carried forward. For these programs, a limited number of informative PCR reactions are run on a very large number of progeny. In genotyping offspring it is likely that the recurrent parent analysis would be done using SSRs on the retained chromosome arms, optimized as described above to allow as much multiplexing as possible. Analysis for introgression could use AFLP reactions preselected to reflect information about the region of interest. Progeny not having the required genotype can be disposed of at an early stage, representing great savings of space and time. Sample preparation will almost certainly become a bottleneck, and it will be worth investing a fair amount of effort in automating this. As numbers of progeny to be screened go up, robustness of the reaction must be very high to reduce the number of reactions repeated. The simpler and more automatic the system the better, since misrecording of data becomes a significant factor as the numbers of samples increase. Software that allows some personalization of routine types of analyses is very worthwhile at this stage.

Conclusion

The increased use of PCR generated markers is significantly affecting the way in which genetic analysis can be automated in plants. DNA marker acquisition methods, combined with advances in high throughput sequencing strategies, allow researchers, plant breeders, and environmentalists to realize automating the steps required for DNA marker assisted selection, bulk segregant analysis, linkage analysis, varietal identification, and taxonomic classification. Increases in sample throughput coupled with decreases in the cost of sample analysis can be driven by a number of factors. The use of robotics can allow sample volumes to be reduced to the sub-miroliter level resulting in decreased costs with increased sample density. For markers that are polymorphic due to differences in molecular weight, high throughput electrophoresis systems can resolve over 5000 SSR genotypes per day or 5000 polymorphic AFLP polymorphic data points per day. Substantial efforts are being made to increase the throughput of electrophoresis systems by either developing multiple array capillaries or micro-channeled slab gel systems. Sequence based polymorphic linked markers or specific gene detection can be analyzed in a gel free fashion. PCR reactions can be performed in multiwell plates by robotics, and through the use of fluorescent labeling, the conformation of the PCR product can be measured on-line or at end point. This approach could be miniaturized resulting in cost savings and increased throughput. Sequence based poymorphisms may also be detected by the use of DNA probes that are attached to a microchip. This technology has been applied to high throughput sequencing and has been given the acronym SBH (sequencing by hybridization) (Cantor et al, 1992, Drmanac et al, 1993). There are two basic forms of SBH. In the first format, a single oligonucleotide probe is used to probe an array of samples immobilized to a silica chip. In the second format, an array of short oligonucleotides, usually 8-mers, are immobilized, and the sample hybridized to the array. Specific gene sequences or a sequence based polymorphisms could be rapidly identified for plant identity, gene identity, trait selection, germplasm identification, and trueness to type. Arrays of oligonucleotides could be bound to silica chips and a sample hybridized. Through the use of either fluorescent labeling and microscopic detection or the use of electrochemical detection, positive hybridizations could be scored. This would allow the rapid screening of many samples on a very small scale with all of the liquid deliveries automated. The bottleneck in the scenario would be a computational one. The weakness from a developmental perspective would be in the design of optimal hybridization conditions to minimize false positives and false negatives. The benefits would be the mass production of microchip oligonucleotide arrays that could be placed in automated instruments that would control the reaction conditions, detection, and data analysis. Oligonucleotides used in this way have been called Geosensors (Cubicciotti, 1993).

REFERENCES

Botstein D, White RL, Skolnick M, Davies RW (1980): Construction of a genetic linkage map in man using restriction fragment length polymorphisms. *Amer J Hum Genet* 32:314–331

Caetano-Anolles G, Bassam BJ, Gresshof PM (1992): Primer-template interactions during DNA amplification fingerprinting with single arbitrary oligonucleotides. *Mol Gen Genet* 235:157–165

Caillat-Zucman S, Garchon H-J, Costantino F, Cot S, Bach J-F (1993): Automation of large-scale HLA oligotyping using a robotic workstation. *Biotechniques* 15:526–531

Cantor CR, Mirzabekov A, Southern E (1992): Report on the sequencing by hybridization workshop. *Genomics* 13:1378–1383

Cubicciotti R (1993): Genosensors: the next step in biosensor technology? *The Genesis Report.*® Montclair, New Jersey: The Genesis Group

Drmanac R, Drmanac S, Strezoska Z, Paunesku T, Labat I, Zeremski M, Snoddy J, Funkhouser WK, Koop B, Hood L, Crkvenjakov R (1993): DNA sequence determination by hybridization : a strategy for efficient large-scale sequencing. *Science* 260:1649–1652

Earley JJ, Kuivaniemi H, Prockop DJ, Tromp G (1994): Robotic automation of dideoxyribonucleotide sequencing reactions. *Biotechniques* 17:156–165

Garner HR, Armstrong B, Lininger DM (1993): High-throughput PCR. *Biotechniques* 14:112–115.

Harrison D, Baldwin C (1993): Use of an automated workstation to facilitate PCR amplification, loading agarose gels and sequencing of DNA templates. *Biotechniques* 14:88–97

Holland, PM, Abramson RD, Watson R, Gelfand DH. (1991): Detection of specific polymerase chain reaction product by utilizing the 5′ to 3′ exonuclease activity of Thermus aquaticus DNA polymerase. *Proc Natl Acad Sci USA* 88:7276–7280

Holmberg A, Fry G, Uhlén M (1994): Automatic preparation of DNA templates for sequencing on the ABI catalyst robotic workstation. In: *Automated DNA Sequencing and Analysis*, Adams MD, Fields C, Venter CJ, eds. London: Academic Press

Knorr D, Otteson K (1995): Use of DNAs to purify specific plant DNA sequences. Unpublished data.

Lee LG, Connell CR, Bloch, W (1993): Allelic discrimination by nick-translation PCR with fluorogenic probes. *Nucl Acid Res* 21(16):3761–3766

Livak KJ, Flood SJA, Marmaro J, Giusti W, Deetz K (1995a): Oligonucleotides with fluorescent dyes at opposite ends provide a quenched probe system useful for detecting PCR product and nucleic acid hybridization. *PCR Methods and Applications* (in press)

Livak KJ, Marmaro J, Todd JA (1995b): Towards fully automated genome-wide polymorphism screening. *Nat Genet* 9:341–342

Michelmore RW, Paran I, Kesseli RV (1991): Identification of markers linked to disease resistance genes by bulked segregant analysis: A rapid method to detect markers in specific genomic regions using segregating populations. *Proc Natl Acad Sci USA* 88: 9828–9832

McCollum C, Chakerian V, Kaufman J, Wentz M, Andrus A (1994): Rapid and efficient oligonucleotide synthesis with low reagent consumption via a new synthesis column design: preparation of fluorescent dye labeled primers for application in PCR. *Biomed Pep Prot Nucl Acids* 1 (1):25–30

Mischiati C, Fiorentino D'A, Feriotto G, Gambari R (1993): Use of an automated laboratory workstation for isolation of genomic DNA suitable for PCR and allele-specific hybridization. *Biotechniques* 15:146–151

Nickerson DA (1992): Identification of clusters of biallelic polymorphic sequence-tagged sites (sSTSs) that generate highly informative and automatable markers for genetic linkage mapping. *Genomics* 12:377–387

Nielson PE, Egholm M, O Buchardt (1994): Peptide nucleic acid (PNA). A DNA mimic with a peptide backbone. *Bioconjug Chem* 5:3–7

Olsen AS, Combs J, Garcia E, Elliot J, Amemiya C, de Jong P, Threadgill G (1993): Automated production of high density cosmid and YAC colony filters using a robotic workstation. *Biotechniques* 14:116–123

Perkin Elmer Corporation (1993): Sequencing of DNA by Gel Electrophoresis in Micro Machined Channels, by Balch JW, Davidson C, Gingrich J, Sharof M, Brewer L, Koo J, Smith D, Albin M, Carrano A. Abstract C-2, Genome Sequencing and Analysis Conference VI, Hilton Head, S.C., Sept 17–21, 1994. CRADA no. TC-486-93

Reed PW, Davies JL, Copeman JB, Bennett ST, Palmer SM, PritchardLE, Gough SCL, Kawaguchi Y, Cordell HJ, Balfour KM, Jenkins SC, Powell EE, Vignal A, Todd JA (1994): Chromosome-specific microsatellite sets for fluorescence-based, semi-automated genome mapping. *Nat Genet* 7:390–395

Schwengle DA, Jedlicka AE, Nanthakumar EJ, Weber JL, Levitt RC (1994): Comparison of fluorescence-based semi-automated Genotyping of multiple microsatellate loci with autoradiographic techniques. *Genomics* 22:46–43.

Thomas MR, Scott NS (1993): Microsatelllite repeats in grapevine reveal DNA polymorphisms when analyzed as sequence tagged sites (STSs). *Theor Appl Genet* 86:985–990

Thomas MR, Cain P, Scott NS (1994): DNA typing of grapevines: A universal methodology and database for describing cultivars and evaluating genetic relatedness. *Plant Mol Bio* 25:939–949

Thomas MR, Matsumoto S, Cain D, Scott NS (1993): Repetitive DNA of grapevine: classes present and sequence suitable for cultivar identification. *Theor Appl Genet* 24:121–124

Tingey S (1995): Automation Technology for Plant Genetic Diagnostics. Lecture given Jan 18, 1995 at the Plant Genome III Meeting, Jan 15–19, 1995, San Diego, CA. Organized by Scherago International, Inc., New York, NY

Versluis LF, Rozemuller S, Tonks S, Marsh SGE, Bouwens, JG, Bodmer JG, Tilanus MGJ (1993): High-resolution HLA-DPB typing based upon computerized analysis of data obtained by fluorescent sequencing of the amplified polymorphic exon 2. *Hum Immun* 38:277–283

Weller JW, Ziegle J, Bates SB (1995): Automated High-Throughput AFLP Using Multicolor Fluorescent Labelling. Poster P254 presented at Plant Genome III Meeting, Jan 15–19, 1995, San Diego, CA. Organized by Scherago International, Inc., New York, NY

Welsh J, McClelland M (1990): Fingerprinting genomes using PCR with arbitrary primers. *Nucl Acids Res* 18:7213–7218

Williams JG, Kubelik AR, Livak KJ, Rafalski JA, Tingey SV (1990): DNA polymorphisms amplified by arbitrary primers are useful as genetic markers. *Nucl Acids Res* 18:6531–6535

Zabeau M, Voss P (1992): Selective restriction fragment amplification : a general method for DNA fingerprinting. European Patent Application 92402629.7

Ziegle JS (1992): Application of automated DNA sizing technology for genotyping microsatellite loci. *Genomics* 14 (4):1026–1031

Zimmerman J, Dietrich T, Voss H, Erfte H, Schwager C, Stegemann J, Hweitt N, Ansorge W (1992): Fully automated sanger sequencing protocol for double-stranded DNA. *Meth Mol Cell Biol* 3:39–42

Part V

The Experience of Molecular Marker-Assisted Breeding

15

Molecular Marker Assisted Breeding in a Company Environment

STEPHEN SMITH AND WILLIAM BEAVIS

Introduction

The Need and Genetic Basis for Improved Productivity

As the world's population continues to rise, and it becomes increasingly desirable to avoid excessive use of chemical inputs, plant breeding can provide biological solutions to the necessities of food production and help conserve environmental resources. Continued improvements in the genetic basis of agronomic productivity must be achieved. Applied genetic research has been instrumental in progressing varietal performance and breeding efficiency. Productivity has risen as breeders have brought together in single varieties chromosomal regions exerting positive effects on traits that have a favorable impact on agronomic performance in specific and general environments. Russell (1984), Duvick (1987), Anderson et al (1988), and Tollenaar (1989) have shown that newer varieties outperform older varieties even in low input environments. Older varieties more readily exhibit sterility, lodging, and increased damage from pests when confronted with stress conditions (Patch et al, 1941; Russell, 1974; Austin et al, 1980; Duvick, 1984).

Most significant improvements in agronomic productivity have been contributed by characters that are controlled by many genes (Duvick, 1986). It is generally believed that each gene affecting these characteristics is of relatively small effect and final expression can be significantly modified by the environment. Therefore, it is not surprising that loci and alleles affecting most important agronomic traits are anonymous with respect to their genetic locations, relative magnitude of effects, and modes of expression. Consequently, the information and the means to more efficiently bring together alleles contributing positive effects for

The Impact of Plant Molecular Genetics
BWS Sobral, Editor

complexly inherited agronomic traits have remained unavailable. Further improvements for traits such as pest resistance and fruit, seed, or tuber quality that are under simpler oligogenic control are also desirable.

The Role of the Private Sector and the Company Environment

There has been a shift toward the use of privately bred proprietary inbred lines in United States maize hybrid production during the last two decades. Darrah and Zuber (1986) reported that 92% of U.S. hybrids were made with one or more proprietary inbred lines. New technological tools, such as DNA markers and transformation coupled with improved technical and legal abilities to protect proprietary genotypes, have made plant breeding and product development more attractive for private investment. In contrast, public investment in plant breeding has declined (Collins and Phillips, 1991). Budgetary constraints combined with a climate favoring basic research using high profile technologies have resulted in public research efforts in many countries turning away from plant breeding.

While the specifics of a company environment will differ according to circumstances, some generalities can be summarized as follows: (1) a very high priority on commercial product development; (2) funding from seed sales; (3) accountability to customers and stock holders; (4) focus on applied research; (5) fundamental need to protect intellectual property to conserve return on investment and promote further investment; (6) ability to compete effectively to realize returns on investment; (7) ability to effectively manage an entire process ranging from research and product development through to seed production, distribution, sales, marketing, and agronomic and informational services; (8) the need to conduct research on a scale larger than most public researchers; and (9) the ability to breed varieties adapted to a range of environments and customers.

Privately funded plant breeding organizations depend upon the availability and integration of resources from research to generate new and more productive genotypes. Organizations that have sufficient resources to fund integrated research and breeding programs can be more competitive than groups that are dependent upon common publicly available genotypes. The ability to utilize new and existing genetic, intellectual, and technological resources is the primary feature that determines productive and competitive capabilities of a plant breeding organization. Success results in the availability of improved products for agriculture and an opportunity from the sale of these products to fund further breeding and research.

Each organization is limited by its resources in regard to the size and depth of research that it can fund. In the publicly funded arena, research proposals are evaluated by peers, and they usually do not have as a goal the development of an improved product or process that will have a direct impact on the production cycle in the short to medium term. In the private sector, the researcher must communicate and explain to business managers the need for research that has an impact on product development, usually within a three to fifteen year time frame.

Many business managers lack a technical or scientific background. Success of a research driven private organization is highly dependent upon effective communications between business and research managers.

Funding of research and product development from seed sales does not provide sufficient revenues to fund long term explorative projects that would provide the basic infrastructure for applied research. Basic technological development to acquire fundamental new knowledge is best conducted by diverse talents in programs that extend in scope, time, and funding beyond that which can be provided by the private sector. For example, various genome sequencing and mapping projects currently in progress are very demanding of financial and other resources. Many of the benefits of these initiatives may not be immediately obvious, especially to smaller privately funded organizations. Yet practical applications in the medical field are already evident from the human genome project. New basic knowledge and emerging technologies could have a similar impact on plant breeding and pay dividends to investors, farmers, and the consumer. Unfortunately, public sector botanical research in the United States has not resulted in comprehensive support for such an infrastructure. The private sector must encourage public funding of basic infrastructural needs.

The Role of Molecular Markers in Helping to Provide for the More Effective Management of Germplasm Diversity

The technologies, intellectual skills, and resources that are required to deliver genotypes to meet customer demands are key components that enhance the performance of a specific breeding organization and that benefit agricultural productivity in general. Management of genetic diversity to radically improve genotypic potential underlying agronomic traits provides immense intellectual challenges and requires long term concerted effort and risk. A critical factor determining progress in advancing the genetic basis of productivity is the ability to characterize germplasm in both agronomic and genetic detail.

Information on performance ability and genetic relatedness among inbred lines or varieties from known pedigrees has provided the basic framework within which the parents of breeding populations and testers of heterosis are selected. However, pedigree, morphological, and agronomic performance data fail to provide the optimum detail and precision of genetic characterization (Dudley, 1993, 1994; Graner et al, 1994, Nienhuis et al, 1994). Further challenges are offered by exotic germplasm which may be of unknown pedigree and sufficiently unadapted to allow even preliminary evaluation of its potential usefulness. Molecular markers can help to characterize the genotypic-agronomic potential of chromosomal regions, a capability that has hitherto been impossible. Mapping studies to reveal Quantitative Trait Loci (QTL) have been used most frequently to identify agronomically important chromosomal regions. Variety profiles or fingerprints could also be used to identify QTL (Melchinger et al, 1994). Molecular marker profiles could help improve the heritability or power of selection and thereby help to more efficiently and effectively organize, combine, and select new genotypic combinations.

Transfer of germplasm from one region of the world to other regions is a general rule for most cultivated species. Most agriculturally important plant species are now grown most productively in regions far removed both geographically and ecologically from locations where they were first domesticated or from other centers of genetic diversity for landraces and wild or weedy relatives. In many cases, migration has occurred over hundreds or thousands of years during which time genotypes have evolved that are better adapted to other regions. Movement of genetic resources, whether they be the products of scientific plant breeding or farmer landrace varieties, can continue to benefit consumers worldwide. In either case, exotic germplasm usually would be unadapted to new agroecological environments thereby presenting significant challenges to its evaluation, breeding, and utilization in agriculture. These factors, together with numerous uncharacterized linkages of molecular marker alleles with QTL, present special challenges for identifying and introgressing useful exotic germplasm.

Molecular markers can provide additional ancillary and important support for agriculture. Plant breeding and agriculture benefit from increased funding of research and product development due to effective Plant Breeders' Rights, the support of which can be aided by varietal profiles from molecular marker technology (Melchinger et al, 1994). Also, molecular marker variety profiles can be used to promote a broader genetic base in agriculture. In new legislation, a variety would be classified as an Essentially Derived Variety (EDV) when it is similar to an already existing variety. Widespread attacks by pests and pathogens with the consequent threat of erosion of genetic resources can be guarded against provided an overly narrow EDV boundary is avoided. Molecular marker data could also help farmers identify genetically different varieties as part of a management strategy to hedge risks that might be associated with a narrow genetic base.

It is approximately ten years since intense research began using molecular markers to characterize and manipulate genetic diversity for major crop plants. Molecular markers are being used in some applications, but for others there is hesitancy. It is important to determine the obstacles that need to be overcome to allow molecular markers to aid in achieving improved productivity.

Review of Molecular Marker Studies

Development and Selection of Technologies

Restriction fragment length polymorphisms (RFLPs) first revealed an abundance of genetic marker variants among elite inbred lines and varieties such that genetic distances reflective of pedigree and detailed genetic maps suitable for subsequent mapping of regions associated with agronomic traits could be obtained for many cultivated species.

For purposes of varietal identification and the protection of Plant Breeders' Rights, RFLP technology will maintain a preeminent role in the provision of data until a simpler, more standardized methodology is used routinely. Microsatellite technology, which reveals polymorphisms for numbers of simple sequence repeat (SSR) units at mappable sites through the genome, offers significant practical improvements over RFLPs for varietal identification. SSR technology could provide a standardized and highly accurate set of descriptors once relatively high development costs have been met.

In trait mapping and marker based selection applications, emphasis must be placed upon attaining reduced time and costs of data acquisition. If breeders are restricted to using a technology that can accommodate fewer than 100–200 individuals for each of only a few populations, then only a limited array and number of marker assisted programs can be accomplished. Chromosomal segments associated with agronomic traits of fairly moderate genetic complexity cannot be readily identified with sufficient precision or power by RFLPs. Throughput using RFLPs is limiting, especially during progeny selection with time constraints imposed prior to pollination or harvest. Plant breeders are increasingly using arbitrary primer PCR based methods such as randomly amplified polymorphic DNA (RAPD) (Williams et al, 1990) or amplified fragment length polymorphisms (AFLP) (Zabeau and Vos, 1993) because these methods, in some instances (Ragot and Hoisington, 1993), can provide speedier and more cost effective marker assisted progeny selection. Arbitrary primer methods, because of the competitive nature of the primer-target annealment process, also can be used to very quickly identify, map, and introgress chromosomal regions contributing insect or disease resistance. Profiling of near isogenic lines and bulk segregant analysis can be used to quickly map, characterize, and introgress important chromosomal regions (Michelmore et al, 1991; Paran et al, 1991).

All gel based marker technologies are inherently constrained for sample throughput efficiency and cost of data acquisition. These constraints restrict the number of marker assisted projects of even moderate genetic complexity that can be simultaneously conducted and effectively preclude marker assisted selection for traits of moderate to greater genetic complexity. No marker based scheme of selection will be useful unless it can advance the speed and efficiency of the already rapid and significant progress made using conventional parent and progeny selection methods. However, a more serious and fundamental limitation to the mapping approach, regardless of mapping technology, is the basic difficulty in obtaining reliable performance data for traits that are controlled by more than three to five genes.

Genetic Distance Studies

RFLP profiles of inbred lines and varieties of many cultivated species have been analyzed for their discriminational power and their ability to reveal associations among genotypes that are reflective of pedigree or of heterosis. It is important to identify not only a sufficient number of probe-restriction enzyme combinations that can be discriminative but also to achieve a reasonably thorough sampling of the genome. It has been shown repeatedly that molecular marker data reveal associations among inbred lines or varieties that are reflective of pedigree (Dudley, 1994; Melchinger et al, 1994; Mumm and Dudley, 1994; Nienhuis et al, 1994; O'Donoughue et al, 1994). Correlations between distances calculated from comparative marker profiles and those calculated from expectations based on pedigrees often are high but only when genotypes are 50% or more related by pedigree. Molecular marker data can be used to identify inbred lines or varieties that have common or similar pedigrees. Therefore, the technologies can be useful in the support of Plant Breeders' Rights (Melchinger et al, 1994). However, with regard to agronomic performance, genetic distance data per se are not practically useful predictors of heterosis. Bernado (1992) has indicated that sufficient qualitative information linking genetic markers to QTL affecting heterosis must be available; quantitative distance data do not meet these criteria.

Agronomic Trait Mapping Studies

Initially, isozymes were used to identify and map QTL in tomato and maize (Tanksley et al, 1982, Weller, 1987; Edwards et al, 1987). Since the development of RFLP markers, many studies have been devoted to the identification of QTL for oligogenic traits (Melchinger, 1990) such as insect and disease resistance (Godshalk et al, 1990; Lee et al, 1991; Bubeck et al, 1993; Webb et al, 1994) and for complex or polygenic traits such as soluble solids in tomato (Paterson et al, 1987) and grain yield in maize (Stuber and Sisco, 1991; Stuber et al, 1992; Beavis et al, 1994), and barley (Hayes et al, 1993). All of these studies have used the same basic design in which linkage disequilibrium has been maximized among small numbers of lines (<500) derived from bi-parental crosses of inbred lines. With about one marker locus per 20 cM, each progeny was evaluated for the quantitative trait(s) of interest in replicated field trials at several environments. Three to ten independently segregating QTL usually have been identified. There has been little evidence for QTL × environment interactions, and each QTL has usually been estimated to account for 5% to 25% of the phenotypic variability (Beavis et al, 1994). However, there has been little consistency in the genomic locations of QTL (Beavis et al, 1991) even when the same inbred parents were used (Beavis et al, 1994). The lack of congruent QTL is best explained by recognizing that there is little power to identify QTL unless the trait exhibits high heritability (Van Ooijen, 1992; Carbonell et al, 1992) (Table 1). Thus, for polygenic traits, an experiment that uses a small number of segregating progeny will identify only a small proportion of the QTL. Furthermore, if the QTL are not independently segregating, there will be problems in accurate and precise estimations of QTL effects and genomic position (Haley and Knott, 1992; Martinez and Curnow, 1993). Precision in the estimated QTL effects and genomic position has been improved (Jansen, 1993; Zeng, 1993) but with an associated loss in power to identify QTL (Zeng, 1994). The breeder can improve the precision of QTL estimates without loss of power by relaxing linkage disequilibrium through random sibmating (Beavis et al, 1992). However, it costs more time to develop these

Table 1. Power to Identify Simulated QTL in a Diploid Genome[1]

Simulated Experimental Conditions			
Number of Simulated QTL[2]	Magnitude of Simulated Effects[3]	Sample Size[4]	Power[5]
10	30	100	0.117
10	63	100	0.327
10	95	100	0.391
10	30	500	0.574
10	63	500	0.864
10	95	500	0.935
10	30	1000	0.845
10	63	1000	0.988
10	95	1000	0.999
40	30	100	0.025
40	63	100	0.044
40	95	100	0.059
40	30	500	0.112
40	63	500	0.294
40	95	500	0.458
40	30	1000	0.253
40	63	1000	0.594
40	95	1000	0.774

[1]The simulated diploid genome consisted of 75 independently segregating linkage groups. Genetic markers were placed at the ends of each linkage group. Simulated QTL were assigned to a random set of 20cM linkage groups and placed in the middle of each linkage group. Each simulation consisted of:

[2]Ten or 40 QTL with equal additive effects that explained cumulatively

[3]30, 63 or 95% of the phenotypic variability among

[4]100, 500 or 1000 F_2 progeny.

[5]The power was calculated as the percentage of correctly identified QTL from 200 independent simulations that were conducted for each set of experimental conditions.

populations, and they then require a greater saturation of genetic markers. Nevertheless, despite low power for QTL identification using RFLPs, associations for grain yield have been identified and successfully selected (Stuber and Sisco, 1991; Stuber, 1994), although not all attempts to improve the expression of quantitative traits through marker aided selection have been successful (Tanksley and Hewitt, 1988; Edwards and Johnson, 1994).

Results of Selections Made Using Molecular Marker QTL Data

Reports of results from selection experiments that have used QTL information are few. The reasons are several. Marker technology is still evolving. Selection studies are only possible after detailed QTL mapping or after profiles of many inbred lines with known pedigree relationships to the genotypes being field tested are obtained. Publicly funded programs do not always have available numerous inbred lines that are related in a pedigree breeding scheme. Nor do public programs usually have the field space and support to generate and acquire data from numerous replicated field plots equal to those that are available to many in the private sector. Performance results obtained within privately funded breeding programs are primarily of use in product development rather than for scholarly academic publication.

Molecular marker profiles of inbred lines and varieties including relatives of lines under selection increasingly will be routinely available because these data can help provide security of proprietary germplasm. Goddard (1992) has suggested that fingerprint based marker information could be effective in QTL identification. Performance data of these inbreds are also often obtained in conventional breeding programs. Best linear unbiased predictor (BLUP) procedures are extensively used in animal breeding (Henderson, 1988; van Arendonk et al, 1994). In plants, BLUP has been reported for forest trees (White and Hodge, 1989) and maize (Bernado, 1994). In stimulation studies, Zhang and Smith (1992) found that selection using RFLP data and BLUP analysis was the most effective means of achieving progress. Bernado (1994) showed that maize hybrid yields were predictable using BLUP and RFLP profiles from relatives of the lines under selection.

Backcrossing is widely used to transfer chromosomal regions controlling simply inherited traits into more desirable genetic backgrounds. Molecular markers can be used to effectively speed and improve the precision of backcrossing (Young and Tanksley, 1983; Openshaw et al, 1994) but effective expressions of the desirable trait depend on the genetic background (Tanksley and Hewitt, 1988). Heritable traits that are under oligogenic control such as maturity can also be successfully transferred by backcrossing (Rinke and Sentz, 1961; Shaver, 1976). Such traits can be mapped by molecular markers: for example, maturity (Godshalk et al, 1990); dry matter content of tubers (Freyre and Douches, 1994); soluble solids of tomato (Tanksley et al, 1982); insect resistance (Lee et al, 1991); nematode resistance (Webb et al, 1995); and virus resistance (Bubeck et al, 1993). Oligogenic traits that are not normally highly heritable can become more heritable, at least in the population within which they have been mapped, once they have been molecularly tagged by mapping using inoculation and effective scoring procedures (Webb et al, 1995). Marker assisted transfer of oligogenic factors and exotic germplasm delivered by transformation into elite varieties, therefore, could be expanded in practical breeding programs.

Stuber (1994) reported results of more complex genetic changes effected with marker aided QTL selection. Stuber and Sisco (1991) identified QTL in the inbred lines B73 and Mo17 that could be improved after replacement with alleles from the lines TX303 and Oh43, respectively. Thirty-two parents of enhanced B73/Mo17 crosses and 44% of B73/enhanced Mo17 crosses significantly outyielded B73/Mo17. Subsequent crosses among only the top performing enhanced lines resulted in 5% of the hybrids outperforming the unenhanced B73/Mo17 by 2SD. The two top yielding enhanced hybrids surpassed the check hybrid by 15% or 22–24 bu/acre.

Edwards and Johnson (1994) reported a study to improve performance in sweetcorn. Four successive cycles of recurrent selection based on an index using QTL data were performed with 100–300 individuals. In one population, positive responses were found for six of eleven traits. However, responses were not significantly greater than for the unimproved lines, but data were limited. In the second population, QTL with wholly positive effects were difficult to find. Many chromosomal regions had disruptive effects on yield versus quality traits. Although eight of twelve performance traits responded positively to marker assisted selection, improvement was only achieved by direct phenotypic selection; marker selection resulted in negative progress for five of eight quality traits. Reasons for inconsistent results in the use of marker assisted selection (MAS) for quantitative traits are related to the samples of progeny that were used for these studies and possibly to the selection index.

Greatest gains in efficiency from marker aided selection would be for those traits that exhibit low heritabilities (Lande and Thompson, 1990). The paradox is that there is little power to identify QTL for such traits (Table 1). In order to realize the benefits and efficiencies from marker aided selection, it is important to identify as many QTL as possible and thus large population sizes would be needed (Gimelfarb and Lande, 1994) for consistent results. The need to evaluate large numbers of progeny (>1000) would require new marker technologies and additional perspectives from research. New breeding paradigms would be needed as well to fully realize the power of MAS.

Conclusions

During the past 10–15 years, greater sums have been invested by the private sector in plant breeding and associated support technologies because scientific advances in molecular biology and enhanced Plant Breeders' Rights have created new opportunities in product development. When these investments are successful, they pay dividends first to farmers, industrial customers, and the public consumer and second to shareholders and employees. Most importantly, revenues from seed sales provide additional resources for further research, product development, and genetic resource maintenance. Privately funded organizations usually can only commit resources to programs of research that are projected to have a direct impact on products within 5–20 years. The private sector alone cannot fund basic infrastructure. Therefore, it must provide economic, intellectual, and political support to the conduct of basic research in the public sector.

Molecular marker assisted breeding is becoming a routine component of backcrossing and the mapping and selection for oligogenic traits of high heritability and/or economic worth. Molecular profiles are used to help protect Plant Breeders' Rights in cases of disputed identity and in the establishment of boundaries of ownership around initial varieties. It is essential that EDV boundaries be determined that provide strong disincentives to reverse engineering and varietal plagiarism. Investment and opportunity for improved productivity would be wasted, and the spectre of genetic uniformity and vulnerability could arise if such disincentives are not in place.

With regard to the use of molecular markers to improve multigenic agronomic traits, BLUP procedures could be integrated into plant breeding using currently available profiling methodologies. The state of knowledge with regard to the genetic control and allelic diversity for multigenically inherited traits remains primitive. Such an evaluation should not be surprising due to the immense complexity of genetic and environmental effects underlying these traits. For example, Stromberg et al, (1994) were unable to show that the use of QTLs identified in an early generation selection program were effective in improving yield, a conclusion due in large part to the complex biology they were attempting to investigate. New marker technologies that can assay 100-fold more quickly and less expensively than RFLPs are one prerequisite to allow further progress. However, other limitations in fundamental knowledge exist. Improved criteria of significance and selection indices will be required (Stuber and Edwards, 1986; Lande and Thompson, 1990; Johnson, 1991; Lande, 1992; Zehr et al, 1992; Zhang and Smith, 1992). Physiological and biochemical information is also lacking. Basic progress in research involving geneticists, molecular biologists, statisticians, physiologists, and biochemists will be necessary before more progress can be made in marker assisted selection for multigenically inherited traits. Challenges that will need to be met include: (1) an understanding of genotype x environment effects; (2) improved detection of epistasis; (3) improved resolution of QTL linkage phase; and (4) optimization of selection taking into account population size, selection intensity, size of QTL segment, and the ability to select for multiple traits. Most of these endeavors extend well beyond the scope of funding or accomplishment for a single privately funded organization. However, technological development in genome scanning, targeting mutations, and physically manipulating chromosome segments continually opens up new opportunities as investigative powers thereby improve.

Molecular markers have already made positive contributions to plant breeding. For genetically complex traits, it is only candid to conclude that little of practical use has been learned to date, beyond a reaffirmation that various heritable factors are situated on chromosomes. Such a statement in no way belittles the efforts that have been achieved. It rather reacquaints us with the complex realities that are involved in making progress for agricultural productivity since it is largely founded upon traits with a quantitative genetic basis. Molecular markers have provided a focus and opportunity for an infusion of other technologies and new intellect into research supporting plant breeding. It is crucial for the long term benefit of agriculture to continue and extend this research. Privately funded companies must be in the forefront of supporting these endeavors. Companies then can integrate products of basic research into applied programs for the benefit of their customers. The whole plant research community should strongly advocate research to build an infrastructure that allows an improved ability to productively manage a greater diversity of genetic resources.

REFERENCES

van Arendonk JAM, Tier B, Kinghorn BP (1994): Use of multiple genetic markers in prediction of breeding values. *Genetics* 137:319–329

Anderson JR, Herdt RW, Scobie GM (1988): *Science and Food: The CGIAR and Its Partners*. Washington, DC: World Bank

Austin RB, Bingham J, Blackwell RD, Evans LT, Ford MA, Morgan CL, Taylor M (1980): Genetic improvements in winter wheat yields since 1900 and associated physiological changes. *J Agric Sci* 94:675–680

Beavis WD, Grant D, Albertsen M, Fincher R (1991): Quantitative Trait Loci for plant height in four maize populations and their associations with quantitative genetic loci. *Theor Appl Genet* 83:141–145

Beavis WD, Lee M, Grant D, Hallauer AR, Owens T, Katt M, Blair D (1992): The influence of random mating on recombination among RFLP loci. *Maize Genet Coop Newsletter* 66:52–53

Beavis WD, Smith OS, Grant D, Fincher R (1994): Identification of Quantitative Trait Loci using a small sample of topcrossed F_4 progeny from maize. *Crop Sci* 34:882–896

Bernado R (1992): Relationship between single-cross performance and molecular marker heterozygosity. *Theor Appl Genet* 83:628–634

Bernado R (1994): Prediction of maize single-cross performance using RFLPs and information from related hybrids. *Crop Sci* 34:20–25

Bubeck DM, Goodman MM, Beavis WD, Grant D (1993): Quantitative trait loci controlling resistance to gray leaf spot in maize. *Crop Sci* 33:838–847

Carbonell EA, Gerig TM, Balansard E, Asins MJ (1992): Interval mapping in the analysis of nonadditive quantitative trait loci. *Biometrics* 48:305–315

Collins WW, Phillips RL (1991): Plant breeding training in public institutions in the United States: A survey conducted by the National Plant Genetic Resources Board. *Diversity* 7:28–32

Darrah LL, Zuber MS (1986): 1985 United States farm maize germplasm base and commercial breeding strategies. *Crop Sci* 26:1109–1113

Dudley JW (1994): Comparison of genetic distance estimators using molecular marker data. In: *Analysis of Molecular Marker Data Symposium*; 1994 Aug 5–6; Corvallis, OR. Madison, WI: Crop Science Society of America

Dudley JW (1993): Molecular markers in plant improvement: Manipulation of genes affecting quantitative traits. *Crop Sci* 33:660–668

Duvick DN (1984): Genetic contributions to yield gains of U.S. hybrid maize, 1930 to 1980. In: *Genetic Contribution to Yield Gains of Five Major Crop Plants*, Fehr WR, ed. Madison, WI: Crop Science Society of America

Duvick DN (1987): Sources of genetic advance for the future. In: Burris JS, ed. *Proceedings of the Ninth Annual Seed Technical Conference*; Feb 24–25, 1987. Ames, IA: Seed Science Center, Iowa State University

Duvick DN (1986): Past achievements and expectations for the future. *Econ Bot* 40:289–297

Edwards M, Johnson L (1994): RFLPs for rapid recurrent selection. In: *Analysis of Molecular Marker Data Symposium*; 1994 Aug 5–6; Corvallis, OR. Madison, WI: Crop Science Society of America

Edwards MD, Stuber CW, Wendel JF (1987): Molecular-marker facilitated investigations of quantitative trait loci in maize. I. Numbers, genomic distribution and types of gene action. *Genetics* 116:113–125

Freyre R, Douches DS (1994): Development of a model for marker-assisted selection of specific gravity in diploid potato across environments. *Crop Sci* 34:1361–1368

Gimelfarb A, Lande R (1994): Simulation of marker assisted selection in hybrid populations. *Genet Res Camb* 63:39–47

Goddard ME (1992): A mixed model for analyses of data on multiple genetic markers. *Theor Appl Genet* 83:878–886

Godshalk EB, Lee M, Lamkey KR (1990): Relationship of restriction fragment length polymorphisms to single-cross hybrid performance of maize. *Theor Appl Genet* 80:273–280

Graner A, Ludwig WF, Melchinger AE (1994): Relationships among European barley germplasm: II. Comparison of RFLP and pedigree data. *Crop Sci* 34:1199–1205

Haley CS, Knott SA (1992): A simple regression method for mapping quantitative trait loci in line crosses using flanking markers. *Heredity* 69:315–324

Hayes PM, Liu BH, Knapp SJ, Chen F, Jones B, Blake T, Franckowiak J, Rasmusson D, Sorrels M, Ullrich SE, Wesenberg D, Kleinhofs A (1993): Quantitative trait locus effects and environmental interaction in a sample of North American Barley germplasm. *Theor Appl Genet* 87:392–401

Henderson CR (1988): Progress in statistical method applied to quantitative genetics since 1976. In: *Proceedings of the 2nd International Conference on Quantitative Genetics*; 1987 May 31–June 5; Raleigh, NC. Sunderland, MA: Sinnauer Assoc

Jansen RC (1993): A general model for mapping quantitative trait loci by using molecular markers. *Theor Appl Genet* 85:252–260

Johnson GR (1991): RFLP assisted early generation selection for advanced generation testcross performance. In: *Proceedings of the Twenty-seventh Illinois Corn Breeders School*, 1991 Mar 4–5; Champaign, IL: University of Illinois

Lande R (1992): Marker-assisted selection in relation to traditional methods of plant breeding. In: *Plant Breeding in the 1990s*, Stalker HT, Murphy JP, eds. Wallingford, UK: CAB International

Lande R, Thompson R (1990): Efficiency of marker-assisted selection in the improvement of quantitative traits. *Genetics* 124:743–756

Lee M, Melchinger AE, Guthrie WD (1991): Molecular marker analysis of host-plant resistance to European Corn Borer in corn. In: *Proceedings of the Twenty-seventh Illinois Corn Breeders School*; 1991 Mar 4–5; Champaign, IL: University of Illinois

Martinez O, Curnow RN (1992) or (1993): Estimating the locations and sizes of quantitative trait loci using flanking markers. *Theor Appl Genet* 85:480–488

Melchinger AE (1990): Use of molecular markers in breeding for oligogenic disease resistance. *Plant Breed* 104:1–19

Melchinger AE, Graner A, Singh M, Messmer MM (1994): Relationships among European barley germplasm: I. Genetic diversity among winter and spring cultivars revealed by RFLPs. *Crop Sci* 34:1191–1199

Michelmore RW, Paran I, Kesseli RV (1991): Identification of markers linked to disease resistance genes by bulked segregant analysis: A rapid method to detect markers in specific genomic regions using segregating populations. *Proc Natl Acad Sci USA* 88:9828–9832

Mumm RH, Dudley JW (1994): A classification of 148 U.S. maize inbreds: I. Cluster analysis based on RFLPs. *Crop Sci* 34:842–851

Nienhuis J, Tivang J, Skroch P (1994): Analysis of genetic relationships among genotypes based on molecular marker data. In: *Analysis of Molecular Marker Data Symposium*; 1994 Aug 5–6; Corvallis, OR. Madison, WI: Crop Science Society of America

O'Donoughue LS, Souza E, Tanksley SD, Sorrells ME (1994): Relationships among North American oat cultivars based on restriction fragment length polymorphisms. *Crop Sci* 34:1251–1258

Openshaw SJ, Jarboe SG, Beavis WD (1994): Marker-assisted selection in backcross breeding. *Analysis of Molecular Marker Data Symposium*; 1994 Aug 5–6; Corvallis, OR. Madison, WI: Crop Science Society of America

Paran I, Kesseli R, Michelmore R (1991): Identification of restriction fragment length polymorphisms and random amplified polymorphic DNA markers linked to downy mildew resistance genes in lettuce using near isogenic lines. *Genome* 34:1021–1027

Patch LH, Still GW, App BA, Crooks CA (1941): Comparative injury by the European corn borer to open-pollinated and hybrid field corn. *J Agr Res* 63:355–368

Paterson AH, Lander ES, Hewitt JD, Peterson S, Lincoln SE, Tanksley SD (1988): Resolution of quantitative traits into Mendelian factors using a complete linkage map of restriction fragment length polymorphisms. *Nature* 335:721–726

Ragot M, Hoisington D (1993): Molecular markers for plant breeding: Comparisons of RFLP and RAPD genotyping costs. *Theor Appl Genet* 86:975–984

Rinke EH, Sentz JC (1961): Moving corn-belt germplasm northward. In: *Proceedings of the Annual Corn and Sorghum Industry Research Conference*; Dec 6–7, 1961, Chicago, IL. Washington, DC: American Seed Trade Association

Russell WA (1974): Comparative performance of maize hybrids representing different eras of maize breeding. In: *Proceedings of the 29th Annual Corn and Sorghum Research Conference*; 1974 Dec 10–12; Chicago, IL. Washington, DC: American Seed Trade Association

Russell WA (1984): Agronomic performance of maize cultivars representing different eras of breeding. *Maydica* 29:375–390

Shaver DL (1976): Conversions for earliness in maize inbred. *Maize Genet Coop Newslett* 50:20–23

Stromberg LD, Dudley JW, Rufener GK (1994): Comparing conventional early generation selection with molecular marker assisted selection in maize. *Crop Sci* 34:1221–1225

Stuber CW (1994): Enhancement of grain yield in maize hybrids using marker facilitated introgression of QTLs. In: *Analysis of Molecular Marker Data Symposium*; 1994 Aug 5–6; Corvallis, OR. Madison, WI: Crop Science Society of America.

Stuber CW, Sisco PH (1991): Marker-facilitated transfer of QTL alleles between elite inbred lines and responses in hybrids. In: *Proceedings of the 46th Annual Corn and Sorghum Industry Research Conference*; Dec 11–12, 1991, Chicago, IL. Washington, DC: American Seed Trade Association

Stuber CW, Edwards MD, Wendel JF (1987): Molecular marker-facilitated investigations of quantitative trait loci in maize. II. Factors influencing yield and its component traits. *Crop Sci* 27:639–648

Stuber CW, Lincoln SE, Wolff DW, Helentjaris T, Lander ES (1992): Identification of genetic factors contributing to heterosis in a hybrid from two elite maize inbred lines using molecular markers. *Genetics* 132:823–839

Tanksley SD, Hewitt J (1988): Use of molecular markers in breeding for soluble solids content in tomato: a re-examination. *Theor Appl Genet* 75:811–823

Tanksley SD, Medina-Filho H, Rick CM (1982): Use of naturally occurring enzyme variation to detect and map genes controlling quantitative traits in an interspecific backcross of tomato. *Heredity* 49:11–25

Tollenaar M (1989): Genetic improvement in grain yield of commercial maize hybrids grown in Ontario from 1959 to 1988. *Crop Sci* 29:1365–1371

Van Ooijen JW (1992): Accuracy of mapping quantitative trait loci in autogamous species. *Theor Appl Genet* 84:803–811

Webb DM, Baltazar DM, Rao-Arelli AP, Schupp J, Clayton K, Keim P, Beavis WD (1995): Genetic mapping of soybean-cyst-nematode race-3 resistance loci in the soybean. *Theor Appl Genet:* in press

Weller JI (1987): Mapping and analysis of quantitative trait loci in Lycopersicon (tomato) with the aid of genetic markers using approximate maximum likelihood methods. *Heredity* 59:413–421

White TL, Hodge GR (1989): *Predicting Breeding Values with Applications in Forest Tree Improvement*. Dordrecht, The Netherlands: Kluwer Academic Press

Williams JGK, Kabelik AR, Livak KJ, Rafalski JA, Tingey SV (1990): DNA polymorphisms amplified by arbitrary primers are useful as genetic markers. *Nucl Acids Res* 18:6531–6535

Young ND, Tanksley SD (1989): Restriction Fragment Length Polymorphism Maps and the concept of graphical genotypes. *Theor Appl Genet* 77:95–101

Zabeau M, Vos P, inventors (1993): Selective Restriction Fragment Amplification: a general method for DNA fingerprinting. European Patent Application No. 0534858

Zehr BE, Dudley JW, Chojecki J, Saghai-Maroof MA, Mowers RP (1992): Use of RFLP markers to search for alleles in a maize population for improvement of an elite hybrid. *Theor Appl Genet* 83:903–911

Zeng Z-B (1993): Theoretical basis of precision mapping of quantitative trait loci. *Proc Natl Acad Sci USA* 90:10972–10976

Zeng Z-B (1994): Precision mapping of quantitative trait loci. *Genetics* 136:1457–14680

Zhang W, Smith C (1992): Computer simulation of marker-assisted selection utilizing linkage disequilibrium. *Theor Appl Genet* 83:813–820

16

Molecular Biology and Traditional Breeding Applied to the Improvement of Maize Nutritional Quality

MAURÍCIO A. LOPES AND BRIAN A. LARKINS

Introduction

Maize (*Zea mays* L.) provided the agricultural basis for the development of early civilization in the American continent, and after the arrival of Columbus in 1492, it spread to virtually all continents of the world. Today, this cereal is recognized as a strategic food crop that provides an enormous amount of protein and energy for humans and livestock. Due to its amenity to genetic manipulation, high productivity, wide range of uses, and remarkable adaptability to diverse environments, much research has been dedicated to characterize and improve the biosynthetic pathways leading to the accumulation of starch and protein, the major storage reserves in the maize seed. In particular, much effort has been devoted to understanding the process of storage protein synthesis in maize endosperm.

The storage proteins of maize and most cereals are characterized by low contents of certain amino acids that are essential for human and monogastric animal nutrition. This is the primary cause for the poor nutritional quality of most cereal seeds. Many people in developing nations are dependent on a few or in some cases a single staple food. Maize is an important cereal in many of these countries, especially in Africa, where its consumption as the principal staple is frequently associated with protein deficiency. The improvement of maize protein quality can be an effective means of reducing malnutrition whether consumed directly in the human diet or indirectly consumed as a consequence from greater availability of animal products resulting from on-farm use of maize with improved nutritional quality.

The advances in biotechnology over the past decade have expanded the hori-

The Impact of Plant Molecular Genetics
BWS Sobral, Editor

zon for developing biochemical and genetic markers that can be used as breeding tools to help speed up the process of cultivar development. In particular, molecular biology has provided new insights into the process of protein biosynthesis in the endosperm of agriculturally important cereal species. Through a better understanding of gene regulation, synthesis, and accumulation of storage proteins, we may be better able to develop novel strategies to help maize breeders improve the nutritional quality of the seed while maintaining the agronomic attributes of the plant. Using these strategies to increase and broaden high quality maize consumption can help improve the nutritional situation of many people in developing countries, since the potential for improvement and utilization of this cereal is far from realized.

This chapter describes our current knowledge on the biosynthesis and genetic regulation of storage protein accumulation and on the use of molecular markers to help unravel processes that define physical and biochemical properties of the maize endosperm. We also discuss approaches of genetics and molecular biology that can help breeders improve the nutritional quality of maize while preserving the physical properties of the seed and the agronomic performance of the plant.

Maize Endosperm Development

Pollination in maize leads to a phenomenon known as double fertilization, a unique biological process in which one haploid sperm nucleus fuses with the egg to produce the embryo, while a second sperm nucleus fuses with the two female polar nuclei to form a triploid endosperm. The maternal ovule cells surrounding the embryo sac give rise to the outer layer of the mature caryopsis, the pericarp. Therefore, the embryo has one genome from each parent, the triploid endosperm receives two genomes from the female and one genome from the male, and the pericarp has maternal constitution only.

Normally, endosperm development requires fertilization, even though this tissue does not have the means to transfer its genetic information to another organism (Sargant, 1900). Its major function is to serve as a specialized seed storage tissue filled with diverse compounds that provide nitrogen, sulfur, and energy for germination and early stages of seedling growth (Lopes and Larkins, 1993). Components such as storage and nonstorage proteins, phytin, oils, carotenoids, polysaccharides, and free amino acids are known to exist in varying proportions in the endosperm.

Reserve carbohydrate is critical for providing the embryo with a source of energy and carbon until it is photosynthetically competent. In maize, the principal storage carbohydrate is starch, which is composed of two alpha-glucan polymers, amylose and amylopectin, that are packed as crystalline granules in amyloplasts. Maize endosperm contains also a variety of proteins. The most abundant of these are storage proteins. Many other proteins also accumulate to high levels within the endosperm. Since endosperm proteins are the major determinants of maize nutritional quality they are described in more detail in the topics that follow.

Maize Endosperm Proteins

Mature maize endosperm contains a complex of starch granules and protein bodies embedded in a proteinaceous network or matrix (Wolf et al, 1952a, b; Duvick, 1951, 1955, 1961). It is implied that the structural characteristics of the endosperm are a consequence of the interactions between these components. Duvick (1961) attributed the opaqueness of the soft, starchy regions that are usually found in the inner part of the endosperm tissue to light refraction resulting from air spaces formed by ruptures of the cytoplasmic matrix upon drying. Dombrink-Kurtzman and Bietz (1993) suggest that maize protein composition may influence endosperm texture and physical properties of the kernels. They show that hard endosperm regions appear to accumulate more storage proteins than soft regions.

Proteins correspond to approximately 10% of the maize endosperm dry weight. For practical purposes, we divide them into two main categories: storage and nonstorage proteins (Wallace et al, 1990). The most abundant type (approximately 65% of the total) are storage proteins known as prolamines or zeins. By definition, storage proteins accumulate for the sole purpose of storing nitrogen and sulfur for the seedling when it germinates. Storage proteins typically have high amide contents and in some cases are rich in sulfur containing amino acids. They generally are found within vacuoles or endoplasmic reticulum membranes as insoluble accretions. Many other proteins also accumulate to high levels within the endosperm. Nonstorage proteins found in maize endosperm are collectively known as nonzeins and will be described later.

Maize Storage Proteins: Zeins

Zeins consist of a complex mixture of polymorphic alcohol-soluble polypeptides. The onset of their synthesis occurs approximately 8–10 days after pollination, and the proteins are accumulated throughout seed development. They are synthesized as preproteins by membrane-bound polyribosomes, cotranslationally transported into the lumen of the endoplasmic reticulum (RER) and assembled into insoluble, membrane-bound accretions termed protein bodies (Figure 1) (Larkins and Hurkman, 1978; Lending and Larkins, 1989). Accumulation of storage proteins in discrete cellular bodies is an adaptation that probably prevents their exposure to enzymes responsible for turnover of metabolic proteins. Other potential advantages of sequestering these proteins in membrane-bound organelles are the facilitated packaging provided by an environment of proteins with similar biochemical properties and interactive capabilities, and the fact that the proteins are deposited in a relatively nonhydrated condition, which facilitates seed desiccation.

Expression of zein genes is primarily limited to endosperm, although trace

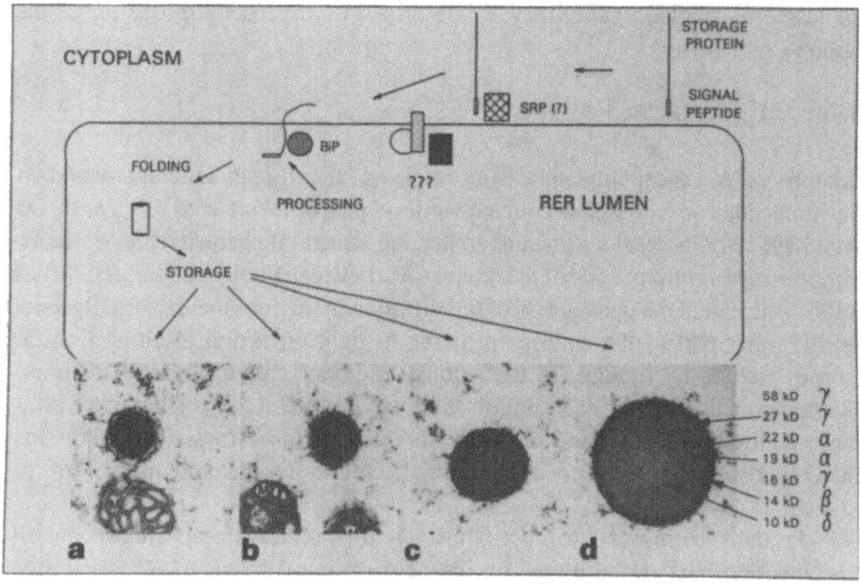

Figure 1. Pattern of zein deposition during protein body formation. Zeins are synthesized as preproteins by membrane bound polyribosomes and cotranslationally transported into the lumen of the endoplasmic reticulum (RER) where they are processed, folded, and stored. Zeins are assembled into insoluble, spherical membrane-bound accretions termed protein bodies (a to d) (Larkins and Hurkman, 1978; Lending and Larkins, 1989). Immunocytochemical studies have revealed that protein bodies in the outer endosperm layers (a) are small and accumulate mostly beta- and gamma-zeins, while those in the central, developmentally more mature endosperm cells are filled with alpha-zeins (d). Early in development, the alpha-zeins are present in small amounts and form small, distinct globules within a matrix of beta- and gamma-zeins (b and c). As alpha-zein synthesis and accumulation increase, the beta-, delta-, and gamma-zeins are progressively displaced to more peripheral parts of the protein body. The protein bodies are surrounded by rough endoplasmic reticulum. M = mitochondria (After Lending and Larkins, 1989).

amounts can be found in the embryo. There is no evidence for transcription of any zein gene subfamily in leaf, root, or shoot tissue. Zeins have no enzymatic activity, and their only known function is to provide nitrogen, sulfur, and carbon skeletons for germination and early phases of seedling growth. They survive desiccation for long periods of time. The diversity of zeins in regard to structure and solubility properties makes it possible to divide them into four distinct types, called alpha-, beta-, gamma-, and delta-zeins (Figure 2) (Esen, 1986; Larkins et al, 1989). Alpha-zeins, which typically account for about 70% of the total storage protein fraction, are composed of polypeptides of apparent molecular weight 19- and 22-kD. These proteins can be further separated into many different groups by IEF (Righetti et al, 1977; Hagen and Rubenstein, 1981), indicating that they correspond to a large number of polypeptides with similar solubility properties but considerable charge heterogeneity. In fact, alpha-zeins have been showed to be encoded by a large family of related genes (Hagen and Rubenstein, 1981;

Pedersen et al, 1982). The 14-kD beta-zein, 16-kD and 27-kD gamma-zeins, and 10-kD delta-zein require reducing agents, such as 2-mercaptoethanol, to be soluble in alcoholic solvents. These proteins have been found to be encoded by genes present in only one or two copies in the genome (Pedersen et al, 1986; Prat et al, 1987; Kirihara et al, 1988).

All zeins are deficient in lysine and tryptophan, amino acids that are essential for human and livestock nutrition. The alpha-zein proteins have high contents of glutamine (25%), leucine (20%), alanine (15%), and proline (11%) and range in size from 210 to 245 amino acids. The beta-zein protein is 160 amino acids long and contains less glutamine (16%), leucine (10%), and proline (9%), than the alpha-zeins, but has significantly more methionine (4%) and cysteine (7%). The 27-kD gamma-zein is a cysteine-rich (7%) protein of 180 amino acids, with high content of proline (25%) (Prat et al, 1987). The delta-zein is a small protein of 130 amino acids. It has a very high content of the sulfur amino acid methionine (23%) (Kirihara et al, 1988).

Immunocytochemical techniques have been used to determine the distribution of the various types of zeins within protein bodies (Ludevid et al, 1984; Lending et al, 1988; Lending and Larkins, 1989), and the distribution of the various types of protein bodies in regions of the developing maize endosperm (Figure 1) (Lending and Larkins, 1989). The distribution of the different zein types within the protein body appears to be determined by the temporal and spatial synthesis of the proteins during endosperm development. Protein bodies in the youngest endosperm cells, those located in the outer endosperm layers, are small

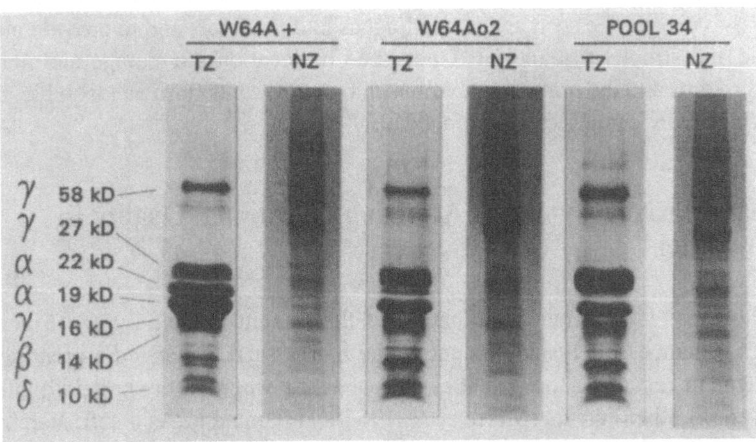

Figure 2. Total zein (TZ) and nonzein (NZ) proteins from maize endosperm. Total endosperm proteins were obtained and fractionated according to the method described by Wallace et al, (1990). SDS-polyacrylamide gradient gels (7.5%–18%, w/v) were prepared according to Laemmli (1970) and stained with Coomassie blue. Description of the components of both fractions is presented in the text. Lanes labelled TZ were loaded with protein corresponding to 1.5 mg of mature endosperm flour, and lanes labelled with NZ were labelled with protein corresponding to 3.0 mg of endosperm flour. W64A+ e W64A*o2* are wild-type and *opaque*-2 isogenic versions of the inbred line W64A. Pool 34 is a modified *o2* genotype developed by CIMMYT/Mexico.

and accumulate mostly beta- and gamma-zeins, while those in the central, developmentally more mature endosperm cells, are filled with alpha-zeins. Early in development, the alpha-zeins are present in small amounts and form small, distinct globules within a matrix of beta- and gamma-zeins. As alpha-zein synthesis and accumulation increase, the beta- and gamma-zeins are progressively displaced to more peripheral parts of the protein body (Lending and Larkins, 1989). Recently, the delta-zein has been found to predominantly occur in the center of the protein body, within the alpha-zein core (Esen and Stetler, 1992).

None of the zein polypeptides contain signals for RER retention. It is possible that the interactive properties of these proteins is an important factor controlling their accumulation in this organelle. Recent studies suggest that hydrophobic peptide repeats play a role in the retention of wheat proteins within the RER (Altschuler et al, 1993). Perhaps, the repeated peptides found in zeins are involved in RER retention as well.

Other Endosperm Proteins: NonZeins

Nonzeins, or the alcohol-insoluble proteins (albumins, globulins, and glutelins), are a good source of the essential amino acids that zeins lack (Habben et al, 1993). However, their low amount relative to the zeins diminishes their contribution to the overall nutritional quality of the maize grain. These proteins perform housekeeping and seed-specific functions during endosperm development. Among these are structural proteins, protease inhibitors, and biosynthetic enzymes. This diverse group of proteins (Figure 2) may act secondarily as a source of nitrogen and sulfur for the seedling during germination, but its primary function is to protect the seed against pathogens and predators and to provide biosynthetic and structural functions. Upon seed desiccation the storage and nonstorage proteins and the remaining cytoplasmic components form an insoluble matrix in which starch granules are embedded.

Altered Protein Synthesis Affecting Nutritional Quality in Maize Endosperm

In the early 1960s, the identification of the mutant *opaque*-2 (*o2*) (Figure 3), which enhances the lysine content of the maize seed, generated a great deal of excitement regarding the potential to genetically improve the protein quality of maize and other cereals (Mertz et al, 1964). This mutant has a soft, starchy endosperm that does not transmit light (opaque) when placed on a light box. In addition, the mutation reduces the synthesis of the storage protein fraction, zein, and causes a partial increase in the synthesis of nonzein proteins, that contain higher amounts of lysine (Figure 2) (Habben et al, 1993).

In addition to lysine and tryptophan, the amounts of histidine, arginine, aspartate plus asparagine, and glycine are markedly increased in *o2* mutants. The glutamate plus glutamine, alanine, and leucine contents are decreased, with

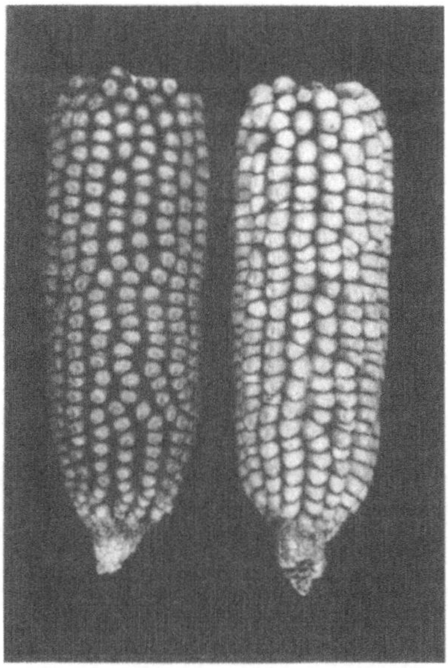

Figure 3. Mature ears of the normal (left) and *o2* (right) isogenic versions of the inbred line W64A. Normal maize endosperms usually have a vitreous, hard outer layer and only a small internal region with soft, starchy texture. The *o2* mutation causes reduction in storage protein accumulation (Figure 2) and development of a homogeneous, starchy endosperm at maturity. The loose packing of endosperm components appears to lead to formation of air spaces that block light transmission through the mutant seed. For this reason, seeds of this mutant have a dull appearance and are called opaque.

leucine being decreased the most (Figure 4) (Nelson, 1980). Apparently, there are no significant changes in the amino acid composition of endosperm proteins from *o2* and normal maize. The differences between the content of amino acid residues are due to quantitative and not qualitative changes in protein composition (Sodek and Wilson, 1971; Habben et al, 1993).

Studies of zein synthesis during endosperm development indicate that *o2* mutants have a lower rate of zein accumulation, compared to normal genotypes. In addition, *o2* mutants accumulate little or no zeins in the 35 days after pollination (DAP), in contrast to normal genotypes which continue to accumulate zeins until late in development (Tsai and Dalby, 1974; Dalby and Tsai, 1974). Kodrzycki et al. (1989) has shown that the synthesis of zein proteins in *o2* mutant genotypes is delayed and reduced. The 19-kD alpha-zein proteins are not detected until 14 DAP, and the 22-kD alpha-zeins are found only in trace amounts. The 14-kD beta-zein and 16- and 27-kD gamma-zeins are detected around 12 DAP, and their accumulation is not as dramatically affected by the mutation.

Ultrastructural analysis has shown that reduction of alpha-zein synthesis causes the protein bodies in *o2* endosperm to be small. Generally they are 0.1 to 0.3 μm in diameter in developing endosperm, while the wild-type protein bodies are between 1 and 2 μm in diameter. Immunocytochemistry has revealed that protein bodies in the mutant contain primarily beta- and gamma-zeins, with only small locules of alpha-zein (Lending and Larkins, 1989; Geetha et al, 1991). The small size of the protein bodies has been postulated to cause the formation of air spaces around the starch grains as the endosperm desiccates, causing the soft phenotype of the mature mutant seed (Robutti et al, 1974).

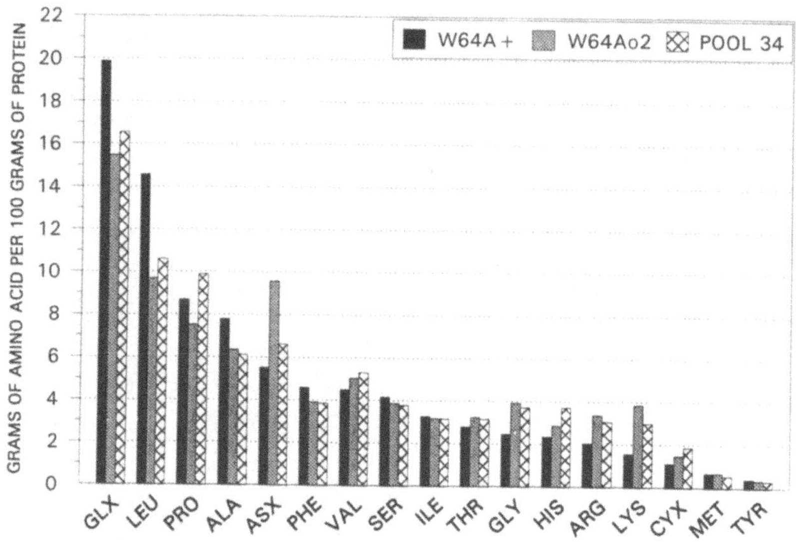

Figure 4. Comparative amounts of amino acids from mature endosperm flour of W64A, W64A*o2*, and the modified *o2* Pool 34.

Several studies have shown that the reduction of zein proteins in opaque mutants is correlated with reduced amounts of the corresponding mRNAs (Pedersen et al, 1980; Burr and Burr, 1982; Marks et al, 1985; Boston et al, 1986). These results suggest that the mutations alter zein gene transcription. This has been demonstrated for *o2* using nuclear run-on transcription analysis (Kodrzycki et al, 1989). A substantial reduction has been shown in alpha-zein gene transcription in the *o2* mutant relative to the normal genotype. *o2* differently affects the transcription of genes encoding the 22- and 19-kD alpha zeins, with the former being almost totally blocked. This result indicates that although the alpha-zein genes share common sequence elements responsible for their coordinate expression (Thompson and Larkins, 1989), there must also be somewhat different regulatory factors that control their transcription.

Since the *O2* locus is located on the short arm of chromosome 7, the transcription data are consistent with the gene acting as a positive, transacting transcriptional activator of the zein coding sequences, which occur on chromosomes 4, 7, and 10 (Soave and Salamini, 1984). Schmidt et al, (1987) used the transposable element *Spm* as an insertional mutagen to tag the locus and obtained three mutable *o2* alleles. DNA sequences flanking the *Spm* insertion in one of them were verified as *o2*-specific. Following cloning by recombinant DNA procedures, this sequence provided a probe to clone the wild-type allele. Northern blots indicated that the *O2* gene is expressed in wild-type endosperm but not in leaf tissues or endosperm homozygous for the mutant allele. Transcripts were detected in endosperms homozygous for *o7* and *fl2*, indicating that *O2* expression occurs independently of the other two putative regulators of zein gene expression. Another successful attempt to clone the *O2* locus was reported by Motto et al, (1988). They isolated the *O2* gene by transposon tagging with the element *Activator* (*Ac*), using an approach similar to that of Schmidt et al (1987).

O2 has been shown to encode a basic domain/leucine zipper transcriptional activator (Schmidt et al, 1990) that regulates 22-kD alpha-zein genes (Schmidt et al, 1990; Lohmer et al, 1991; Schmidt et al, 1992), as well as a ribosome inactivating protein (RIP) gene (Bass et al, 1991). Genetic and molecular aspects of *O2* regulation have been reviewed by Schmidt (1993). Another gene encoding a protein with a similar regulatory domain has been isolated and the protein it encodes shown to interact with *O2* in vitro (Pysh et al, 1993). The fact that transcriptional activator complexes are potentially involved in regulation of genes encoding proteins with diverse functions indicates that very complex regulatory processes may take place during endosperm development. Transcription factors containing a similar activator domain, such as v-fos and v-jun, form heterodimeric complexes (Rauscher et al, 1988; Smeal et al, 1989; Jones, 1990), so it is possible that the *O2* protein interacts with other transcription factors to regulate zein gene expression in the endosperm tissue. Further evidence for this comes from the multiple pleiotropic effects caused by *o2* and other opaque mutations believed to affect regulatory genes (Damerval and De Vienne, 1993; Motto et al, 1989).

Genetic Modification of o2 *Mutants*

Modification of the floury endosperm of *o2* is a naturally occurring phenomenon that leads to the formation of hard and vitreous endosperm regions in the mutant seed. Genes that condition this phenotype are generally called *o2* modifiers. Biochemical analyses have shown that the high levels of essential amino acids associated with the *o2* mutation persists even after the physical modification of the endosperm (Paez et al, 1969).

The ability of modifier genes to alter the soft endosperm of *o2* mutants while maintaining high nutritional quality was successfully utilized to convert several maize cultivars to genotypes known as Quality Protein Maize (QPM) (Vasal et al, 1980; Bjarnason and Vasal, 1992; Villegas et al, 1992; Gevers and Lake, 1992). The breeding approach consisted of sequential cycles of selection in *o2* stocks that have genetic variability for seed modification. The objective of the breeder was to accumulate favorable modifier genes, thus promoting a progressive increase in seed hardness. Every cycle of selection included careful analysis of diverse plant traits, as well as physical and biochemical analysis of seed traits, so that selected material had desirable agronomic properties as well as high nutritional quality.

The action of endosperm modifier genes in maize has been the subject of a considerable number of reports during the last 25 years (Glover and Mertz, 1987). However, due to the complexity of this trait, no detailed research has been done to unravel the inheritance and mode of action of the modifier genes. The need for complex breeding strategies and many cycles of selection to achieve a reasonable degree of endosperm modification (Vasal et al, 1980) indicates that this trait has complex inheritance.

Phenotypic variation ranging from completely unmodified to completely modified kernels with a plethora of intermediate, partially modified types, is found in single ears of F_2 progenies obtained of F_1 hybrids from modified and nonmodified *o2* mutants (Lopes and Larkins, 1991). The existence of such a wide range of segregation in small seed samples suggests that the number of genetic factors responsible for endosperm modification is not large, although their interaction must be complex. In fact, extensive variability in the manifestation of endosperm modification can be detected, depending on genotype. In segregating kernels involving certain combinations of modified and opaque parents, the opaque and vitreous regions occur either at the tip or at the crown of the endosperm but are never interspersed. However, other parental combinations generate kernels that segregate as mosaics with irregularly interspersed opaque and vitreous regions (Figure 5). These dramatic variations in the nature of phenotypic segregation suggest that the genetic background affects the spatial expression of endosperm modifiers (Lopes and Larkins, 1991), thus complicating genetic analysis of this trait.

Some improved modified *o2* varieties have been released by national research programs, but the adoption of these improved materials by farmers and returns to society in both developed and developing countries have been minimal to date. Although QPM germplasm constitutes a rich genetic resource that consumed many million dollars and more than two decades of research effort, hope for its further development has been diminishing year after year. Unfortu-

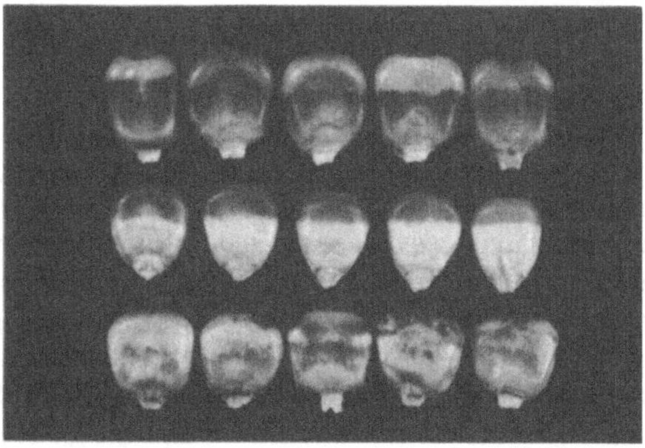

Figure 5. Seeds from F_2 ears obtained from crosses of *o2* by modified *o2* genotypes. Formation of modified endosperm lead to alterations in physical structure of the mutant seed that appear to act independently of the normal clonal mode of development. Extensive variability in the manifestation of endosperm modification can be detected, depending on genotype. In segregating seeds involving certain genetic backgrounds, the opaque and vitreous regions occur either at the tip (middle row) or at the crown (top row) of the endosperm. However, certain combinations generate seeds that segregate as mosaics with irregularly interspersed opaque and vitreous regions (bottom row). These dramatic variations in the mode of phenotypic segregation indicate that the genetic background affects the spatial expression of endosperm modifiers (Lopes, 1993).

nately, doubt about QPM has been based on sparse knowledge of its potential and great magnification of its problems. Apparently, the major challenge faced by QPM is the lack of knowledge about the mechanisms that define nutritional quality and modification of the soft texture of the mutant endosperm. Without a better knowledge of these mechanisms it will be difficult to solve the few remaining problems that plague QPM development and utilization. A better knowledge of the mechanisms involved in modification of the soft texture of the endosperm is necessary in order to facilitate efficient transfer of this trait to elite materials and to develop highly productive QPM cultivars. Also, modification of the vitreous phenotype of the kernel is incomplete in some genetic backgrounds, resulting in endosperm that sectors into vitreous and starchy regions. The basis for the incomplete modification is unknown, but it diminishes the physical quality of the grain. The necessity of introgressing multiple genetic traits like the *o2* mutation, modifier genes, and genes responsible for lysine containing proteins add significant complexity to the process of developing elite QPM germplasm. The lack of biochemical markers for each of these traits has slowed the rate with which germplasm can be developed and has limited the possibilities for identifying elite materials. However, the feasibility of such a task has been vitalized by biotechnological approaches that provide innovative solutions for helping breeders manipulate complex traits. Genetic markers such as RFLPs and biochemical markers such as endosperm proteins can be used in conjunction with traditional selection procedures to develop QPM genotypes with improved nutritional quality and agronomic characteristics equal to the best normal maize cultivars.

Developing Biotechnological Tools to Help Improve Modified *o2* Maize.

Mapping Endosperm Modifier Genes

The poor understanding of modifier gene action has slowed efficient transfer of these genes into elite materials and has delayed development and commercial utilization of this germplasm. Genetic mapping of modifier genes could accelerate their transfer to commercially valuable germplasm and would facilitate their isolation and molecular characterization.

A gradual increase in the proportion of hard to soft endosperm is consistently observed in reciprocal F_1 hybrids developed by crossing *o2* and modified *o2* genotypes (Figure 6). Also, a continuous phenotypic variation ranging from completely opaque to completely modified seeds is detected in ears of F_2 progenies segregating for endosperm modification (Lopes and Larkins, 1991). Selfing of F_2 individuals with extreme phenotypes (modified and opaque) has shown that the paternal phenotypes persist in the F_3, indicating that modifier genes function in an additive fashion. Careful analysis of F_2 progenies indicate that it is possible to recover similar amounts of parental phenotypes at a pro-

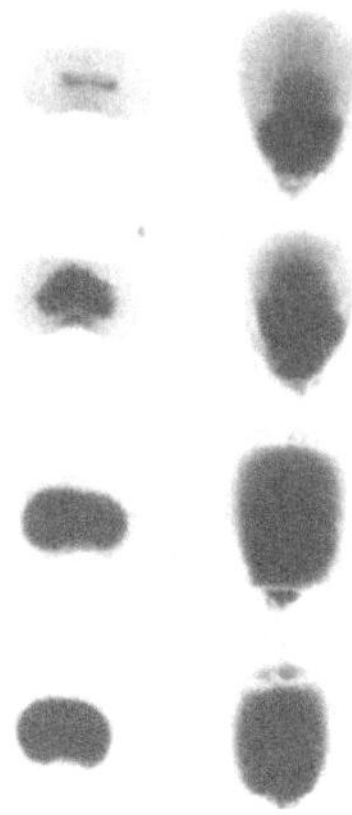

Figure 6. *o2* modifier genes alter seed phenotype in an additive fashion. Seeds from representative samples of modified *o2* (Pool 33 QPM), *o2* (W64A*o2*), and from their reciprocal F_1 hybrids were photographed on a light box. Seeds on the left were cut in cross section with a razor blade to better show internal endosperm structure. Identification of each pair of seeds, from top to bottom is Pool 33 QPM (P1), P1 × P2, P2 × P1, and W64A*o2* (P2), respectively. Note a gradual decrease in the proportion of hard to soft endosperm with the reduction of copies of the modified genome in the triploid endosperm (from top to bottom) (Lopes and Larkins, 1991).

portion that approximates 1/16 of the total progeny (Figure 7). This proportion of parental phenotypes suggests the involvement of two additive loci with the trait (Lopes and Larkins, 1995; Lopes et al, 1995).

A modified *o2* individual with two independent additive modifier gene loci would have six active modifier alleles in the triploid endosperm (Figure 8). Active modifier alleles (identified by A and B) are taken here as effectors of modification, in contrast to the inactive counterparts (a and b) that do not contribute to the trait. Reciprocal crosses of modified *o2* with *o2* should result in F_1 hybrids with four and two modifier alleles in the endosperm, depending on how the

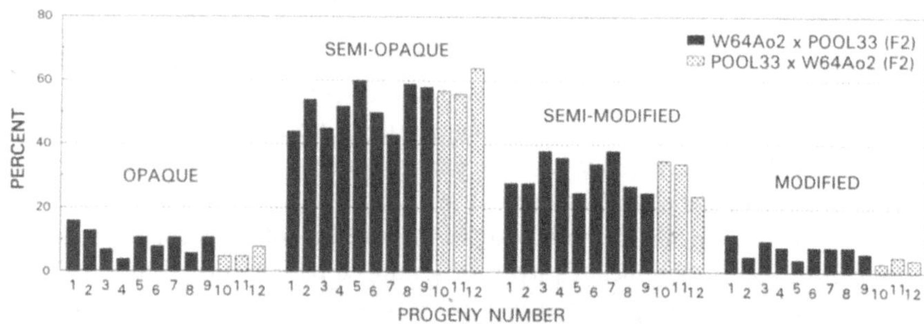

Figure 7. Analysis of seed phenotype in F_2 progenies from crosses between *o2* and modified *o2* genotypes. A. Seeds from twelve F_2 ears were separated into four phenotypic classes according to the degree of endosperm light transmission. Progeny of individual F_2 ears were analyzed separately. Hatched bars represent the number of seeds in each class for progenies 1–9 from the cross W64A*o2* × Pool 33; filled bars are progenies 10–12 from the reciprocal cross Pool 33 × W64A*o2*. (Lopes and Larkins, 1995).

cross is made (Figure 8a). The dosage variation in the number of modifier alleles in the endosperm, considering both parents and their progeny (from 6 to 0), would explain the stepwise variation in the seed phenotype of these crosses (Figure 6) (Lopes and Larkins, 1991).

Extending the 2-gene model, any reciprocal F_1 cross between modified *o2* and *o2* should generate F_2 progeny with a continuous pattern of segregation for seed modification (Lopes, 1993; Lopes and Larkins, 1995). This is due to triploidy and additivity, factors that contribute to generate complex patterns in the endosperm of segregating seeds. In addition to having individuals with 0 to 6 mod-

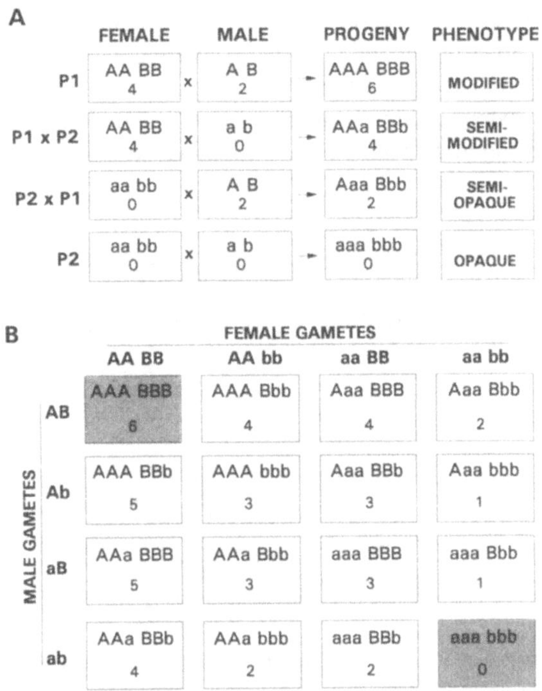

Figure 8. Prediction of endosperm genotypes in reciprocal crosses and F_2 populations considering the hypothesis of two additive modifier loci (A and B). A. P1 and P2 indicate modified *o2* and *o2* parents, respectively. The endosperm is triploid with two contributions from the female parent and one contribution from the male parent. For additive loci, the number of active alleles, indicated by capital letters, affect the phenotype in a dosage-dependent fashion. Therefore, endosperm progeny with 6, 4, 2, and 0 doses of active modifier alleles show a gradation in the degree of endosperm modification, from completely modified to completely opaque (Figure 7). B. A prediction of endosperm genotypes generated in F_2 progenies derived from crosses between o2 and modified o2 genotypes. This is an extension of the 2-gene model shown in A. Selfed F_1's should generate progeny with 16 genotypic classes. Phenotypic classes can be predicted by the number of active modifier alleles (A and B) for each genotype. Opaque (0 active alleles), semiopaque and semimodified (1 to 5 active alleles) and modified (6 active alleles) classes are expected. Each parental genotype (opaque and modified) is expected to be recovered in 1/16 of the segregating progeny (shaded squares).

ifier alleles, F_2 progenies may vary in phenotype depending on differential contribution of modifier genes to the phenotype and/or epistatic relationships between the genes involved. For instance, individuals with three modifier alleles could have different phenotypes, depending on their genotypes (AAAbbb, aaaBBB, AaaBBb, AAaBbb). Due to difficulty in precisely evaluating small variations in seed phenotype, assessment of these effects are possible only after molecular characterization of the modifier genes.

RFLP analysis has been used to determine map location of putative modifier genes and their possible association with the gamma-zein locus, using a Bulked Segregant Analyses (BSA) strategy (Michelmore et al, 1992) with pooled F_2 individuals from the two extreme parental phenotypes (Lopes et al, 1995). Approximately 90 combinations of RFLP markers/restriction enzymes were used to search for genomic regions associated with endosperm modification. The results showed that a modifier gene map to the centromeric region of chromosome 7, where gamma-zein genes are located (Figure 9) (Lopes et al, 1995). Another modifier gene has been mapped to the long arm of chromo-

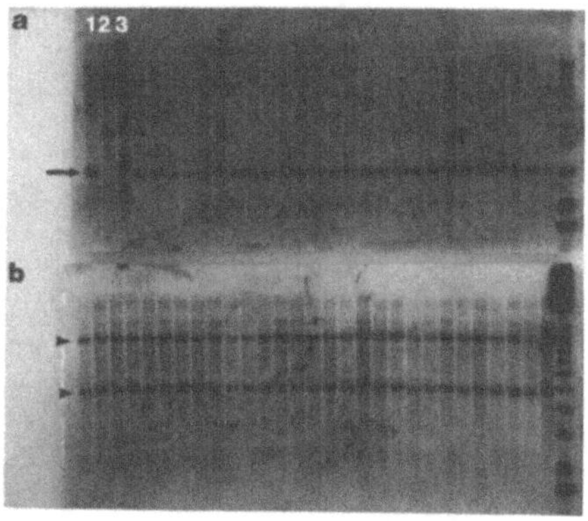

Figure 9. Colocalization of a modifier gene and the gamma-zein locus. RFLP mapping by bulked segregant analysis indicated a modifier gene locus in the centromeric region of chromosome 7, where gamma-zein genes are located (Lopes et al, 1995). To evaluate the relationship between allelic variation at the gamma-zein locus and endosperm modification, Southern blots with a gamma-zein gene sequence were used. The population was obtained from the cross W64A*o2* × Pool 33 QPM. F_2 seeds displaying parental phenotypes (opaque and modified) were selected, planted, and leaf tissue was collected from each individual for analysis. Each F_2 plant was selfed, and the phenotype of the F_3 seeds evaluated. Leaf tissue from F_2 plants that bred true for the selected phenotype was used for DNA extraction and Southern blots. DNA samples from the *o2* parent (W64A*o2*) (blot a, lane 1), modified *o2* parent (Pool 33) (blot a, lane 2), F_1 (blot a, lane 3), F_2 individuals with *o2* phenotype (blot a, remaining lanes), and F_2 individuals with modified *o2* phenotype (blot b, all lanes) were digested with HindIII. The blots were probed with a 1050 bp sequence corresponding to the 5′ end of a gamma-zein gene, described in Figure 10. The arrow in blot a indicates the band corresponding to the Ra gamma-zein locus, and the arrowheads in blot b indicate the two bands corresponding to the alternative locus that contain the A and B genes (Das and Messing, 1987). The results show that there is a clear correlation between endosperm phenotype and allelic composition at the gamma-zein locus.

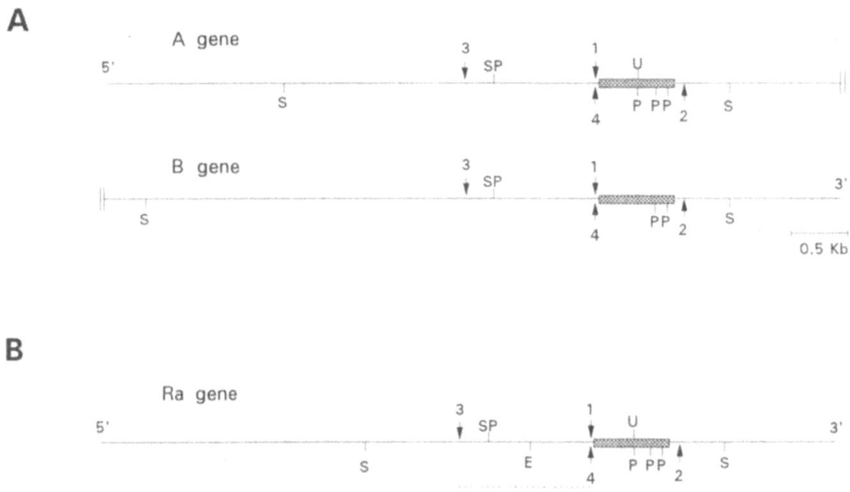

Figure 10. Targeting, amplification, and analysis of gamma-zein sequences by the polymerase chain reaction (PCR). (A) Detailed analysis of the gamma-zein locus has been reported by Das et al, (1990). Rectangles represent sequences bound by the ATG and TGA codons. Oligo nucleotides (indicated by arrows) were designed for amplification of the coding (oligos #1 and #2) and flanking (oligos #3 and #4) sequences of the tandemly repeated A and B genes (A), and Ra gene (B). Restriction enzyme sites are S, Sal I; E, Eco RI; U, Pvu II; P, Pst I; and SP, Spe I. Variation in PstI, PvuII, and EcoRI restriction sites at the locations indicated in the map allow identification of PCR-amplified sequences corresponding to each of the alternative alleles at the locus. Dotted line in B indicates the 5′ fragment used as a probe in the blots of Figure 9. (Lopes et al, 1995)

some 7 (Lopes et al, 1995) where *o15*, a newly described opaque mutation maps (Dannenhofer et al, 1995).

The gamma-zein protein has been shown to be encoded by only one or two genes (Figure 10) (Das and Messing, 1987; Das et al, 1991). Some inbreds, such as A188, W23, and W22, contain two tandemly linked genes, designated A and B (Das and Messing, 1987; Das et al, 1990). In spite of their extensive sequence identity, these two genes can be easily distinguished because the A gene contains three Pst I sites, while the B gene contains only two. Also, the A gene has a Pvu II site in the middle of the coding region. This site is absent in the B gene (Figure 10). Most of the commonly used inbred lines, such as W64A, Oh43, Mo17, and B73 contain a single rearranged A gene (Ra) that has been shown to be generated by a recombinational event that led to excision and loss of the B gene (Das et al, 1990). Colocalization of a modifier gene and the gamma-zein locus could be an indication that the gamma-zein locus itself is involved in the process of endosperm modification, or it could indicate that gamma-zein genes are located close to a modifier gene locus. Obviously, the mapping data presented are insufficient to validate either of these alternatives.

Another evidence for the linkage between a locus that carries two gamma-zein genes and endosperm modification has been provided by analysis of recombinant inbred lines. These lines were developed from a cross between a modified genotype that has the A and B genes (Pool 33 QPM) and an *o2* line that has only the rearranged A gene (W64A*o2*). Amplification and restriction analysis of gamma-zein gene sequences indicate that recombinant inbred lines that have low levels of endosperm modification (seed density) and accumulate lower levels of the gamma-zein protein, inherit the gamma-zein gene from the *o2* parent (rearranged A alleles) (Figure 11) (Lopes et al, 1995).

After analysis of many modified *o2* materials (33 different genotypes), no individual has been found to carry only one gamma-zein gene (Lopes, 1993). Therefore, it appears that seed modification and overproduction of the gamma-zein protein, are definitely linked to the presence of the A and B alleles at this locus. An intriguing aspect of this relationship is the fact that a few inbred lines (W22, W23, and A188) are known to have the A and B genes at the gamma-zein locus (Das et al, 1987). Introduction of the *o2* mutation into these genotypes does not result in modified seeds. Therefore, it appears that the gamma-zein locus alone does not affect *o2* seed phenotype. In addition, the DNA sequence of the alleles cloned from modified genotypes reveals no major differences in the coding and 5′ untranslated regions when compared to alleles of nonmodified opaque and normal genotypes (Lopes, 1993; Lopes et al, 1995). These results are not surprising, since genetic analysis clearly indicates the existence of additional modifier genes (Dannenhofer et al, 1995; Lopes et al, 1995; Lopes and Larkins, 1995; Lopes, 1993). Therefore, it appears that a gamma-zein locus with the A

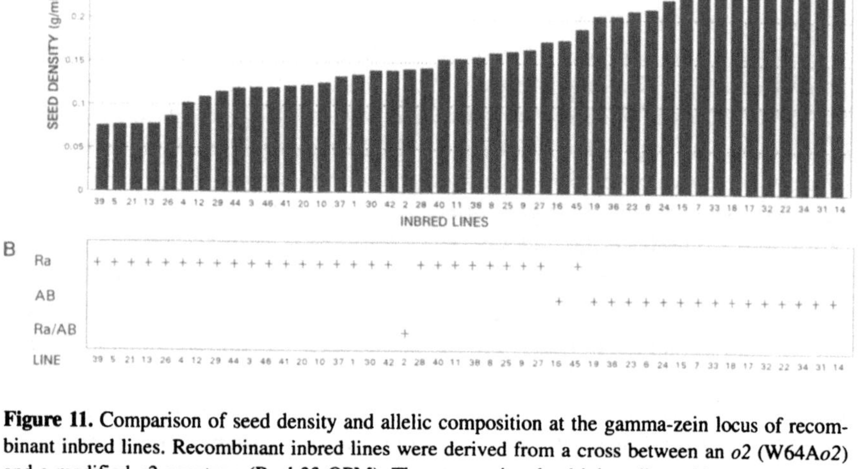

Figure 11. Comparison of seed density and allelic composition at the gamma-zein locus of recombinant inbred lines. Recombinant inbred lines were derived from a cross between an *o2* (W64A*o2*) and a modified *o2* genotype (Pool 33 QPM). The process involved inbreeding with selection to fix a wide range of seed phenotypes (Lopes, 1993; Lopes and Larkins, 1995; Lopes et al, 1995). (A). Seed density was determined for each line as described by Lopes (1993) and plotted in ascending order. (B). Allelic composition at the gamma-zein locus was determined for each line using PCR amplification, as described in Figure 10. Presence of the Ra, A and B, Ra/A and B is indicated by a "+" sign. (Lopes et al, 1995).

and B genes is necessary but not sufficient to effect modification of the *o2* phenotype.

Biochemical Changes Associated with o2 *Modification*

The development of improved techniques for extraction and measurement of endosperm storage proteins has revealed that endosperm modification is associated with a two to threefold increase in the accumulation of the 27-kD gamma-zein (Wallace et al, 1990; Lopes and Larkins, 1991). No other significant change occurs in any protein component of the modified seed compared to nonmodified *o2*. Further analysis has shown that in reciprocal crosses between modified and nonmodified genotypes, as well as in F_2 progenies segregating for seed modification, the degree of endosperm modification and increased synthesis of gamma-zein are both dosage dependent and directly correlated (Lopes and Larkins, 1991). Furthermore, analysis of the steady state level of gamma-zein mRNA indicated a correlation between gene dosage and mRNA levels in the developing endosperm. The increased level of gamma-zein mRNA in modified endosperm could be a result of increased gene transcription or more stable mRNA, but it cannot be a consequence of gene amplification, since gamma-zein genes are found in only two copies per modified genome (Geetha et al, 1991). Inbred lines that have no modifier genes and one or two gamma-zein genes in the genome do not develop modified endosperm when converted into *o2*. Therefore, the process of seed modification appears to lead to accumulation of factors that act on gamma-zein gene transcription and/or mRNA stability in order to increase the net amount of mRNA and protein in the endosperm (Geetha et al, 1991; Or et al, 1993).

To further characterize the changes in gamma-zein protein accumulation conditioned by *o2* modifiers, the subcellular location and spatial distribution of this protein has been determined in developing endosperms by immunocytochemical techniques (Geetha et al, 1991). Interestingly, the increased synthesis of gamma-zein in modified *o2* mutants does not lead to larger protein bodies. Protein bodies in modified genotypes have about the same diameter as those in nonmodified *o2* mutants (about 0.4 μm). However, there is an alteration in spatial distribution of this protein. In modified *o2* mutants, high concentrations of the protein extend for multiple cell layers toward the center of the endosperm, while in the wild-type and *o2* endosperms gamma-zein is located mostly in the outer cell layers of the endosperm tissue (Lending and Larkins, 1989)

Despite the fact that the number of protein bodies per unit endosperm volume in modified and nonmodified *o2* types has not been compared, higher accumulation of gamma-zein must lead to formation of many more of these structures in the cytoplasm of modified endosperm than in unmodified types (Lopes and Larkins, 1991). As the seed matures, the desiccation of the endosperm may lead to disintegration of RER membranes, exposing the gamma-zein protein to the cytoplasmic matrix. The gamma-zein protein has an extensive repeat structure, a high proline and cysteine content. The fact that this protein has the ability to crosslink, forming high molecular weight oligomers in vitro (Lopes and

Larkins, 1991), in addition to the necessity of strong reducing/denaturing conditions to quantitatively extract it from maize endosperm (Wallace et al, 1990), suggest it is extensively complexed in vivo. During desiccation, the protein bodies become closely packed (Torrent et al, 1989) and possibly crosslinked by the gamma-zein protein. This could cause the vitreous phenotype of the mature modified *o2* kernels (Lopes and Larkins, 1991).

Accumulation of alpha-zein proteins has been shown to be positively associated with endosperm hardness (Dombrink-Kurtzman and Bietz, 1993). This conclusion is consistent with the still unproven association between the soft, floury phenotype of *o2* mutants and reduced accumulation of alpha-zeins. If this association is correct, it can be imagined that development of seed hardness in *o2* mutants would be possible through increased accumulation of a storage protein capable of offsetting the defect caused by reduced alpha-zein. Since gamma-zein genes are not regulated by the *O2* protein, increased accumulation of this protein can be achieved in *o2* backgrounds, provided that the necessary regulatory processes are amenable to selection. This appears to be the case in modified *o2* genotypes.

The gamma-zein protein in modified *o2* genotypes does contain lysine (Lopes et al, 1995). Therefore, it is difficult to imagine that selection for seed modification would cause increased accumulation of a gene product unrelated to the modification process, thus antagonizing efforts to maintain protein nutritional quality. This is especially true because the selection procedures used for *o2* modification placed a high priority on maintenance of seed nutritional quality (Vasal et al, 1980; Bjarnason and Vasal, 1992; Gevers and Lake, 1992). Therefore, it appears that increased accumulation of the lysine-poor gamma-zein protein is a necessary route to alter seed phenotype. Apparently, the cost of this process is a slight reduction of seed nutritional quality relative to the unmodified *o2* (Figure 4) (Ortega and Bates, 1984; Lopes, 1993), due to a decrease in the lysine-rich nonzein proteins that occurs upon increased storage protein accumulation (Lopes, 1993). The mechanisms determining the quantitative balance between these protein fractions remain unknown.

Summary, Conclusions and Future Directions

Although a great deal of research has been devoted to the nutritional improvement of seed storage proteins, to a large extent this work has not been very successful. While it has been possible to increase the protein content of seeds through breeding, this approach has had little impact on the composition of essential amino acids. This is simply because genes encoding storage proteins that contain the most limiting essential amino acids have not been found. In 1964, the identification of a mutation in maize that enhances the lysine content of the seed generated a great deal of excitement regarding the potential to genetically improve the protein quality of maize and other cereals. This mutation, *o2*, reduces the content of zein storage proteins and increases the proportion of other proteins that contain lysine. However, the soft starchy phenotype of the mutant seeds, as well as their reduced protein content and yield, prevented *o2* from becoming agronomically useful.

The discovery of a system of modifier genes that cause formation of hard and vitreous endosperm regions in *o2* mutant seeds led to a series of efforts to develop mutant genotypes with normal seed phenotype. The development of breeding methods that rely on progressive accumulation of *o2* modifier genes led to the development of germplasm with normal kernel phenotype and enhanced nutritional value. However, these modified *o2* mutants have not yet reached widespread commercial production because the components and mode of action of the genetic system involved in seed modification are poorly understood. Problems related to stability of modification, to transfer of modifier genes to elite inbred lines, and to slightly reduced nutritional quality in comparison to the original *o2* mutant indicate that further development and future usefulness of modified *o2* may depend upon deciphering the genetic, biochemical, and molecular mechanisms acting in these genotypes.

A combination of genetic and biochemical analyses has been used to investigate the number and mode of action of modifier genes and possible association between seed modification and gamma-zein deposition. These analyses indicate the possible existence of two additive modifier loci in the populations studied. Phenotypic analysis have been confirmed by RFLP mapping that identifyed two modifier loci on chromosome 7. One of these modifier loci maps to the same chromosomal region where gamma-zein genes are located.

Confirming our genetic analysis, biochemical characterization of storage proteins from normal, *o2*, and modified *o2* genotypes indicate that independently developed modified *o2* genotypes have seeds with higher than normal contents of gamma-zein, a cysteine-rich storage protein. Even though no proof is available for the direct involvement of gamma-zein in *o2* endosperm modification, the consistent association between these traits is supportive of the hypothesis that during seed desiccation protein bodies may become closely packed and possibly crosslinked by the gamma-zein protein and other cysteine-rich proteins, thus causing the vitreous phenotype of the mature modified *o2* kernels.

Our results represent the first important steps toward the dissection of a complex modifier system. The understanding of *o2* modifier gene action promises to shed light on important aspects of seed development, seed physical structure, and nutritional quality. Knowledge of the genomic location of modifier genes will greatly facilitate the use of marker-assisted selection for their rapid transfer to elite inbred lines. Also, the finding of the antagonistic relationship between endosperm modification and nonzein accumulation (Lopes, 1993) can be used to develop selection strategies to enhance nutritional quality in highly modified *o2* materials. Monitoring of the nonzein content is an inexpensive and rapid procedure compared to expensive and cumbersome monitoring thorough amino acid analysis (Lopes et al, 1995).

Although modification of *o2* maize represents a significant step in the development of high quality maize, further increases in essential amino acid content may be possible through genetic engineering of seed protein genes. However, introducing an engineered gene into the genome may prove insufficient to significantly increase the nutritional value of the seed, since the endosperm accumulates great amounts of low quality storage proteins. The understanding of the *o2* modifier system could provide the mechanisms to reduce the synthesis of low quality endogenous storage proteins while allowing for high levels of expression of the engineered genes.

REFERENCES

Altschuler Y, Rosemberg N, Harel R, Galili G (1993): The N- and C-terminal regions regulate the transport of wheat gamma-gliadin through the endoplasmic reticulum in *Xenopus* oocytes. *Plant Cell* 5:433–450

Bass HW, Webster C, O'Brian GR, Roberts JKM, Boston RS (1992): A maize ribosome-inactivating protein is controlled by the transcriptional activator *opaque-2*. *Plant Cell* 4:225–234

Bjarnason M, Vasal SK (1992): Breeding Quality Protein Maize (QPM). *Plant Breed Rev* 9:181–216

Boston RS, Fontes EBP, Shank BB, Wrobel RL (1991): Increased expression of the maize immunoglobulin binding protein homolog b-70 in three zein regulatory mutants. *Plant Cell* 3:497–505

Burr FA, Burr B (1982): Three mutations in *Zea mays* affecting zein accumulation: a comparison of zein polypeptides, *in vitro* synthesis and processing, mRNA levels, and genomic organization. *J Cell Biol* 94:201–206

Dalby A, Tsay CY (1974): Zein accumulation in phenotypically modified lines of *opaque*-2. *Cereal Chem* 51:821–825

Damerval C, De Vienne D (1993): Quantification of dominance for proteins pleiotropically affected by *opaque-2* in maize. *Heredity* 70:38–51

Dannenhoffer JM, Bostwick DE, Or E, Larkins BA (1995): opaque-15, a maize mutation with properties of a defective opaque-2 modifier. *Proc Natl Acad Sci USA*: (in press)

Das OP, Messing JW (1987): Allelic variation and differential expression at the 27-kilodalton zein locus in maize. *Mol Cell Biol* 7:4490–4497

Das OP, Minzi SL, Koury M, Benner M, Messing J. (1990): A somatic gene rearrangement contributing to genetic diversity in maize. *Proc Natl Acad Sci USA* 87:7890–78

Das OP, Poliak E, Ward K, Messing J (1991): A new allele of the duplicated 27 kD zein locus of maize generated by homologous recombination. *Nucl Acids Res* 19:3355–3330

Dombrink-Kurtzman MA, Bietz JA (1993): Zein composition in hard and soft endosperm of maize. *Cereal Chem* 70:105–108

Duvick DN (1951): Development and variation of the maize endosperm (Ph.D Dissertation). St. Louis, MO: Washington University

Duvick DN (1955): Cytoplasmic inclusions of the developing and mature maize endosperm. *Am J Bot* 42:7171–725

Duvick DN (1961): Protein granules of maize endosperm cells. *Cereal Chem* 38:374–384

Esen A (1986): Separation of alcohol-soluble proteins (zeins) from maize into three fractions by differential solubility. *Plant Physiol* 80:623–627

Esen A, Stetler DA (1992): Immunocytochemical localization of delta-zein in the protein bodies of maize endosperm cells. *Am J Bot* 79:243–248

Geetha KB, Lending CR, Lopes MA, Wallace JC, Larkins BA (1991): *opaque*-2 modifiers increase gamma-zein synthesis and alter its spatial distribution in maize endosperm. *Plant Cell* 3:1207–1219

Gevers HO, Lake JK (1992): Development of modified *opaque*-2 maize in South Africa. In: *Quality Protein Maize*, Mertz ET, ed. Saint Paul, MN: American Association of Cereal Chemists

Glover DV, Mertz ET (1987): Corn. In: *Nutritional Quality of Cereal Grains: Genetic and Agronomic Improvement*, Agronomy Monograph, Olson RA, Frey KJ, eds. Madison, WI: American Society of Agronomy

Habben JE, Kirleis AW, Larkins BA (1993): The origin of lysine-containing proteins in *opaque*-2 maize endosperm. *Plant Mol Biol* 23:825–838

Hagen C, Rubenstein I (1981): Complex organization of zein genes in maize. *Gene* 13:239–249

Jones N (1990): Transcriptional regulation by dimerization: two sides to an incestuous relationship. *Cell* 61:9–11

Kirihara JA, Hunsperger JP, Mahony WC, Messing JW (1988): Differential expression of a gene for a methionine-rich storage protein in maize. *Mol Gen Genet* 211:477–484

Kodrzycki R, Boston R, Larkins BA (1989): The *opaque*-2 mutation of maize differentially reduces zein gene transcription. *Plant Cell* 1:105–114

Laemmli UK (1970): Cleavage of structural proteins during assembly of the head of bacteriophage T4. *Nature* 227:680–685

Larkins BA, Lending CR, Wallace JC, Galili G, Kawata EE, Geetha KB, Kriz AL, Martin DN, Bracker CE (1989): Zein gene expression during maize endosperm development. In: *The Molecular Basis of Plant Development*, Goldberg RB, ed. New York: Alan R. Liss

Larkins BA, Hurkman WJ (1978): Synthesis and deposition of zein in protein bodies of maize endosperm. *Plant Physiol* 62:256–263

Lending CR, Larkins BA (1993): Effect of the *floury-2* locus on protein body formation during maize endosperm development. *Protoplasma* 171:123–133

Lending CR, Kriz AL, Larkins BA, Bracker CE (1988): Structure of maize protein bodies and immunocytochemical localization of zeins. *Protoplasma* 143:51–62

Lohmer S, Maddaloni M, Di Fonzo N, Hartings H, Salamini F, Thompson RD (1991): The maize regulatory locus Opaque-2 encodes a DNA-binding protein which activates transcription of the b-32 gene. *EMBO J* 10:617–624

Lopes MA (1993): Genetic and biochemical characterization of the maize mutants floury-2 and modified opaque-2 (PhD Dissertation). Tucson, AZ: University of Arizona

Lopes MA, Larkins BA (1991): Gamma-zein content is related to endosperm modification in quality protein maize (QPM). *Crop Sci* 31:1655–1662

Lopes MA, Larkins BA (1993): Endosperm origin, development and function. *Plant Cell* 5:1383–1399

Lopes MA, Larkins BA (1995): Genetic Analysis of opaque-2 modifier gene activity in maize endosperm. *Theor Appl Genet*: in press

Lopes MA, Takasaki K, Helentjaris T, Larkins BA (1995): RFLP mapping indicates the involvement of gamma-zein genes with the formation of vitreous endosperm in modified opaque-2 maize. *Mol Gen Genet*: in press

Ludevid MD, Torrent M, Martinez-Izquierdo JA, Puigdomenech P, Palau J (1984): Subcellular localization of glutelin-2 maize (*Zea mays* L.) endosperm. *Plant Mol Biol* 3:227–234

Marks MD, Lindell JS, Larkins BA (1985): Quantitative analysis of the accumulation of zein mRNA during maize endosperm development. *J Biol Chem* 260:16445–16450

Mertz ET, Bates LS, Nelson OE (1964): Mutant gene that changes protein composition and increases lysine content of maize endosperm. *Science* 145: 279–280

Michelmore RW, Paran I, Kesseli RV (1991): Identification of markers linked to disease-resistance genes by bulked segregant analysis: A rapid method to detect markers in specific genomic regions by using segregating populations. *Proc Natl Acad Sci USA* 88:9828–9832

Motto M, Di Fonzo N, Hartings H, Maddaloni M, Salamini F, Soave C, Thompson RD (1989): Regulatory genes affecting maize storage proteins. *Oxf Sur Plant Mol Cell Biol* 6:87–114

Motto M, Maddaloni M, Ponziani G, Brembilla M, Marotta R, Di Fonzo N, Soave C, Thompson R, Salamini F (1988): Molecular cloning of the *o2-m5* allele of *Zea mays* using transposon marking. *Mol Gen Genet* 232:488–494

Nelson OE (1980): Genetic control of polysaccharide and storage protein synthesis in the endosperm of barley, maize, and sorghum. *Adv Cereal Sci Tech* 3:41–71

Or E, Boyer SK, Larkins BA (1993): opaque-2 modifiers act post-transcriptionally and in a polar manner on gamma-zein gene expression in maize endosperm. *Plant Cell* 5:1599–1609

Ortega EI, Bates LS (1983): Biochemical and agronomic studies of two modified hard-endosperm *opaque-2* maize (*Zea mays* L.) populations. *Cereal Chem* 60:107–111

Paez AV, Helm JL, Zuber MS (1969): Lysine content of *opaque*-2 maize kernels having different phenotypes. *Crop Sci* 9:251–252

Pedersen K, Argos P, Narayana SLV, Larkins BA (1986): Sequence analysis and characterization of a maize gene encoding a high-sulphur zein protein of Mr 15,000. *J Biol Chem* 261:6279–6284

Pedersen K, Bloom KS, Anderson JN, Glover DV, Larkins BA (1980): Analysis of the complexity and frequency of zein genes in the maize genome. *Biochemistry* 19:1644–1650

Pedersen K, Devereux DR, Wilson E, Sheldon E, Larkins BA (1982): Cloning and sequence analysis reval structural variation among related zein genes in maize. *Cell* 29:1015–1026

Prat S, Perez-Grau L, Puigdomenech P (1987): Multiple variability in the sequence of a family of maize endosperm proteins. *Gene* 52:41–49

Pysh LD, Aukerman MJ, Schmidt RJ (1993): *OHP1*: A maize basic domain/leucine zipper protein that interacts with *Opaque-2*. *Plant Cell* 5:227–236

Rauscher FJ, III, Voulalas PJ, Franza BR Jr, Curran T (1988): Fos and Jun bind cooperatively to the AP-1 site: reconstitution in vitro. *Genes Dev* 2:1687–1699

Righetti PG, Gianazza E, Viotti A, Soave C (1977): Heterogeneity of storage proteins in maize. *Planta* 136:115–123

Robutti JL, Hoseney RC, Wassom CW (1974): Modified *opaque*-2 corn endosperms. II. Structure viewed with a scanning electron microscope. *Cereal Chem* 51:173–180

Sargant E (1900): Recent works on the results of fertilization in angiosperms. *Ann Bot* 14:689–712

Schmidt RJ (1993): *Opaque-2* and zein gene expression. In: *Control of Plant Gene Expression*, Verma DPS, ed. Boca Raton: CRC Press

Schmidt RJ, Burr FA, Burr B (1987): Transposon tagging and molecular analysis of the maize regulatory locus *opaque-2*. *Science* 238:960–963

Schmidt RJ, Burr FA, Aukermam MJ, Burr B (1990): Maize regulatory gene *opaque-2* encodes a protein with a "leucine zipper" motif that binds to zein DNA. *Proc Natl Acad Sci USA* 87:46–50

Schmidt RJ, Ketudat M, Aukerman MJ, Hoschek G (1992): *Opaque*-2 is a transcriptional activator that recognizes a specific target site in 22-kD zein genes. *Plant Cell* 4:689–700

Smeal T, Angel P, Meek J, Karin M (1989): Different requirements for formation of Jun:Jun and Jun:Fos complexes. *Genes Dev* 3:2091–2100

Soave C, Salamini F (1984): Organization and regulation of zein genes in maize endosperm. *Phil Trans R Soc Lond B* 304:341–347

Sodek L, Wilson CM (1971): Amino acid composition of proteins isolated from normal, *opaque-2*, and *floury-2* corn endosperms by a modified Osborne procedure. *J Agric Food Chem* 19:1144–1150

Thompson GA, Larkins BA (1989): Structural elements regulating zein gene expression. *BioEssays* 10:108–113

Torrent M, Geli MI, Ludevid MD (1989): Storage-protein hydrolysis and protein-body breakdown in germinated *Zea mays* L. seeds. *Planta* 180:90–95

Tsai CY, Dalby A (1974): Comparison of the effect of *shrunken*-4, *opaque*-2, *opaque*-7 and *floury*-2 genes on the zein content of maize during endosperm development. *Cereal Chem* 51:825–829

Vasal SK, Villegas E, Bjarnason M, Gelaw B, Goertz P (1980): Genetic modifiers and breeding strategies in developing hard endosperm *opaque-2* materials. In: *Improvement of Quality Traits of Maize for Grain and Silage Use*, Pollmer WG, Phillips RH, eds. London: Matinus Nijhoff

Wallace JC, Lopes MA, Paiva E, Larkins BA (1990): New methods for extraction and quantitation of zeins reveal a high content of gamma-zein in modified *opaque*-2 maize. *Plant Phys* 92:191–196

Wolf MJ, Buzan CL, MacMasters MM, Rist CE (1952a): Structure of the mature corn kernel. I. Gross anatomy and structure relationships. *Cereal Chem* 29:321–333

Wolf MJ, Buzan CL, Rist CE (1952b): Structure of the mature corn kernel. III. Microscopic structure of the endosperm of dent corn. *Cereal Chem* 29:349–361

Part VI

Examples of Social and Economic Impact of New Technologies

17

Impacts of Biotechnology in International Agriculture and Forestry

VICTOR M. VILLALOBOS

Introduction

Biotechnology is emerging as one of the most innovative achievements in life sciences of this century and is influencing almost every aspect of human life. The achievements of modern biotechnology are due to scientific discoveries made in the past two decades during which advances in genetics, biochemistry, molecular and cell biology have allowed development of genetic engineering or recombinant DNA techniques. Molecular biology has produced a group of biotechnological tools, based on the use of living organisms, which creates or modifies products, plants, or animals, and develops microorganisms for specific uses.

The results generated by the use of biotechnology, even in this initial stage, can be applied to several disciplines and industries such as:

- pharmaceuticals;
- microbiology for engineering microorganisms to produce compounds (like antibiotics, vitamins, and aminoacids);
- protein industry for food and feeds; and
- medicine for therapeutics and diagnostics of a wide range of diseases.

In other areas biotechnology facilitates the production of a higher quality and greater quantity of compounds than would be possible through the use of conventional techniques. Among all the fields in which biotechnology has been applied, agriculture is undoubtedly the sector that has benefited most from this technological innovation. It can already be foreseen that modern biotechnology will bring about new changes related to sustainable agriculture that are now unimaginable. Prevention of plant disease through transformation, genetic improvement through new gene combinations unavailable in nature, improvement of the qual-

The Impact of Plant Molecular Genetics
BWS Sobral, Editor

ity of agricultural products, and reduction of environmental degradation are but a few of the contributions that this new technology can offer to agriculture.

Biotechnology when applied to plants, includes the following two major groups of tools, which are, to a certain extent, complementary: (1) tissue culture, and (2) genetic engineering or recombinant DNA. Both techniques provide new ways to overcome problems in plant production and protection, restriction of plant adaptation to a marginal environment, abiotic stresses, and resistance to pests and diseases. Furthermore, breakthroughs in biotechnology allow the achievement of agricultural development by producing unique gene combinations through genetic engineering.

Despite the optimism with which these emerging techniques are regarded, there is apprehension and concern over the social, economic, and technical implications of biotechnology in agriculture. This paper analyzes the benefits to be achieved by applying biotechnology and the negative implications for those who choose not to be part of this revolution.

Current Agriculture Situation

Agricultural systems, empirical or specialized, have met the constantly increasing demand for food since early days. Increasing demographic and market needs require new developments in agriculture. Technology, therefore, has been developed for extensive agriculture, involving the use of genetically uniform varieties, mechanization, post-harvest technology, processing, and development of the plant protection technology. Although these improvements have increased the total yield of crops, as demonstrated during the present century, and more recently by the Green Revolution during the 1970s and early 1980s, there is concern that the excessive uniformity could result in plant vulnerability and, therefore, unsustainable agricultural production. Crop uniformity might foster genetic erosion, reduce the genetic base of a plant population, and increase the vulnerability of cultivated plants to pests, diseases, and adverse environmental factors such as salinity and drought. Furthermore, extensive agriculture reduces food diversification due to the concentration on a limited number of crops (Villalobos, 1995).

An important factor influencing agriculture is undoubtedly the dramatic increase in human population which creates an increased pressure over the fragile equilibrium of the ecosystems. Only 30 years ago, the world population was 3 billion. In 1990 it was 5.3 billion and is expected to grow to 7.2 billion by the year 2010. Some 94 percent (1.8 billion) of the total increase in world population will occur in the developing countries which have the highest incidence of poverty and malnutrition. How has agriculture responded to these increases in human world population? Until now, production has grown faster than population. Per capita production is today about 20% above that of 30 years ago. However, for much of the developing world, food availability is far from adequate for all people to have access to sufficient food to provide food security at all times.

It is clear that to decrease or eliminate hunger involves more than expanding crop production. What are the real choices in this intricate situation? Part of the solution for producing enough food for Earth's burgeoning population, without destroying the environment, is to increase potential yields per unit of area, not just only of major crops but also of local species. This will require further scientific advances in food production and thereafter the transferring of technology and improved farming practices to farmers in the developing countries. In this respect, plant biotechnology, along with conventional plant breeding, will play an important and increasing role.

Economic and Social Impacts of Modern Plant Biotechnology

Plant biotechnology could have a considerable influence on the economic relationship between developed and developing nations. In developing countries, biotechnology can contribute by increasing the productivity of local crops and enhancing diversification opportunities and food sources. Simple tissue culture techniques allow the massive multiplication of superior genotypes, mainly from vegetative propagated species like root, tubers, fruits, medicinal plants, and spices. Through tissue and apical meristem culture, diseases (primarily virus) can be eradicated. Virus elimination in planting stock can increase yields by up to 40%. Negative effects could involve substitution of commodities. The threat of the biotechnological substitution of products such as vanilla and cocoa butter is frequently mentioned, as is the substitution of some pharmaceutical compounds, scents, flavorings, and spices. It is true that biotechnologically-derived substitutes for these products are objects of research, but the higher production cost of this methodology is still a limiting factor. If, in addition, we consider society's tendency to prefer natural products, the significance of this threat is greatly diminished. A good example is the case of the vanilla industry; the use of a very simple biotechnology tool such as tissue culture techniques for cloning superior (more resistance to disease and high yielding) individuals, could be a strategy to reduce the risk of vanilla from Madagascar being replaced by that from large-scale fermenters producing vanillin in synthetic form, since natural vanilla will be more competitive.

Another type of risk should be taken more seriously in developing countries in particular: lack of knowledge of the scope and the limitations of modern biotechnology. Countries should be able to identify those technologies that can and cannot be useful to them. They should also recognize that the development of biotechnology is a long and costly process, and that its benefits may not be evident in the short term. Governments mostly rely on picked teams of scientists, who serve as interlocutors between science and policy-makers throughout the process. Above all, it must be realized that biotechnology is only desirable as a tool for the development of agriculture and not as an end in itself.

Other negative effects are related to the potential risk from the deliberate release into the environment of genetically modified organisms (GMOs). Developing countries, particularly those identified as centres of dispersion of plant species, run greater risks compared to developed countries, since the former have more genetic diversity and more marginal lands where the balance of living organisms, soil, and climate is delicate (Commandeur et al, 1995)

Regardless of the potential of plant biotechnology, developments differ from crop to crop. Consequently, breakthroughs to increase the production of one crop may be at the expense of another, particularly where species are closely related, e.g., among cereals, *Solanaceae* and *Cucurbitaceae*. Furthermore, the diffusion will also be unequal among extensive crops like cotton, tobacco, tomato, soya, maize, rice, and wheat, which are gaining more attention from basic research and capital investment than small-scale crops for local consumption.

Commandeur et al (1995), have analyzed the impact of biotechnology on developing countries and indicate that although it is not yet clear to what extent the overall impact of modern biotechnology on the economies of developing countries will be, it is clear that these countries will all be affected differently by these developments. They consider that:

- Countries that combine a net import of agricultural products with a high technological potential will probably gain from biotechnology, since they could be oriented towards self-sufficiency (for example India and China);
- Countries that combine a net export of agricultural products with a high technological potential could gain from biotechnology if it helps to diversify their exports (for example Brazil and Malaysia);
- Countries that combine a net export of agricultural products with a low technological potential will be damaged by such substitutive developments, and it is highly improbable that they will use biotechnology (many African and Caribbean countries);
- Countries that combine a net import of agricultural products with a low technological potential might in the short-run profit from low world market prices. In the long run, however, low prices might hamper their domestic food production (for example Bangladesh and Ethiopia).

The most important agricultural export commodities (coffee, banana, and soya beans) of Latin American countries, will probably not be negatively affected by biotechnology in the foreseeable future. Brazil will probably benefit from biotechnology, since it could be applied to increasing export production faster than other countries. Also some Asian countries, like Malaysia, have a tradition of running large-scale plantations and have built up the capacity to use technological changes (Commandeur et al, 1995).

The danger of biotechnology with the more traditional techniques of plant improvement is not the displacement of traditional crops. The danger is that developing countries could miss the opportunities it provides.

Crop production per unit area may be a poor indicator of the social benefit of biotechnology applications. Especially in the case of small farmers, yield increase may not result in income gain if it is achieved at higher input costs. Small farmers are very sensitive to the external costs arising from new technologies in regard to access to inputs and credit. They benefit more from biotechnology applications that reduce production costs (eg. biofertilizers and biopesticides) than from those that increase yield potential. Small farmers may turn to agricultural practices such as the improvement of crop resistance to combat disease, pests, and drought if the benefits biotechnology provides are too expensive (Commandeur et al, 1995).

Forestry Production and Trade

There is a clear distinction between agricultural crops and forest species in regard to exploitation and improvement. These differences also reflect the potential benefits of biotechnology. Forests are complex ecosystems capable of providing a wide range of economic, social, and environmental benefits. Forests and woodlands are essential for human life, but their benefits and services are valued differently by different people and different groups. Local, national, and international interests in forest resources also vary. Furthermore, the numerous roles that forests are expected to play in local, national, and global development change dramatically over time. The long history of selection and improvement of many agricultural crops is in contrast to that of forest species, where the shift from exploitation to domestication is much more recent (FAO, 1994a).

According to the Food and Agricultural Organization of the U.N. (FAO) (1994b), the roles of forestry are changing. The changes began in the 1970s when growing awareness of how local communities control and depend on forests prompted efforts to strengthen local stakes in forest management, programs, and activities. By the 1980s, countries began to recognize that forests have a global role in the stability of biodiversity and in the protection of threatened indigenous and traditional cultures. In the 1990s, sustainable development became a primary focus of policy discourse (FAO, 1994). Concern about the evolving roles of forestry moved onto a crowded stage at the United Nations Conference on Environment and Development (UNCED) in June 1992. UNCED highlighted forestry development and environmental issues by developing a set of forest principles, devoting a chapter of its document, Agenda 21, to combating deforestation (Chapter 11) and focusing on the importance of nonwood functions of forests in the biodiversity and climate change conventions. A number of countries have launched specific international programs to follow up on UNCED forestry recommendations. This broad consensus on principles of sustainable forest management represents the first commitment of responsibilities beyond national boundaries. Turning this principle into practice, however, presents a more formidable task (FAO, 1994b).

The State of Forest Resources

In its most recent resource assessment, FAO estimates the world forest area at be 3.4 billion ha, or 26% of the earth's land area. FAO's definition of forest comprises ecological systems with a minimum of 10% crown coverage of trees. The regional distribution of world forest cover is presented in Figure 1. Four countries account for more than 50% of the world's forests: the Russian Federation (22%), Brazil (16%), Canada (7%) and the United States (6%). Table 1 lists the most forested countries in the world according to total area. Some of the least forested countries have extremely arid climates, for example Algeria, Egypt, and Saudi Arabia. Others include countries which were once heavily forested but which cleared most of their original forests for agriculture, human settlement, and infrastructure. Examples include Bangladesh, Haiti, and the United Kingdom.

Temperate and boreal forests occupy 1.64 billion ha, just less than half of the world's forest cover. More than 70% of these are located in the Russian Federation, while the rest is in Canada (15%) and the United States (13%). In general, the aggregate size of temperate forests in industrial countries is stable or even slightly increasing through afforestation efforts. In Europe, the forested and wooded land area increased by 2 million ha between 1980 and 1990. Temperate zone forests are widely recognized for their enormous contribution of global industrial timber supplies as well as for their nonwood products, recreation, and environmental services. However, they are often less appreciated for their plant and animal life than the tropical rain forests, even though they contain some of the tallest and oldest trees in the world (FAO, 1994b). The pharmaceutical value

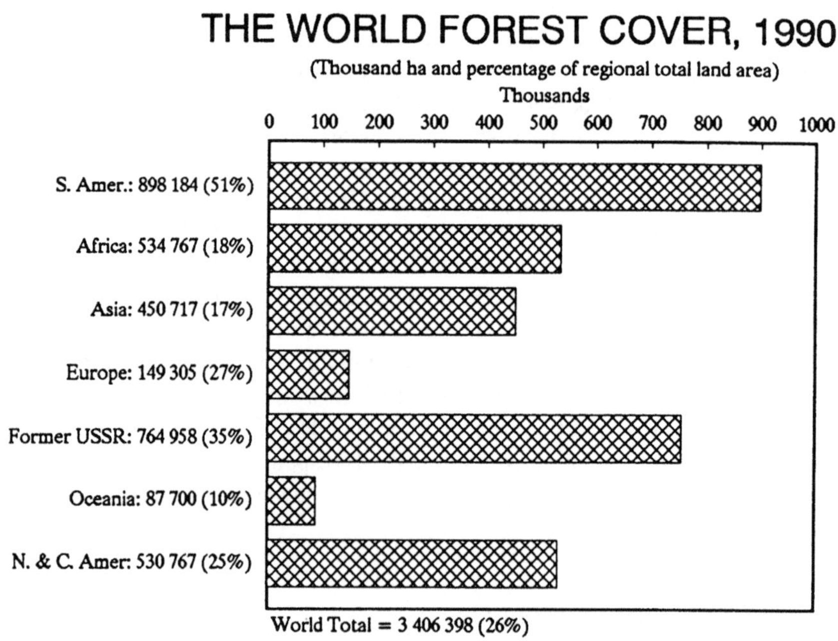

Figure 1. The state of world forest resources, the world forest cover, 1990 (thousand ha and percentage of regional total land area) (FAO, 1994c).

Table 1. Total Forest Area (1990)

Country	Total Forest Cover*	Percentage of Land Area	Hectares Per Caput
Former USSR	754 958	35	2.6
Brazil	561 107	66	3.7
Canada	247 164	27	9.3
United States	209 573	23	0.8
China	127 780	14	0.1
Zaire	113 275	50	3.1
Indonesia	109 549	61	0.6
Peru	67 906	53	3.0
Colombia	54 064	52	1.7
India	51 729	17	0.1

*'000 ha (FAO, 1994c)

of biodiversity in the temperate zone is also substantial. A recent study found that 28% of Canadian trees have medicinal properties.

The tropical forest zone contains 1.76 billion ha. More than half of the world's 718.3 million ha of rain forests are located in two countries: Brazil (41%) and Indonesia (13%). The rate of causes and effects of deforestation differ greatly from one country or region to another. During the past decade, the total area of rain forest cut was 46 million ha.

The available data on tree plantation areas suggests that the 100 million ha of plantations in the world provides for 7%–10% of the world's present commercial wood consumption. An additional 14 million ha of rubber and coconut oil palm plantations is not included in the area of forest plantations. Plantations cannot provide the full range of goods and services supplied by a natural forest. They are tree crops, analogous to agricultural crops, with a simplified ecology of one or at most, a few, species usually chosen for their yield and ease of management. The primary purpose of most plantations is to produce wood or other products quickly and cheaply. Their role, which is a highly valuable one, complements national or global forestry management strategies. The area dedicated to plantations is growing at an average rate of around 2.6 million ha per year. A recent review of tropical plantations concluded that planning is generally poor, particularly for vital issues such as matching species and sites. Temperate countries also have numerous examples of plantations that have failed, or sites that have been degraded, due to the establishment of large blocks of single-species plantations introduction of unsuitable species, and planting of even-aged plantations. In the developing countries, further development of plantation forestry is constrained by the shortage of land. With expanding farming populations using all the unforested land for food production, the areas available for plantations are becoming increasingly restricted. The experience of the past two decades demonstrates that degraded land may be the only resource available to the landless poor. There are, however, large areas where the natural forest has been badly degraded or where soil fertility has been lost through overcultivation, which could be used for plantations.

Tree Improvement

For industrial forestry species, of which there are 100 million hectares of plantation worldwide (Gauthier, 1991), the most common approach to improvement is recurrent selection in genetically quite broad breeding populations. Four generations have been completed in *Eucalyptus grandis* and three in some *Pinus* species, but very few programs are as advanced as these (Gauthier, 1991). The use of open-pollinated seed orchards is the most common approach for the capture of genetic gain, while a few advanced programs are characterized by the development of superior full-sib families or clones. Vigor, form, and wood quality are the priority criteria for selection in most programs, with disease or insect-resistance of importance only in a few cases.

FAO (1994c) prepared an extensive study of the potential contributions and limitations of biotechnology in forest tree improvement. He indicated that the major limitations to rapid improvement for most of the established industrial species are:

- long generation intervals, related to poor juvenile-mature correlations, and the long juvenile phase before flowering;
- low effectiveness of selection for many characters, due to low heritability or difficulty in assessment; and
- the use of open-pollinated seed orchards when only a part of the genetic variation is available.

Major research priorities for these established industrial species should be the development of methods for the propagation of full-sib families and clones, the development of methods for early and more accurate selection, and the promotion of precocious flowering.

Application of Biotechnology to Forestry

Tissue Culture

One of the main aims of vegetative propagation of trees is to multiply individuals old enough to have expressed their superior characteristics. The traditional methods for vegetative propagation of forest species are: rooted cuttings, grafting, or rooted needle fascicles for pine trees. Although these methods have been practiced for centuries, rooting capacity, survival and growth decline in direct proportion with the age of the parent plant. Furthermore, in many conifers, rooted branch cuttings tend to grow plagiotropically. The main advantage of using tissue culture for tree cloning is its potentially unlimited multiplication rates (Thorpe and Biondi, 1984). Thus, while a rooted cutting can produce a single plant from which, several years later, further cuttings are available, with today's techniques of tissue culture, it can produce several auxiliary as well as adventitious shoots (Villalobos et al, 1985). Both these types of shoots can in turn be induced to form additional axillary and/or adventitious shoots. The intent of such clonal propagation is to reproduce large quantities of uniform plants of selected genotypes. Over 1000 plant species have been micropropagated, including over 100 forest tree species. With sufficient research effort, successful protocols could be developed for most tree species. In general, a research project of this type may be described as short to medium term with a moderately high expectation of success.

Asexual multiplication of both hard and softwoods using tissue culture methods can be achieved through three approaches (Thorpe et al, 1991) namely: (1) enhancement of axillary bud breaking; (2) production of adventitious buds; and (3) somatic embryogenesis. The first two approaches lead to plantlet formation via organogenesis through the production of unipolar shoots, which must then be rooted in a multistage process. Somatic embryogenesis leads to the differentiation of a bipolar embryo through differentiation steps that are often similar to zygotic embryogenesis.

Cost of In Vitro *Plantlet Production*

The cost of generating plants in vitro varies according to species and is dependent on the techniques used. In all cases, in vitro multiplication has a higher unit cost than conventional propagation. Currently, the most advanced micropropagation technology available is based on the multistaged organogenesis process, which is extremely labor and cost intensive (Thorpe and Hasnain, 1988). Therefore this process can be profitable only for species that cannot be multiplied conventionally, in the cases in which there is a very limited number of individuals, in which the cost of conventional clonal propagation is high, or in cases in which a high return is expected from the resulting genetic gain, disease-free stock, etc.

Studies made on production costs in New Zealand (Smith, 1986), France (Boulay, 1985) and Canada (Hasnain et al, 1986) indicated that micropropagated conifer plants cost three to ten times as much as seedlings, and labor costs represent 60%–80% of the total. Therefore, for in vitro techniques, namely, shoot multiplication by axillary or adventitious budding, and somatic embryogenesis, to be cost effective need to be improved or implemented in countries with comparative advantages such as low labor and production costs. At present, there are no studies on production costs for forest species in laboratories in developing countries. However, there are many countries that have developed an efficient system for the micropropagation of several plant species including ornamentals, fruit trees, horticultural, and industrial crops. Countries like China, Brazil, Costa Rica, India, Thailand, and Mexico could be good alternatives for implementing massive multiplication programs of forest species since the technical capacity exists and yet the labor cost is still very competitive.

Biotechnology in Tree Improvement

Biotechnology offers little to meet the urgent priorities in forest tree improvement in comparison with annual crops, although forest tree species are the subject of very active biotechnology research programs. Some possibilities of technology substitution or supplementation for tree improvement in the short term exist. In the short term, the technologies are:

- The use of molecular markers in quality control in advance breeding programs;
- The use of micropropagation techniques in the integrated clonal propagation systems featuring the commercial planting of cuttings harvested from rapidly multiplied, micropropagated stool plants of the selected clones; and
- The use of markers in essential taxonomic studies and investigations of mating systems.

Long term priority areas are:

- Genetic engineering for cone sterility;
- The use of molecular markers and DNA transformation techniques to investigate genetic processes at the molecular level, in particular those relating to complex and multigenic traits such as growth, adaptation, and stem and wood quality;
- Molecular studies of the maturation state;
- Development of somatic embryogenesis protocols in combination with artificial seed technology as a clonal propagation method for industrial plantation species; and
- Genetic engineering of useful traits.

Recently, FAO published a review of the present state and future potential of biotechnologies for forest tree improvement, with special reference to their application in developing countries. The document confirms that new technologies offer possibilities which will greatly facilitate the work of the tree breeder in the future. It stresses, however, that the potential of the powerful new tools that biotechnology offers can only be realized if applied within existing, sound tree improvement programs and if based on basic, biological knowledge of the species concerned. FAO (1994c) critically analyzed the benefits and limitations of the available biotechnologies for tree improvement. The following sections summarize the major findings of the study on biotechnology applied to forest species.

In Vitro *Selection*

Many recent publications concerning crop plants have reported useful correlations between in vitro responses and the expression of desirable field traits (most commonly disease-resistance), although positive results are also available for tolerance to herbicides, metals, salt, and low temperatures.

For the selection criteria of major general importance in forest tree species, in particular vigor, stem form, and wood quality, however, poor correlations with field responses will limit the usefulness of in vitro selection. in vitro selection is therefore likely to have very limited application in forest tree species, and only to be of possible interest in a few programs in which there is a disease-resistance selection problem, but it is of no broad strategic value as a research objective.

Somaclonal Variation

Variation induced during cell or callus cultures has been reported for many species. For some crops, variants have been produced showing economically useful levels of traits such as resistance to disease and increased levels of salt. The phenomena are still not clearly understood, and persistence of modifications through subsequent sexual generations has been demonstrated in some cases but not in others.

Assuming stability of the characteristics, the approach is of most value when selection can be done at cell level (e.g. by exposure to a phytotoxin or to high mineral levels), thus enabling the screening of large numbers of genotypes, and when the level of the trait sought lies outside the range available naturally in the species. Cold tolerance in the *Eucalyptus* is an example that may meet these criteria. No immediate applicability is evident to the new tropical hardwoods or to nonindustrial species, for which naturally available genetic variation generally is poorly defined.

Protoplast Fusion

This is a field with a long history of research (Gamborg et al, 1981). There have been several successes, notably with species of the *Brassicaceae* and *Solanaceae*. Severe limitations are imposed, however, by the taxonomic relationship of parents and the requirement for regeneration from protoplasts. Expectations of success with a new pair of species is foreseen to be low.

There is little apparent applicability of protoplast fusion to forest tree species, particularly those (e.g., some tropical hardwoods and nonindustrials) for which investigations of the possibilities for traditional hybridization remain minimal. In the longer term, the objectives of protoplast fusion may be better served by manipulations at the DNA level.

Genetic Engineering

Crops transformed with genes of insect and virus resistance, as well as resistance to various types of herbicides, are being commercialized or are close to commercial application. Plants into which these genes have been inserted include the poplars. Many projects are in progress with forest tree species, in particular for modification of lignin biosynthesis through anticense technology. Insertion into a new species of the insect or herbicide resistance genes currently available would constitute a major research undertaking, and successful application would be dependent on being able to regenerate from the transformed cells. Manipulation of more complex traits would be a much more formidable undertaking, and, although rapid progress is being made in this field, much research remains to be done. The availability of effective transformation techniques remains an obstacle, although improved techniques are being developed. Regeneration is difficult for some tree species, but the problem may be overrated; the noncompetence of mature materials is not necessarily an obstacle to effective application of genetic engineering, provided that the juvenile explants respond satisfactorily. An often overlooked research component is field testing which would be required before a responsible recommendation for large scale deployment of transgenic plants could be made. Such testing could be extensive and prolonged, depending on the species and genes involved. Research projects of this type are necessarily intensive, and must be regarded as long term with only modest expectation of success.

Insect resistance is of potential value, e.g., in the poplars and some pines, *Eucalyptus*, and tropical hardwoods. A single gene may be sufficient to confer resistance for very short rotation cycles, but a much more cautions approach should be applied with long rotation species. The work involved in introducing several resistance genes in sufficient numbers to ensure that insects do not acquire tolerance during a long rotation, should not be underestimated. The reduction of lignin biosynthesis is a valuable objective for the pulp species. The production of herbicide tolerance genes is of some interest, but in many programs the advantage of using unguarded herbicide application may not be sufficient to pay for the research program. The extent to which genetic engineering permits substitution of environmentally friendly herbicides such as glyphosate for currently used residual herbicides could be an important factor. Cold tolerance genes are likely to be of some commercial value in many species, in particular the *Eucalyptus*. Much remains to be done to establish that sufficient tolerance can be conferred using antifreeze proteins and to extend the work to tree species.

It is important that genetically engineered genotypes are of high quality with respect to other traits as well. The clonal test is the most logical basis for integration of genetic engineering into traditional tree improvement programs. For these reasons, genetic engineering is most appropriately conducted with species in which breeding programs are advanced and sound clonal forestry programmes can be realistically contemplated. Research on this subject should not assume a high priority with species for which natural variation available within the taxon remains poorly investigated.

Molecular Markers

Techniques for the analysis of isozymes, Restriction Fragment Length Polymorphism (RFLPs) and Random Amplified Polymorphic DNA (RAPDs), are now well established, and adaptation of procedures to suit new species can be expected to be a realistic short term undertaking.

Markers have important immediate applications in supportive research for advanced breeding programs with industrial species, mainly in relation to quality control, e.g., checking of clonal identification, orchard contamination, and within-orchard mating patterns, by fingerprinting isozymes will be satisfactory for many, but not all, of these purposes. The major current value of molecular markers lies in: (1) support to long term strategic research; (2) advance in the understanding of basic genetic mechanisms; and (3) understanding genome organization at the molecular level. For forest tree species, the study of quantitative traits will be an important aspect of this work in the coming years. This work will be most efficiently concentrated on a few model species, e.g., *Pinus taeda* and in particular hybrids such as *P. elliottii X P. caribaea.*

Biosafety

The prevention of the escape of genes into wild populations is likely to become an important concern, and sterility should be an early target of genetic engineering work with forest tree species. The major factor limiting applications of genetic engineering in forest tree species is the state of knowledge of molecular control of the traits which are of most interest i.e., those relating to growth, adaptation, and stem and wood quality. Genetic engineering of these traits remains a distant prospect.

Germplasm Conservation Using Biotechnology

Since the early 1960s there has been a significant increase in the number of ex situ collections of genetic material and consequently an increase in the number of samples stored in seed banks and field collections.

According to the FAO (1994a), there are 4.41 million accessions currently in ex situ storage. Of these, 50% are maintained in industrialized countries, and 38% in developing countries with the Consultative Group on International Agriculture Research Centres (CGIAR) holding the remaining 12% (FAO, 1994; Villalobos and Engelmann, 1995).

Although genetic variability of a large number of crops are preserved ex situ, many crops that are important at the national and local level are poorly represented in the existing collections. These problems include conservation of recalcitrant seeds, which are unable to withstand desiccation, as in the case of many important economic crops such as mango, rubber, cocoa, coconut, coffee, oil palm, and many others. Moreover, there are practical problems in applying long-term seed storage to perennial plants, including trees, since their juvenile period is very long and they do not produce seeds for several years. On the other hand, some vegetatively propagated species are either sterile or do not have stable sexual reproduction.

Success in germplasm storage is determined by the inherent longevity and physiological storage behaviour of plant species (i.e., orthodox, recalcitrant, or intermediate), by initial quality, such as moisture content of the material storage, and storage methods and conditions (FAO, 1993).

The possibility of developing complete plants originating from isolated cells, tissues or organs, has made it feasible to establish germplasm banks utilizing plant tissue culture approaches.

In vitro storage has been proposed as a means to overcome some of the above-mentioned conservation problems. Tissue culture technology has, in fact, been a strong motivation for the planning, research, and development of alternative sources of conservation. In principle, in vitro techniques are appropriate for conservation, the development of a complete plant being the expected result in all cases. For this reason, it is essential to maintain an elevated level of tissue organization during storage (Villalobos and Engelmann, 1995).

Although tissue culture techniques are used for in vitro conservation, at present only 37,600 accessions are stored in this way (FAO, 1994a). In vitro collections are subject to discussion, since they must be maintained in a carefully controlled environment. Practical maintenance problems, associated with the risks of microbial contamination and critical genetic stability, have to be considered. In vitro techniques imply the substitution of natural with artificial conditions. This has the advantage that light and temperature are controlled in a reduced space, and, in many cases, for plants with short reproductive cycles, such as some roots, tubers, and other vegetatively propagated species, in vitro transfer intervals are less frequent than the harvest cycle in the field. Another important advantage is the possibility of producing virus-free plants in large numbers, independent of climatic conditions. In the modern context of conservation and with a rational use of genetic diversity, in vitro conservation should include the elimination of virus from the stored material and the capability to micropropagate the germplasm in large quantities when necessary (Villalobos et al, 1991). Recently, cryopreservation has been used successfully on plants for long-term conservation. This technique is based on the reduction and subsequent arrest of the metabolic functions, including the cellular division in the explants. This is accomplished when material is brought to a liquid N_2 temperature ($-196°C$). Cryopreservation, however, is at an earlier stage of development. Extensive research is still needed to set up new freezing techniques, to apply existing ones to additional plant species, and to experiment with them on a large scale.

In Vitro *Conservation and Cryopreservation of Forest Species*

Although increasingly used operationally for the storage of threatened germplasm agricultural species, in vitro storage and cryopreservation have little to offer for this purpose in forest tree species. Gene pools of most of the established industrial species are reasonably well-preserved in in situ and ex situ stands and, for the short and medium terms, in seed stores. Undoubtedly, urgent gene conservation problems exist for many tree species, particularly among the tropical hardwoods and nonindustrial species. Distribution of these species is poorly known, as are their biological characteristics. Although seeds of some species will be recalcitrant, many are orthodox. Major impediments to the conservation of forest tree germplasm include the availability of resources sufficient for only a very small fraction of the survey and collection work that would be required before any germplasm could be stored, and the unreliability of many existing seed storage facilities. The replacement of existing facilities with more sophisticated technology is not likely to make a positive impact on the gene conservation problem with tropical tree species. Even for the recalcitrant species, activities should favour the establishment of ex situ planting which will facilitate evaluation of the material. In the longer term, cryopreservation and in vitro storage may have some application as a back-up conservation strategy for populations of well surveyed species which are recalcitrant such as poplars and walnut.

Cryopreservation warrants much more attention as a means of maintaining juvenility and capturing genetic gains offered by clonal forestry with industrial species. The technology is thus applicable mainly where good breeding programs are already in place, clonal forestry is a realistic goal, and rejuvenation is difficult, in particular for the conifers.

With the increasing emphasis on international cooperative breeding of both industrial and nonindustrial forest tree species in tropical countries, international movement is likely to become more important as breeding programs advance, and in vitro approaches could find application in the longer term.

ACKNOWLEDGMENT

The author gratefully acknowledges J. Suttie for the suggestions provided to the manuscript.

REFERENCES

Boulay M (1985): Some practical aspects and applications of the micropropagation of forest trees. In. *In Vitro Propagation of Forest Tree Species. Proc Intl Sump*; 1984 May; Bologna, Italy.

Commandeur P, van Roozendaal G, Junne G (1995): *The Impact of Biotechnology on Developing Countries. Opportunities for Technology-assessment Research and Development Cooperation.* A study commissioned by the Büro für Technikfolgen-Abschätzung bein Deutschen Bundestag (TAB), Amsterdam

FAO (1993): *Ex Situ Storage of Seeds, Pollen and In Vitro Cultures of Perennial Woody Plant Species*, FAO Forestry Paper 113. Rome: FAO

FAO (1994a): *Survey of Existing Data on Ex Situ Collections of Plant Genetic Resources for Food and Agriculture.* CPGR ex 1/94/5

FAO (1994b): *The State of Food and Agriculture. Forest Development and Policy Dilemmas*, FAO Agriculture Series No. 27. Rome: FAO

FAO (1994c): *Biotechnology in Forest Tree Improvement with Special Reference to Developing Countries*, FAO Forestry Paper 118. Rome: FAO

Gamborg OL, Shyluk JP, Shahin EA (1981): Isolation, fusion and culture of plant protoplasts. In: *Plant Tissue Culture Methods and Applications in Agriculture*, Thorpe TA, ed. Orlando, FL: Academic Press

Gauthier JJ (1991): Plantation wood in international trade. Presented at the seminar "Issues Dialogue on Tree Plantations, Benefits and Drawbacks." Geneva: CASIN

Hasnain S, Pigeon R, Overend PP (1986): Economic analysis of the use of tissue culture for rapid forest improvement. *For Chron* 82:240–245

Smith DR (1986): Forest and nut trees. I. Radiata Pine (*Pinus radiata* D. Don). In: *Biotechnology in Agriculture and Forestry*, Vol. 1, *Trees*, Bajaj YPS, ed. Berlin: Springer-Verlag

Thorpe TA, Biondi S (1984): Conifers. In: *Handbook of Plant Cell Culture*, Vol. 2, Sharp WP, Evans DA, Ammirato PV, Yamda Y, eds. New York: Macmillan

Thorpe TA, Hasnain S (1988): Micropropagation of conifers: Methods, oportunities and costs. In: *Tree Improvement—Progressing Together, Proceedings of the 21st Meeting of the Canadian Tree Improvement Association*, Morgenstein EK, Boyle JB, eds.; 1987 Aug; Truro, NS. ON: Canadian Forestry Service

Thorpe TA, Harry IS, Kumar PP (1991): Application of micropropagation to forestry. In: *Micropropagation*, Debergh PC, Zimmerman RH, eds. Netherlands: Kluwer Academic Publishers

Villalobos VM (1995): Advances in genetics: Food and agricultural products. In: *Medical Science and the Advancement of World Health*, Lanza R, ed. Medford, MA: Tufts University Press, in press

Villalobos VM, Ferreira P, Mora A (1991): The use of biotechnology conservation of tropical germplasm. *Biotechnology Advances* 9:197–215

Villalobos VM, Engelmann F (1995): *Ex situ* conservation of plant germplasm using biotechnology. *World J Microbiol and Biotech*: in press

Villalobos VM, Young EC, Thorpe TA (1985): Origin of adventitious shoots in excised radiata pine cotyledons cultured *in vitro*. *Can J Bot* 63:2172–2176

18

Molecular Anthropology of Cassava Cyanogenesis

HANS ROSLING

Earlier chapters have convincingly shown that molecular genetics has provided science with an amazing new set of tools. It has been suggested that plant genetics using these tools can provide new options for sustainable increase of food crop yields in the least developed countries. This is good news in the decade when the human population is growing more rapidly than ever and when unsustainable use of natural resources has climbed to the top of the global research agenda. The unsustainable technology of the richest nations seems to be matched by an equal lack of sustainability of the farming systems in the poorest nations. Especially in sub-Saharian Africa many populations have entered a vicious circle of poverty due to stagnant economies, rapid population growth, and accelerating environmental deterioration due to unsustainable farming systems (Harrison, 1992). The late general director of UNICEF, James Grant (1994), labeled this vicious down-going spiral, "Poverty, Population and Environment spiral, the PPE-spiral". This chapter reviews the relationship between cassava cyanogenesis and the paralytic disease konzo, a human disease induced by the PPE-spiral. The aim is to use this example for a discussion of how molecular plant genetics can better contribute to the reversal of the down-going PPE-spiral in the least developed tropical communities.

Extending the Upstream and Downstream Allegory

The poorest agricultural communities in the least developed countries are complex both as biological and social systems, and at present they are undergoing rapid and unpredictable changes. These farming systems are poorly documented, and the interlink between the social and biological mechanisms involved in the down-going spiral are poorly understood. It goes without saying that basic plant biotechnology needs to interact with applied agricultural research and various

The Impact of Plant Molecular Genetics
BWS Sobral, Editor

forms of observational studies of the socio-economic realities and farming practices of these poor populations in order to have an impact on their difficulties. Such links between basic and applied science, as well as development projects, are now becoming more dynamic. More interactions occur between investigations at different explanatory levels, i.e., molecule and cell, plant and field, as well as farm and community level. The one-dimensional dichotomy into upstream and downstream research is giving way for a conceptual continuum (Brader, 1994). This means that Delta-research, i.e., observational studies of the end-users farming systems, has been added to the up- and downstream research, and the need for interaction in both directions between the different levels of the research stream has been emphasized. However, the stream allegory can be further expanded. This chapter argues for the addition of what I call rain-research that collects and transmit experiences and knowledge on biological phenomena by observational studies in remote farming communities. Such studies can be highly productive if performed with participation of biological scientists who directly transmit observations to upstream biological and molecular laboratories. The mostly female farmer in the poorest communities can identify problems and possible solutions for farming systems researchers communicating with downstream researchers. She can also contribute to upstream research with a wealth of accumulated knowledge of the biological system she operates if the visiting researchers possess methods for the collection of this biological knowledge. My limited experience of poor agricultural communities in Africa indicates that there, in addition to the research stream continuum is a place for some rain-research. As shown in Figure 1, the rain research acquires biological information from studies of the rural communities and transforms this directly into raindrops over the often dry ambient of upstream molecular laboratories. The rain researchers from plant biology can benefit from the recent development of survey methods for rural communities in the third world, in the same way that health researchers have (Scrimshaw and Gleason, 1992; Melville, 1993).

Widening the Concept of Molecular Anthropology

The combined use of anthropological methods from the social sciences and biochemical methods from science in the same surveys has been called "molecular anthropology" (Rosling et al, 1993). The term was first coined for the use of determination of aminoacid sequence of serum proteins in studies of prehistoric human evolution (Zuckerkandl, 1963). Ethnobotany combines ethnographic surveys of local population's use of herbal drugs with chemical identification of pharmacologically active substances in tropical plants. Such combinations are, of course, complements to, not substitutes for, established research methods. Conventional scientific studies as well as socio-economic studies are crucial for the understanding of the biology and society in resource poor rural communities. However, the experience from investigations of medical effects in humans of toxic exposure from cyanogenic glucosides in cassava roots indicates that bio-

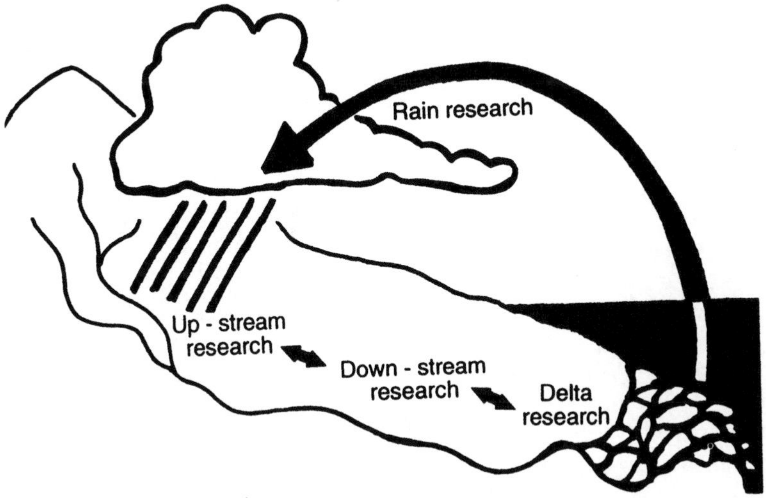

Figure 1. A hydrological stream allegory for agricultural research composed of upstream basic science, downstream applied science, delta research; i.e., interdisciplinary farming system research and rain-research, a direct feed back from rural to upstream laboratories with biological observations not provided through the established counterflow mechanisms.

logical sciences can advance rapidly by borrowing methods from social science and the humanities for elucidation of the local knowledge on biological phenomena. The combined use of research methods from the humanities and biochemistry may also be a cost-effective design option for studies of the biology of farming systems in poor tropical communities. With explorative interview and observational survey methods the researcher can acquire the accumulated biological experiences of poor farming communities (Patton, 1990). Subsequent experiments and biochemical analysis can sort this experience into knowledge and superstition. To label this molecular anthropology is a marketing trick to convey old wisdom already expressed by Louis Pasteur. He noted that in the field of observation, chance favors the prepared mind, i.e., biological understanding is needed for collection and interpretation of biological phenomena in observational studies. It is a well-known principle in medical research that molecular studies can advance more rapidly through a close link to clinical practice.

Konzo, a Newly Identified Neurological Disease

The identification of the paralytic disease konzo (Figure 2) and its relationship to cassava cyanogenesis exemplifies the combined use of explorative social survey methods and conventional biomedical methods. We have given the name konzo to this toxico-nutritional disease (Howlett et al, 1990; Rosling and Tylleskär, 1995). It is the local name among the first affected population, and it means "tied legs" in their language (Trolli, 1938). In collaboration with institu-

Figure 2. Women and children affected with konzo of varying severity in a Savannha village with very poor soil in Bandundu Region, Zaire. Note soaked cassava roots drying on the roof.

tions in Africa and Europe this research odyssey has passed from medicine and epidemiology to biochemistry and food science, and hence to agricultural sciences and plant genetics, thereby reaching the subject of this book.

Cassava, a Major Food Crop

Cassava (*Manhiot esculenta* Crantz) is a high-yielding tropical crop. Its starchy roots provide the main part of the dietary energy for about 400 million people. The roots of the cassava plant can, in fact, provide more dietary energy per hectare and working hour than any other staple crop (Figure 3). With further research and development, this number one converter of solar to dietary energy has great potential for feeding a growing population in tropical countries. The edible cassava leaves are also an important source of nutrients. Cassava ranks fourth among the tropical food crops, and due to its agricultural advantages, cassava has become the main staple crop on marginal land in sub-Saharan Africa (Cock, 1985).

Cassava Cyanogenesis

Cassava roots and leaves contain the cyanogenic glucoside linamarin and smaller amounts of lotaustralin (Figure 4). The levels of cyanogenic glucosides in cassava roots vary considerably among different genetic varieties, among different

Figure 3. Family members in northern Mozambique peeling cassava roots from a high yielding bitter variety.

plants of the same variety, and among roots of the same plant. The amount of glucosides is associated to bitterness (Sunderesan et al, 1987), and the levels of glucosides in roots range from negligible levels to a potentially lethal amount of cyanogenic substances. The glucoside levels also depend on environmental factors, such as drought and soil fertility (Bokanga et al, 1994a).

Populations using bitter and toxic cassava varieties as their main staple food remove the cyanogens by processing of roots before consumption. During domestication of cassava in Amazonia thousands of years ago, people developed safe processing methods which reduce the toxic constituents to negligible levels (Dufour, 1989). These methods first disintegrate plant cells and thereby bring linamarin in contact with an endogenous glucosidase that breaks the glucosides into glucose and acetone cyanohydrin. In a second step, drying or heating will

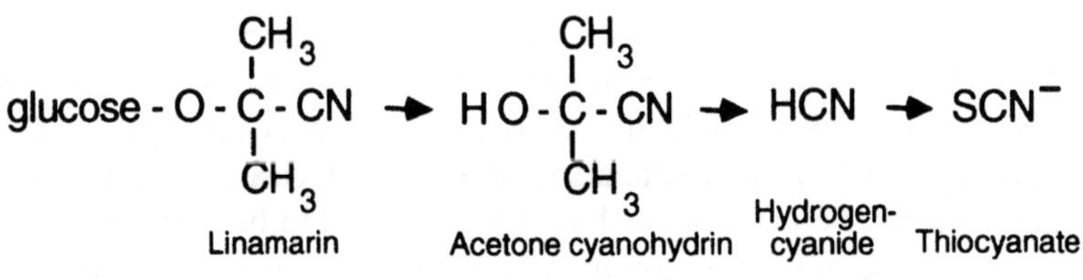

Figure 4. Cyanogenesis of cassava and cyanide detoxification in the human body.

break down cyanohydrin to yield hydrogen cyanide that evaporates into the air at 27°C (Figure 4). Fermentation through soaking in water followed by drying, as well as fermentation of grated root pulp in sacks followed by drying, are effective methods for cyanogen removal. During thousands of years, cassava farmers have preferentially chosen to grow bitter and toxic varieties, but the advantages provided by these varieties have not been elucidated (Bokanga et al, 1994b).

However, the biosynthetic pathway for the cyanogenic glucosides in cassava has been elucidated, and the structural genes of the rate limiting enzymes have recently been identified and sequenced (Koch et al, 1994). This means that plant genetics can start to manipulate the cyanogenic character of cassava before we know why poor farmers prefer to grow bitter and toxic varieties that need processing rather than growing nonbitter varieties that can be eaten fresh or directly boiled. It may be assumed that the task of upstream researchers should be to provide downstreamers with acyanogenic cassava varieties to be released to tropical farmers to avoid one of their many health hazards. However, explorative surveys by delta researchers in African savannah and rain forest areas with poor soils consistently reveal that farmers state that bitter and toxic varieties provide higher yields and that bitterness also provides protection against attacks by vermin and theft by humans (Goosen, 1994; Rosling et al, 1993). A better organoleptic character of the porridge made from flour from cassava roots has also been attributed to bitterness and toxicity of roots. These results indicate that the research priority should be to elucidate the biological and social role of cassava cyanogenesis in farming systems on marginal soils. It seems as though molecular plant genetics rather than removing glucosides should optimize the cyanogenesis of different cassava varieties. For these studies an acyanogetic cassava variety will be an important tool rather than an aim. The studies leading to this conclusion started on 21 August 1981 in a remote dispensary in a drought stricken part of northern Mozambique.

Konzo and Cassava Cyanogenesis in East Africa

Sister Lucia sent a message from the remote missionary dispensary where she had worked for the last 20 years. Serving as district medical officer in the closest town, I received her message. She asked me to come to her dispensary for an immediate investigation of more than thirty women and children who had been admitted during the last week due to sudden onset of paralysis of both legs. Few resources were available to study this epidemic of spastic paralysis in the drought-stricken eastern part of Nampula province in northern Mozambique, one of the most deprived human communities. However, swift support was provided by the Minister of Health, and this enabled our team of local medical officers and specialists from the Ministry of Health to do a classical epidemiology study of the geographical, temporal, and social distribution of 1100 cases of this new form of paralysis. The spastic paralysis of both legs appeared to be a new neurological disease entity. The onset was abrupt, and the disability permanent but not progressive.

We found the 1981 outbreak in Mozambique to be both geographically and temporally associated to monotonous consumption of bitter cassava roots that, due to food shortage, were consumed without the lengthy sun-drying that normally reduced toxicity before consumption. Weeks of discussions with the affected population and observations during visits to many households and fields yielded a consistent description of the diet before and during the drought. In normal years this population cultivated a mixture of more or less toxic and more or less bitter cassava varieties. Roots from the bitter and toxic varieties were only consumed after sundrying for several weeks and a further month of storage in the attic over the fireplace in their small mud houses. During the severe drought in 1981, the population lost all crops except cassava, and the diet was supplemented with a variety of wild plants. This information was obtained through unstructured interviews with open-ended questions, a method that we rapidly found yielded more relevant information than attempts to use structured questionnaires. The open-ended interviews gave a consistent picture of the events triggered by the drought:

> Only the most bitter varieties yielded well during this drought year. The normally nonbitter varieties gave very small roots that this year also were bitter and the formerly bitter varieties tasted more bitter than ever. We tried to live on wild plants to leave the cassava roots in the ground as long as possible. We waited for a rain that would make our cassava roots grow, but when all other food sources were exhausted our hunger made us up-root and eat the bitter roots without prolonged sun-drying. When we did so, many suffered acute poisonings some hours after the meals, most recovered but a few died. Not only did we eat the roots while still toxic, it also seems that people were more susceptible to the poison if they had no fish or beans to eat with the porridge made of cassava flour from bitter roots.

These stories of roots more bitter than ever, shortcuts in processing, and acute poisonings were both geographically and temporally associated to the occurrence of the paralysis, whereas no specific wild plant could be linked to the disease occurrence (Ministry of Health, Mozambique, 1984a,b).

Chemical analysis of blood and urine collected during the epidemic and in follow-up studies verified the highest reported dietary cyanide exposure from cassava. We could also chemically verify an extremely low dietary intake of sulphur aminoacids. In accordance with the farmer's statement we hypothesized an increased susceptibility to cyanide because sulphur is the substrate for cyanide to thiocyanate conversion in the human body (Cliff et al, 1985), as shown in Figure 4.

A total of 4,000 cases of konzo have been documented during the last decade in outbreaks in rural areas with severe agro-ecological problems in Mozambique, Tanzania, Zaire and Central African Republic (Tylleskär, 1994). The same relation between an outbreak of konzo and drought, diet, shortcuts in processing, and cyanide exposure was independently found by Howlett et al (1990) in Tanzania. His careful neurological studies established konzo as a specific human disease that we for reasons given above decided to name konzo. The studies in East Africa made us hypothesize that konzo is induced by a metabolic derangement from the combined effect of high cyanide exposure from insufficiently processed cassava and a very low intake of the sulphur aminoacids providing the substrate for its detoxication.

This etiology was questioned in a report on a konzo outbreak in a cassava eating population in Bandundu region in Zaire. The affected communities had never suffered any drought or food shortage, and the people always soaked bitter cassava roots in water, which is known as an effective method for removal of cyanogens. The epidemic occurrence and familial clustering were regarded as support for an infectious etiology (Carton et al, 1986). However, the absence of signs of infection, and laboratory, as well as advanced radiologic, investigations have ruled out infection as the cause of konzo (Tylleskär et al, 1993).

Konzo and Cassava Cyanogenesis in Zaire

The konzo outbreaks in Zaire have all occurred in very poor rural populations cultivating mainly bitter cassava varieties in collapsing shifting cultivation systems in the southern and central part of Bandundu region (Fresco, 1986). When trying to find an association between famines and konzo in zaire we failed to identify any food shortage periods behind the konzo outbreaks in Zaire. In accordance with Carton et al (1986) we found that all families in the affected communities soaked bitter roots before consumption. In addition the population of Kinshasa consumed large amounts of cassava roots from the affected area, but no konzo case had occurred in the Zairian capital. Furthermore, the bitter and potentially toxic cassava roots had been the main staple food for generations in the affected villages, but konzo outbreaks started abruptly in 1974 and have occurred every year since.

We therefore proceeded with focussed group discussions regarding details of cassava soaking practice over time. When good communication was established with the poor female farmers, they revealed that the outbreaks of konzo in dry seasons coincided with frequent episodes of acute poisoning following cassava meals. With embarrassment they admitted that these poisonings were due to shortcuts in the traditional soaking practices that were common during periods of intensive sales of cassava roots. The construction of a tarmac road from Kinshasa to Bandundu in 1974 coincided with the onset of the epidemics. Following the arrival of trucks purchasing cassava for Kinshasa, the traditional three days soaking was shortened to only 1–2 days as a response to intensive sales. Observations of women's processing practices revealed that the best soaked and dried roots were sold, whereas the families themself ate short-soaked and insufficiently dried roots. The cash earned was mainly used for paying school and health service fees. The sales therefore had high priority. Food chemistry experiments were designed and done in the villages to imitate the traditional and shortcut processing. They were performed under supervision by the local women in the affected communities and revealed that cassava flour from short-soaked roots contained high residues of acetone cyanohydrin that disappear after the transport to Kinshasa (Tylleskär et al, 1991, 1992). This can explain why no cyanide exposure or konzo case has been reported from Kinshasa.

Continued interviews with and observations of cassava producing women made us realize that the period of drying was as important as the period of soaking for avoidance of poisonings from bitter cassava roots. Our studies indicate that remaining cyanohydrins are the major source of dietary cyanide exposure. Cyanohydrins seem to withstand short boiling; however, in the alkaline environment of the gut they can rapidly and completely break down to CN resulting in dietary cyanide exposure. However, if roots are sufficiently dried, almost all cyanohydrins disappear (Tylleskär et al, 1992; Banea et al, 1992). Observations in Tanzania and Cuba suggest that cassava products with high amounts of glucosides can be ingested with limited CN exposure. Experiments verified that this compound pass through the human body unchanged (Hernandez et al, 1995; Mlingi et al, 1993).

Apparently minor differences in the length and sequence of short-cuts in the traditional soaking methods result in considerable variation in remaining cyanogens in the flour used to make the staple porridge. The cause of cyanide exposure is thus not the cassava consumption per se but the inability of poor populations to adhere to the effective traditional processing methods. We found that one further day of sun-drying was enough to render short-soaked roots safe for consumption (Banea et al, 1992). A very high cyanide exposure was also verified in the affected population, and some cases examined at the day of onset had significantly higher blood cyanide levels than controls from the nonaffected population in the same villages.

Cassava Cyanogenesis and the PPE-Spiral

The studies reviewed above indicate that dietary cyanide exposure from cassava may threaten public health in several rural areas of Africa. We have not proven that the cyanide exposure is the main cause, but the evidence is strong enough to urge prevention. The underlying cause is the vicious circle of poverty, population growth, and environmental deterioration known as PPE-spiral (Grant, 1994) that now threatens to create social collapse and unprecedented human misery in several parts of Africa. New outbreaks of konzo are reported from such areas (Tylleskär et al, 1994). However, the most important conclusion is that the toxic varieties of cassava, implicated in the causation of konzo, due to agricultural advantages saved many thousands of people from starvation in konzo affected populations. Paradoxically, the net effect of cassava cyanogenesis on human health may therefore be positive. From a public health perspective konzo may be of local importance, but from a continental perspective it is secondary to health problems like protein-energy malnutrition, tuberculosis, diarrhoeal diseases, malaria, AIDS, Iodine deficiency disorders, and vitamin A deficiency. As it seems that cassava cyanogenesis contributes to food security in the famine-threatened populations affected by konzo, this positive effect is more interesting for public health than the adverse effects resulting from cyanide exposure. Konzo epidemics are warnings of severe agro-ecological crisis with declining soil fertility and increased population pressure on marginal land. The best option appears to be promotion of effective processing of bitter cassava roots. Therefore our results can not be regarded as an argument for provision of acyanogenetic cassava (Tylleskär, 1994). It has even been suggested that modest cyanide exposure may have positive health effects through other mechanisms (Jackson, 1994).

Scientific disciplines tend to propose preventions for cassava toxicity that fit their area of research rather then the realities of the target population. Promotion of other food crops has been the main proposal from medical science, often without considering the agricultural suitability of the areas concerned. Breeding for low cyanide varieties and promotion of varieties before confirmed that they yield well under the harsh conditions in the local intercropping system has been the main action by agricultural science. Biotechnology has proposed acyanogenetic varieties. Food science argues for better processing, and this so far appears to be the most appropriate measure for the communities concerned. Both traditional processing methods and introduction of mechanical milling seem to secure safe cyanogen levels in cassava products.

Reasons for Use of Bitter Cassava

Most communities growing cassava as the major crop in shifting cultivation on poor soil seems to prefer bitter and toxic cultivars. Several studies indicate a west-to-east drift of bitter cassava cultivars from Central to East Africa in the last decades. Drought-threatened cassava communities with poor soil in Mozambique (Essers et al, 1992) and Tanzania (Mlingi et al, 1993) have preferences for bitter varieties. The claim by local farmers that bitter cultivars have higher yields has not been verified by agricultural studies in farming stations. In fact, we do not yet understand to what extent the preferences for bitter and toxic varieties are due to biological or social effects, i.e. due to higher yields or to better protection against theft. Molecular plant genetics can answer many questions related to cassava cyanogenesis. It should elucidate whether the biological advantages of cyanogenesis are due to cyanogenesis per se or to a tightly coupled genetic characteristics. If due to cyanogenesis, is the effect only due to the toxicity of released cyanide, or does cyanogenesis provide other advantages, such as a safe source of nitrogen. In the poor African communities relying on bitter cassava as their main protection against famine, the cyanogenesis may protect the crop against pests and predators, but it may also act only as a marker of an associated preferred characteristic. There is not even a clear understanding of the relation between bitterness and toxicity, since we do not know the molecular source of bitterness. Is the association between bitterness and toxicity an effect of many thousands of years of human selection, i.e. toxic varieties that are not bitter have been systematically destroyed as they constitute a danger? Or is the relation between bitterness and toxicity a much older effect of coevolution between plants and mammals? These questions can be addressed experimentally in upstream laboratories. However, it seems reasonable to try to advance more rapidly by establishing a systematic interlink between these laboratories, wherever they may be situated, and the existing biological knowledge of the poor cassava growing populations. It is reasonable to conclude that this interaction will be easier to establish if the advanced laboratories are situated close to the cassava farmers and if cassava consumers are involved as front line researchers. However, it takes more than geographic setting and ethnic origin to break established scientific traditions.

Conclusions

Needs and knowledge of small scale cassava farmers in areas with severe agro-ecological problems can be effectively identified by modern survey methods, and this information can guide agricultural development programs, basic agricultural science as well as biotechnology at molecular level (Fresco, 1986). This feed-back will mainly pass through the conventional interactions among the different levels of agricultural science. This chapter argues that science sometimes may advance more rapidly if we also give some opportunities for rain-researchers. Rain researchers can be social scientists with specific biological knowledge or biological scientists with trained social abilities. By direct observations and communications with the cassava communities these molecular scientists can offer a cost-effective focused interaction with upstream science. In our future studies on cassava cyanogenesis, we intend to use DNA fingerprinting for varietal identification when collecting farmers' experience of the bitterness, toxicity, and general performance of specific cassava varieties in different environments. To explain the preferred starch character and cooking quality of cassava flour in relation to the cyanogenic character of different cassava varieties, we start with in-depth interviews with women using different varieties for different processing methods. This will help design the studies that will clarify whether released cyanide modifies starch during processing or if cyanogenesis is coupled with special genetic starch qualities. Obviously the result is crucial for breeding programs. What has been said about cassava cyanogenesis applies for most interactions between plants and humans in the harsh life of farming households trying to escape the PPE-spiral. The future well-being of these populations in the least developed countries is of great importance for the global environment as world population will not stop growing until basic needs have been satisfied for all communities of the world. Advanced laboratory research in molecular plant genetics may contribute, and may do so much more effectively if the research methods are critically assessed. Molecular anthropology should often be considered as an option.

REFERENCES

Banea M, Poulter N, Rosling H (1992): Shortcuts in cassava processing and risk of dietary cyanide exposure in Zaire. *Food Nutr Bull* 14:137–143

Bokanga M, Ekanayake IJ, Dixon AGO, Porto MCM (1994a): Genotype-environment interactions for cyanogenic potential in cassava. *Acta Hort* 375: 131–140

Bokanga M, Essers S, Poulter N, Rosling H, Tewe O (1994b): International workshop on cassava safety. *Acta Hort* 375:7–19

Brader L (1994): *Director General's Report. International Institute for Tropical Agriculture, Annual Report, 1993.* Ibadan, Nigeria: IITA

Carton H, Kazadi K, Kabeya, Odio, Billiau A, Martens K (1986): Epidemic spastic paraparesis in Bandundu (Zaire). *J Neurol Neurosurg Psychiat* 49:620–627

Cliff J, Lundquist P, Mårtensson J, Rosling H, Sörbo B (1985): Association of high cyanide and low sulphur intake in cassava-induced spastic paraparesis. *Lancet* ii:1211–1213

Cock JH (1985): *Cassava—New Potentials for a Neglected Crop.* Denver: FA Praeger

Dufour DL (1989). Effectiveness of cassava detoxication techniques used by indigenous peoples in north west Amazonia. *Interciencia* 14:86–91

Essers AJA, Alsén P, Rosling H (1992): Insufficient processing of cassava induced acute intoxications and the paralytic disease konzo in a rural area of Mozambique. *Ecol Food Nutr* 27:17–27

Fresco LO (1986): *Cassava in Shifting Cultivation. A System Approach to Agricultural Technology Development in Africa.* Amsterdam: Royal Tropical Institute

Goosen F (1994): Performance of cassava marketing in Zaire (Thesis). Leuven, Belgium: Catholic University of Leuven

Grant J (1994): *The State of the Worlds Children 1994.* New York: Oxford University Press

Harrison P (1992): *The Third Revolution—Population, Environment and a Sustainable World.* London: Penguin books

Hernández T, Lundquist P, Oliveira L, Pérez Cristiá R, Rodriguez E, Rosling H (1995): The fate in humans of dietary intake of cyanogenic glycosides from roots of sweet cassava consumed in Cuba. In: *Natural Toxins* 3:114–117

Howlett WP, Brubaker GR, Mlingi N, Rosling H (1990): Konzo, an epidemic upper motor neuron disease studied in Tanzania. *Brain* 113:223–235

Jackson FL (1994): The bioanthropological impact of chronic exposure to sublethal cyanogens from cassava in Africa. *Acta Hort* 375:295–309

Koch BM, Sibbesen O, Swain E, Kahn RA, Liangcheng D, Bak S, Halkier BA, Lindberg-Möller B (1994): Possible use of a biotechnological approach to optimize and regulate the content and distribution of cyanogenic glucosides in cassava to increase food safety. *Acta Hort* 375:45–60

Melville B (1993): Rapid rural appraisal: its role in health planning in developing countries. *Trop Doctor* 23:55–58

Ministry of Health Mozambique (1984a): Mantakassa: an epidemic of spastic paraparesis associated with chronic cyanide intoxication in a cassava staple area in Mozambique. 1. Epidemiology and clinical and laboratory findings in patients. *Bull World Health Org* 62:477–484

Ministry of Health Mozambique (1984b): Mantakassa: an epidemic of spastic paraparesis associated with chronic cyanide intoxication in a cassava staple area in Mozambique. 2. Nutritional factors and hydrocyanic content of cassava products. *Bull World Health Org* 62:485–492

Mlingi NV, Assey V, Swai A, McLartey DG, Karlen H, Rosling H (1993): Determinants of cyanide exposure from cassava in a konzo-affected population in northern Tanzania. *Internat J Food Sci Nutr* 44:137–144

Patton MQ (1990): *Qualitative Evaluation and Research Methods*. London: Sage Publications

Rosling H, Mlingi M, Tylleskär T, Banea M (1993): Causal mechanisms behind human diseases induced by cyanide exposure from cassava. In: Roca WN, Thro AM, eds. *Proceedings from the First International Scientific Meeting of Cassava Biotechnology Network*; 1992 Aug 25–28; Cartagena, Columbia. Cali, Colombia: CIAT

Rosling H, Tylleskär T (1995): Konzo, an epidemic upper motoneuron disease associated to cassava and agro-ecological collapse in Africa. In: *Tropical Neurology*, Shakir RA, Newman PK, Poser CM, eds. London: Saunders

Scrimshaw NS, Gleason GR (1992): *Rapid Assessment Procedures—Qualitative Methodologies for Planning and Evaluation of Health Related Programmes*. Boston: International Foundation for Developing Countries

Sunderesan S, Nambisan B, Easwari Amna CS (1987): Bitterness in cassava in relation to cyanoglucoside content. *Indian J Agric Sci* 57:34–40

Trolli G (1938): Paraplégie spastique épidémique, Konzo des indigènes du Kwango. In: *Résumé des observations réunies, au Kwango, an sujet de deux affections d'órigine indéterminée*, Trolli G, ed. Brussels: Fonds Reine Elisabeth

Tylleskär T, Banea M, Bikangi N, Fresco L, Persson L-Å, Rosling H (1991): Epidemiological evidence from Zaire for a dietary aetiology of konzo, an upper motor neuron disease. *Bull World Health Org* 69:581–590

Tylleskär T, Banea M, Bikangi N, Cooke R, Poulter N, Rosling H (1992): Cassava cyanogens and konzo, an upper motoneuron disease found in Africa. *Lancet* 339:208–211

Tylleskär T, Howlett W, Rwiza H, Aquilonius S-M, Stålberg E, Lindén B, Mandahl A, Larsen HC, Brubaker GR, Rosling H (1993): Konzo: a distinct disease entity with selective upper motoneuron damage. *J Neurol Neurosurg Psychiat* 56:638–643

Tylleskär T, Légué F, Peterson S, Kpizingui E, Stecker P (1994): Konzo in the Central African Republic. *Neurology* 44:959–961

Tylleskär T (1994): The causation of konzo. Studies on a paralytic disease in Africa (Thesis). Uppsala, Sweden: Uppsala University

Zuckerkandl E (1963): Perspectives in molecular anthropology. In: *Classification and human evolution*. Washburn SL, ed. New York: Wenner-Gren Foundation

19

Structural Adjustment and Biotechnology Demand in South American Agriculture: The Case of Brazil

CARLOS A. MOREIRA-FILHO, MARCOS E. DA SILVA, GUILHERME L. S. DIAS AND ADRIANA P. A. XAVIER

Introduction

This chapter aims to analyze how the structural adjustment of some "Cone Sur" economies (an ongoing process in Brazil and Argentina and well advanced in Chile) influences the demand for agricultural biotechnologies in those countries. Brazilian agriculture, due to its diversity and size, is treated in depth as a case study. Windows of opportunity for investments in biotechnologies are identified at subsectoral level (e.g., forestry, cash-crops, etc.). The future of South American agricultural research and its prospective economic impact are also considered.

Structural Adjustment Processes and the Agricultural Sector

The performance of the Latin American agricultural sector has been largely determined by policies of general impact, like trade and macroeconomic policies, rather than policies specific to the sector (Garcia, 1989). Since the Great Depression (1929), and especially after World War II, Brazil, Chile, and Argentina have discriminated against agriculture through export taxes, overvalued exchange rates, and protection of the industrial sector.

During the 1960s and 1970s Latin American countries relied too much on external savings to finance import substitution industrialization programs. High oil prices and interest rates halted this growth strategy after 1979. Expansionary fiscal policies (adopted to maintain a large and inefficient public sector), soaring

The Impact of Plant Molecular Genetics
BWS Sobral, Editor

external debt service, and current account deficits led to recession and to the urgent need for structural adjustment policies. The adjustment process, a combination of macroeconomic stabilization and sectoral policy reform, involves reducing the balance of payments and government budget deficits and regaining economic efficiency in order to earn more foreign exchange. What follows is a brief description of the impact of the adjustment process on the agricultural sectors of Brazil, Chile, and Argentina.

The Brazilian agricultural growth process in the 1970s can still be characterized as an extensive system (Dias, 1989). Migration from traditional rural areas to the Center-West and incorporation of new land into cultivation compounded more than two-thirds of output growth. The new land came from cerrado (savannah) areas, where the soil is acid and weak in phosphorous, demanding investments in machinery, lime, and fertilizer. In the early 1970s subsidized rural credit (intended to be a compensation for government policies that discriminated against agriculture) supported these investments charging real interest rates close to zero. After 1973, due to the increase of inflation, these rates became strongly negative. The competition for funds within the government and the financing of the foreign private debt affected the entire agriculture credit system after 1979; investment credit was cut first, then commercial credit, and, finally, working capital (the real change in the volume of rural credit from 1980 to 1989 was minus 76%).

During the 1980s the Brazilian rural sector played an important role in the adjustment process of the Brazilian economy (Dias, 1989). Agricultural annual average growth was 2.6% in the 1980–1992 period (World Bank, 1994), in spite of several unsuccessful economic stabilization plans (Dias, 1992). Grains and oilseeds output growth, for instance, were based almost solely in land productivity (see Table 1). Selective fiscal incentives were given to export oriented activities, mainly to the ones linked to the industrial sector: edible oils; natural fibre textiles; wood industries including cellulose plants; orange juice; and meat industries. Import substitution policies began to be abandoned in 1988–1990. The fertilizer industry was privatized, and fertilizer prices showed a real drop of 27.2% between 1988 and 1993. Import tariffs on agricultural machinery were also substantially reduced between 1990 and 1993. Last, but not least, agricultural en-

Table 1. Brazilian Agricultural Productivity for Selected Crops—1940/94*

	Yield (ton/ha)					
Year	Cotton	Sugar-Cane	Coffee	Maize	Soybean	Wheat
1940	0.6	39.4	0.8	1.2		0.5
1950	0.4	39.5	0.8	1.3		0.8
1960	0.5	42.5	1.0	1.3	1.2	0.6
1970	0.4	46.2	0.6	1.4	1.1	0.9
1980	0.4	57.0	0.9	1.8	1.7	0.8
1988	1.0	62.6	0.9	1.9	1.7	1.6
1994	1.2	66.4	1.2	2.3	2.2	1.7

* IBGE (1994)

trepreneurs understood that land, the dominant production factor in the extensive system, was ceding its strategic role to technology.

The experience of the Argentine agricultural sector was somewhat more dramatic. Agriculture was the staple sector of the economy until the 1930s; the exploitation of the Pampas flat and rich lands pushed economic growth between 1860 and 1930. In the late 20s Argentina and the USA had similar crop yields. Then, from the mid-1930s until the late 1980s, Argentine economic policies taxed agriculture (directly through export taxes and indirectly by protecting the non-agriculture sector), increased government expenditures, and frequently kept an overvalued exchange rate. In 1985 the country was on the verge of hyper-inflation and a severe adjustment program, "Plan Austral", was launched. The program's first two years were disastrous for the agricultural sector; grains and oilseeds output drop from 43.9 million tons in 1984/85 to 32.6 million tons in 1986/87. This bad performance was largely determined by the maintenance of high export taxes on agriculture, fluctuating real interest rates, and the low supply of credit for the agricultural sector (Reca and Garramon, 1989). Agriculture annual average growth was 2.5% in the 1970s and fell to 0.3% in the 1980s (World Bank, 1991). The adjustment program has been substantially modified since 1991; export tax cutoff, gradual opening-up of the economy, and reduction of public expense (ANDIMA, 1993) are expected to revitalize the Argentine agricultural sector.

In the 1990s Argentine agriculture and agribusiness are accounting for about 54% of the country's exports. Trade liberalization encouraged investments in modern farm techniques, counterbalancing the effects of a transitory fall of world agricultural prices (The Economist, 1994). A simulation study for Argentine economy covering the 1930–1984 period (Cavallo, 1989) demonstrated that under a freer trade regime, with monetary and fiscal discipline, trade volume would be increased by 70% and the agriculture output would be 174% higher than the actual level.

Finally, it is worth commenting on Chilean agricultural sector experience. The 1930–1965 period was characterized by government intervention (seeking to buffer external shocks and to reduce current account deficits) and modest growth. The 1965–69 agricultural policies attempted, with some success, to improve farm output and productivity through agrarian reform and better relative prices (Barahona and Quiroz, 1989). The socialist government, 1970–73, disrupted rural production by carrying out a tight control of agricultural prices and a radical land reform. The military junta, 1973–1989, reorganized land reform and land transactions (to the benefit of medium-sized entrepreneurs willing to introduce modern farm technologies) and adopted a neutral pattern of incentives in accordance with a general trade liberalization and a structural adjustment program (Muchnik, 1992). Sectoral government expenditures were centered in agricultural research and infrastructure (ports, roads). After 1990, civilian governments maintained these policies, and the results are presented below.

Chilean agriculture had an average annual growth rate of 5.6% in the 1980–1992 period (World Bank, 1994). The percentage changes on average yields of corn and wheat comparing 1974–83 and 1984–88 periods were +82% and +60% respectively. From 1974 to 1990 Chilean exports grew 3.9 times and agricultural exports grew 11 times. Besides fresh fruit exports (that increased 37 times), other subsectors such as processed fruits, fresh and processed vegetables, and pulp and paper, had improved their export performance. Agricultural imports declined, and a consistent agricultural trade surplus was achieved (Muchnik, 1992).

Significant lessons were learned from the Chilean experience. The aggregate supply response of agriculture to increases in relative prices (part of the adjustment process) was sizable but slow. A simulation study of Coeymans and Mundlak (1991) on Chilean economy (1962–1982) showed that 10% increase in agricultural prices would promote an output growth of 3.2% in the first three years, but as much as 10.2% after ten years, demonstrating how important it is to persist in an economic policy. On the other hand, improvements in yields were due mostly to technological innovations, affecting the quantity and quality of agricultural and ancillary inputs (Muchnik, 1992). A Fundacion Chile report (1989) indicated a strong interaction between liberal reforms and Chilean agricultural technological progress.

In conclusion, these three countries learned that governments can contribute more effectively to economic growth by reducing trade distortions and barriers to innovation. Their agricultural sectors proved to be highly responsive to economic incentives. Under these assumptions, public expenditure on agriculture, in order to secure the welfare of countryside, should emphasize the transferring of appropriate technology to viable farmers. It should also help to bring the small farmers to the modern sector. The following analysis of the Brazilian agriculture is made according to these perspectives.

Brazilian Agriculture Challenges in the 90s

During the 1990s Brazilian agriculture and agribusiness must provide for a growing internal market and remain competitive in external markets plagued by protectionism and an excess of supply. Conservative projections for the year 2000 (considering 1.5% GDP annual average growth and no improvement in the income distribution) show that Brazil should produce: 115 million tons of grains (53% increase over the outstanding 93/94 harvest), 5.6 million tons of bovine meat (almost the double of 1994 output), 3.3 million tons of poultry meat, and 1.6 million tons of swine meat (80% and 34% more than 1990 figures, respectively). In order to reach these targets, the Brazilian agricultural sector needs to incorporate modern production technologies, including biotechnology, combined with efficient transport and storage infrastructure (Megido and Xavier, 1994; Moreira-Filho et al, 1992, 1994).

Another challenge is the progressive closing of the Center-West agricultural frontier; production growth is expected to come mainly from productivity increases and more intensive use of existing farmland. The end of the subsidized frontier expansion will lead to an agricultural growth path that is more economically efficient and regionally balanced, as well as environmentally less damaging (World Bank, 1990). This process will direct investments to activities such as agricultural research and extension, land resource management, and irrigation.

During the 1970s Brazilian agriculture exports became more diversified (processed products increased their relative importance), and the agricultural trade balance experienced a continuous growth (World Bank, 1990). The degree of openness of Brazilian agriculture, i.e., the relation between external sales and domestic sales, showed a sharp increase from 1971 to 1984 (Figure 1) and a tendency to steadiness thereafter. Recently, the domestic market has been absorbing more of the typically exportable agricultural products (soybean derived products, ethanol). On the other hand, some products traditionally sold in the internal market are being displaced by imports due to the fall of relative prices in the international market (rice, maize), and to the end of subsides to consumption (wheat). Therefore, the Brazilian agricultural production growth dynamics is beginning to be determined by the domestic market (Moreira-Filho et al, 1994).

Besides the internal demand, there have been other factors affecting the performance of Brazilian agricultural exports. In the period 1980–1993 the value of these exports had an annual average growth of only 0.5%, whereas the country's exports grew 4.8% annually. In fact, by correcting the 1980–1993 export prices according to the USA Consumer Price Index (CPI-U), Zandonadi (1994) showed that the values of Brazilian agricultural exports had an annual average decline of 3.2%, while total Brazilian exports had a real growth of 0.7% annually. Considering the main agricultural and processing complexes, only soybeans, orange juice, tobacco, and meat exports sustained growth in quantity and value. Taxation of agricultural exports and artificially low international prices (determined by protectionism and excess of supply) were among the causes of this loss in foreign exchange (Zandonadi, 1994).

Nevertheless, during the 1980s Brazil has succeeded in breaking into new export markets for orange juice (being the largest world exporter, with one million tons and sales of US $1.06 billion in 1993), poultry (third in the exporter's rank, just behind EEC and USA), and soybean products (second largest world exporter, with 14 million tons and sales of US $3.07 billion in 1993).

The Brazilian agricultural sector, considering the above account, will depend on technology for its future growth. In the 1970s, new seed varieties and technical packages for soybeans became available through EMBRAPA, the national enterprise for agricultural research, allowing this crop to be grown in the acidic soils of Center-West, the new frontier. In the 1980s, EMBRAPA developed new high-yield varieties of cotton, rice, beans and maize (Brazilian private companies also played an important role in maize genetic improvement), especially for the South and Southeast regions, the old frontier.

For the 1990s EMBRAPA is supposed to reach three main objectives: (1) collaboration with the research efforts of agricultural suppliers, cooperatives, and processors to meet the needs of commercial farmers; (2) research on the wide, strategic issues related to the development of the commercial sector (sustain-

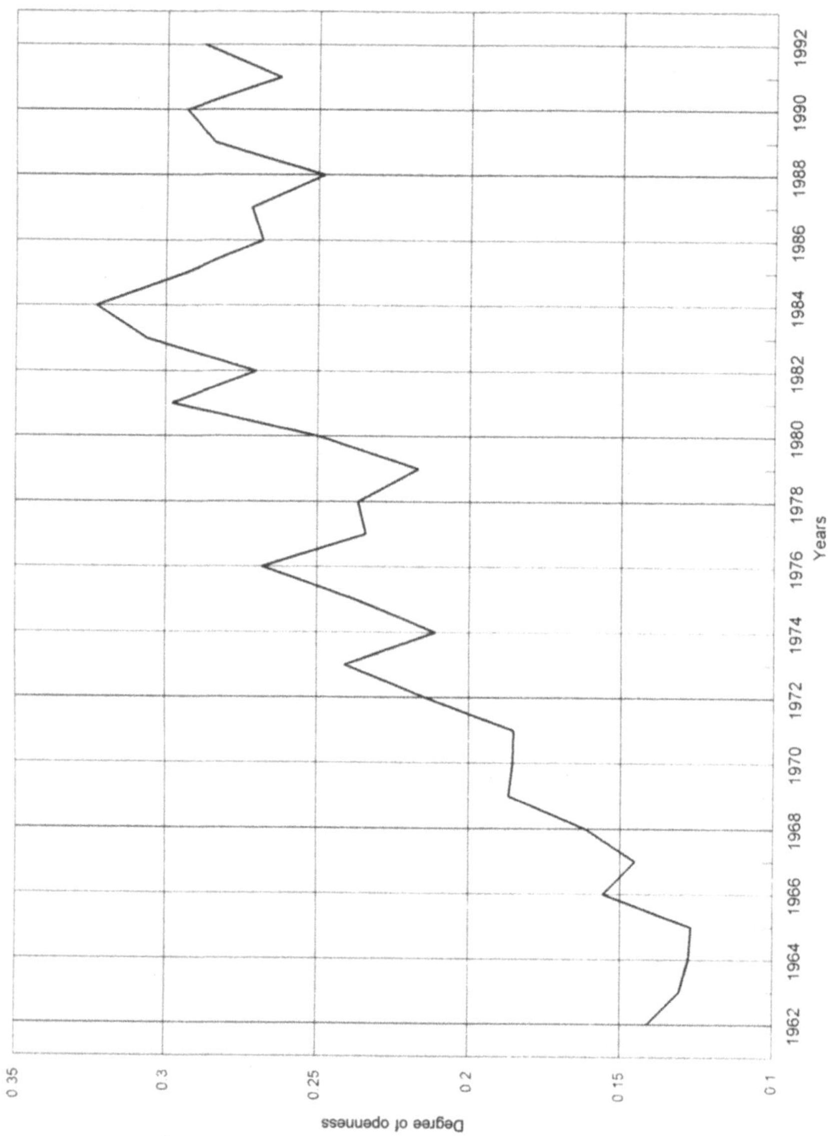

Figure 1. Degree of openness of Brazilian agriculture. The index is the ratio of external sales over domestic sales.

ability, environmental safety) and (3) selection and adaptation of technologies for small-scale farmers. In this context, universities and private companies will have an increasingly important function in promoting technical innovation for commercial agriculture. The main features of this model will be detailed in the following section.

Windows of Opportunity for Agricultural Biotechnology in Brazil

During the adjustment process of Brazilian economy, a few agricultural and industrial complexes emerged as determinants of the agricultural growth dynamics. Considering the export performance and relative importance in the domestic market, soybean, maize, orange juice, sugar-cane, meat, and the forestry industry are the leading segments of Brazilian agribusiness (Moreira-Filho et al, 1994; Burnquist et al, 1994). Technological trajectories and opportunities for investment are identified in these segments in which biotechnologies are relevant for competitiveness.

Biotechnologies for Brazilian Agriculture: An Overview

The scarce knowledge of commercially important plant and animal genes (which could be cloned and transmitted through transgenics), allied to local difficulties on patenting and regulation issues (Possas et al, 1994), render less impressive the short term perspectives for the use of genetic engineering in Brazilian agriculture (Moreira-Filho, 1994). Considering these circumstances, the most promising innovations for immediate application are in the area of genomic genetics, related to gene mapping using molecular markers, like RFLPs and RAPDs (Tanksley et al, 1989; Welsh and McClelland, 1990; Williams et al, 1990; Sobral and Honeycutt, 1993); they link classical breeding programs to modern population genetics and molecular biology, what is called marker-assisted selection.

SOYBEAN AND MAIZE

Brazil, with 18% of world soybean production, is the largest exporter of soybean cake, the second of grain, and the third of oil. Nevertheless, productivity (2.2 tons/ha) is still below that of the USA and Argentina (around 2.5 tons/ha), the country's main competitors in the international market. Maize production reached 29 million tons in the 1993/1994 harvest, but the internal market demand is growing fast (27.5 million tons in 1994), and Brazil faces the risk of becoming an importer of this crop in the late 1990s (Jank, 1994). Brazilian average productivity is low (2.3 tons/ha), although most farmers in the South and Center-West attain productivities well over 3.5 tons/ha (the average productivity in Center-West for the 1993/1994 harvest).

At present, an important biotechnological goal in the Brazilian soybean and maize segments (pursued by consortia between public and private institutions) is the use of marker-assisted selection to increase (or to introduce) pest and environmental stress resistance into commercial plant varieties and, in some cases, to improve their protein quality (FINEP, 1995). EMBRAPA's research on biological nitogen-fixation in soybean varieties is well advanced.

MEAT

Brazil was an important exporter of bovine meat, mainly processed meat, during the 1980s (536,000 tons in 1985). In 1994 about 350,000 tons were exported, 9.3% less than in 1993 and 20% less than in 1992. According to several studies (Agroanalysis, 1986/87), the international market could absorb, during the 1990s, around one million tons yearly of Brazilian bovine meat. The domestic production was 3.9 million tons in 1993 (Desouzart, 1994). The repressed internal demand (due to distortions in income distribution) is estimated in one million ton/yr (Homem de Mello, 1988), just above the total bovine meat exports of EEC in 1994 (950,000 tons).

Brazil is the fourth world producer (3.2 million tons in 1994) and third largest exporter (416,000 tons in 1993) of poultry meat. The technological level is well advanced (feed efficiency, disease control, production of hybrids), supporting international competitiveness. The production of swine meat has been around 1.2 million tons since 1987. Exports are restricted due to strict barriers which are not tariff related (Desouzart, 1994). The internal market offers reasonable prices to producers (predominance of vertical integration to the processing industry), but the stimulus from export demand is needed. The use of genetic resources is modest if one considers the recent developments in swine improvement.

The same approach of teaming up breeding and molecular characterization of germoplasms is beginning to be employed for improving animal productivity and resistance to diseases (Moreira-Filho, 1990). EMBRAPA is conducting a project for improving bovine productivity (meat and milk) in South-East Brazil based on the use of PCR-generated molecular markers. Genetic polymorphisms for growth hormone, kappa-casein, and beta-lactoglobulin are being studied. DNA fingerprinting is being used for better characterization of Nelore, Indubrasil, and Canchin races, which are important in Brazilian cattle.

PULP AND PAPER

In 1994 Brazil was the seventh in the rank of cellulose producers (5.7 million tons) and eleventh in paper production (5.6 million tons). The sector has experienced an accelerated growth since the mid-1970s, mainly due to the expansion of hardwood pulp production (Figure 2). This expansion has been made possible by the development of clonal propagation techniques and genetic improvement programs for *Eucalyptus* (Cruz et al, 1988). These procedures were pioneered by Aracruz (Brandão, 1984), a cellulose company based in the Brazilian state of Espírito Santo, and are now widely used by the major cellulose producers in the country.

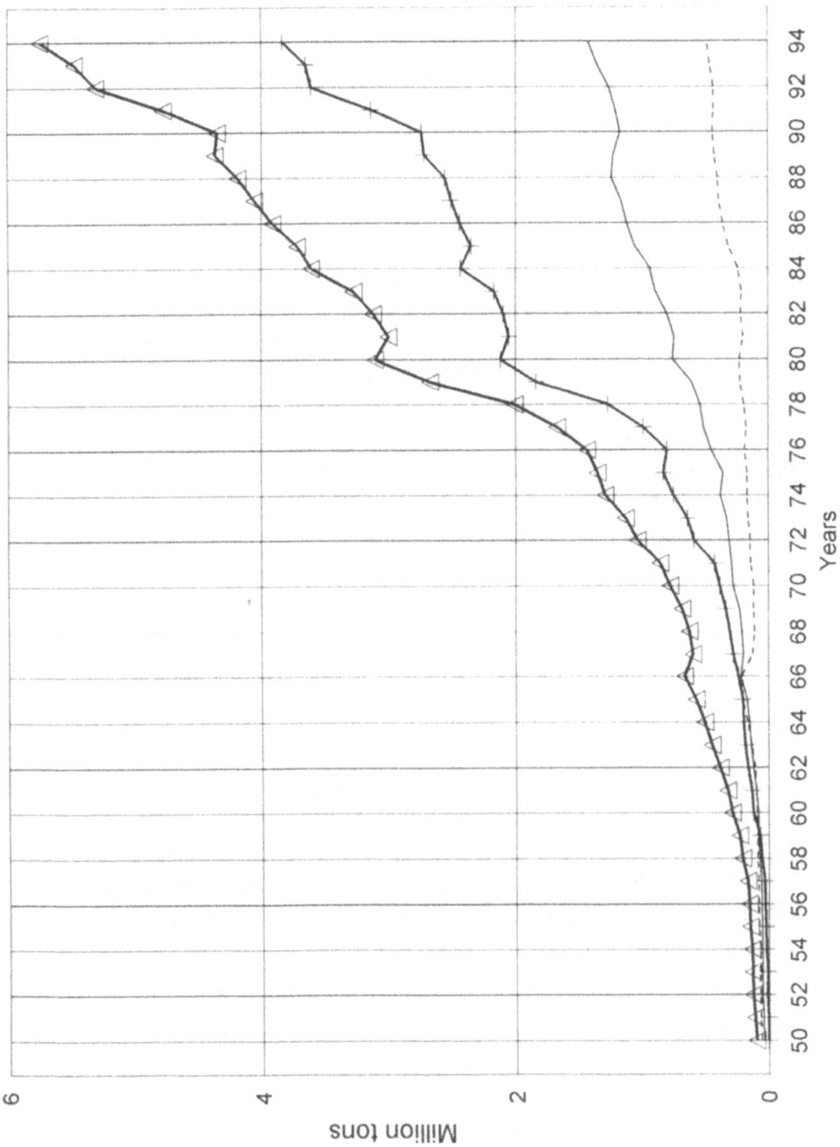

Figure 2. Brazilian production of hardwood (cross), softwood (solid line), and high-yield (dotted line) pulp. Total production is indicated by the triangle-marked line (ANFPC, 1994).

The sector of cellulose and paper is characterized by a high concentration of demand and supply, both in forestry and industrial segments, economies of scale, and a fairly stable technological frontier. The competitiveness of Brazilian companies depends mainly on the possession of a large forestry basis. High genetic quality and proximity to the pulp and bleaching units grant a cheap wood supply (Table 2). Financial strength, credit availability (governmental and private), updated production process, and a sophisticated trade network comprise the set of Brazilian competitive advantages (Jorge, 1994).

The internal market demand still has a good potential for growth. Brazil consumes only 30 kg/year/per capita of paper, compared to 70 kg/year/per capita in Portugal and 300 kg/year/per capita in the U.S. Economic development and improvements in income distribution are supposed to exert a significant impact in paper consumption. Total sales of pulp and paper amounted to US $5.6 billion in 1994, and exports represented 30% of this figure.

The pulp and paper market has a cyclic behaviour which is strongly correlated to the performance of the world economy. The long-run trend for the international market points to a growth around 3.7% per annum until year 2000 (FAO, 1982). Nevertheless, there are uncertainties regarding subsectoral demands; paper for printing is supposed to grow above the average figure, and paper board can have a slower growth due to substituition by recycled materials. For the Brazilian producers the question is how to diversify in a typical commodity market. One option is to increase their participation in the specialty markets. Another is to use their forestry technology, based on clonal propagation, for domesticating new varieties and species and to get a share of the expanding market of wood-based panels and laminated veneer for general construction.

SUSTAINABLE FORESTRY OPERATIONS

Projections for the year 2000 indicate a doubling of net removals of saw and veneer logs in Africa, Asia, and Oceania, and a threefold increase in Latin America. The demand for fine and specialty timber can well outstrip supplies, thus resulting in higher prices for species in great demand, or in the substitution of certain species by other wood or nonwood materials (Moreira-Filho et al, 1992).

Table 2. Cellulose and Pulp Wood Production Costs in Selected Countries—1994 (US$/ton)*

Country	Cellulose	Pulp Wood
Finland	300	171
Brazil	302	100
Portugal	315	131
Sweden	330	166
Spain	330	158
Southern USA	381	113
Eastern Canada	424	115
France	342	178

* ANFPC (1994)

On the other hand, some EEC wood importers, like Holland and the U.K., decided to stop buying forest products from nonrenewable sources after 1995.

The situation described above opens wide opportunities for sustainable forestry operations, mainly for laminated-veneer lumber. In Brazil these operations can adopt the technological trajectory first used by the local pulp and paper industry (Cruz et al, 1988), based on clonal propagation, genetic improvement (using molecular markers), and adaptation of new varieties. Further development of those technologies (and of complementary wood processing techniques) to meet the needs of industries involved in sustainable forestry operations will be necessary.

Sugarcane

The sugar-cane complex has been one of the more heavily controlled industries in Brazil (World Bank, 1990). The controlled prices have very little to do with internal or external market factors and depend on the ambitious goal of substituting sugar-cane derived ethanol for imported petroleum-based fuels, the Proálcool (National Alcohol Program).

Proálcool was planned after the 1973 oil crisis and gained momentum after 1977. In that year, sugar-cane derived ethanol represented only 0.04% of the value of domestic agricultural production, reaching 7.8% in 1991. Petrobras, the state oil company, buys all the anhydrous ethanol (used in gasoline blends) and half of the hydrated (fuel) ethanol, the rest being handled by private distributors. The international market of about one billion liters is dwarfed by the Brazilian's production of about 12–13 billion liters.

According to initial Proálcool projections, ethanol-fuelled vehicles could account for 95% of the new cars produced in Brazil from the mid-1980s to early 1990s. If this figure was attained, the demand for ethanol would be 40 billion liters in 1995 (Barros and Mello, 1988), almost 2.5 times the present country's production capacity, requiring huge investments in distilleries and an increase of 16.8 million ha in the sugar-cane cultivated area. Plans began to change in late 1980s, and the production of ethanol-fuelled vehicles decreased sharply, being adjusted to an ethanol supply around 13 billion liters per year.

The Achille's heel of Proálcool is the governmental price policy that subsidizes producers and consumers. As shown in Table 3, there is an enormous disparity of costs and productivity among ethanol producers. Official prices keep in business even the most inefficient producers, and the efficient ones have extraordinary profits, in a typical Ricardian model. The annual economic losses associated with this program were in order of US $1.0–1.3 billion during the 1980s (World Bank, 1989), or 0.3%–0.4% of GDP.

Competition could erase the original sins of Proálcool, since there are no technological barriers impeding gains of efficiency in the program (Campanario and Pizysieznig-Filho, 1988). The largest Brazilian producer of ethanol and sugar, COPERSUCAR, is involved in research programs for mapping the genome of sugar-cane (Al-Janabi et al, 1993; Silva et al, 1993, 1995; Sills et al, 1995) and

Table 3. Productivity Indices and Operational Costs for Ethanol According to Producer's Classes in the State of São Paulo

Classes	Agricultural Yield ton/ha	Industrial Yield l/ton	Global Yield l/ha	Operational Cost US$/l
1	89,3	78,9	7045,8	0,120
2	73,2	80,1	5863,3	0,118
3	86,2	68,9	5939,2	0,128
4	70,7	73,6	5203,5	0,128
5	70,9	71,6	5026,4	0,134
6	87,3	67,7	5910,2	0,126
7	72,8	68,5	4986,8	0,129
8	58,2	62,3	3625,9	0,141

* Campanário and Pizysieznig Filho (1988)

for developing new products of high added value from sugar, such as biodegradable plastics. Brazil is one of the lowest-cost sugar producers in the world (3 million tons were exported in 1993). With the end of government intervention in the sector, Brazilian producers would tend to rely more on biotechnologies than on subsides to increase their efficiency and develop new products and markets.

ORANGE JUICE

Brazil is the largest world exporter of frozen concentrated orange juice. Exports have grown continuously since 1976 and represented 18.5% of agricultural exports in 1992. Brazilian productivity is 50% below Florida's, but production costs are 48% less, leading to the competitiveness of Brazilian product in the international market (Burnquist et al, 1994).

In order to increase productivity Brazilian producers must control diseases caused by the bacteria *Xyella fastidiosa* and the cytric tristeza virus (CTV). At present, intensive research is being conducted on these subjects, involving genetic improvement assisted by molecular markers (Moreira-Filho et al, 1994). Other developments are being pursued in agricultural techniques (seedling, grafting, cloning). The industry has well-articulated linkages with the public research sector, mainly in the state of São Paulo which is responsible for 83% of orange production and 96% of the orange juice processing capacity.

Technological Trajectories for Brazilian Agriculture

A ranking of biotechnological trajectories for Brazilian agriculture, considering their relative importance in the next few years, is shown in Table 4. The preponderance of genomic genetics is due to the short term goals of important subsectors (like cash-crops, livestock and pulp and paper), involving better characterization of genetic resources. This effort will pave the way for the wide employment of genetic engineering in the future (which also depends on political and economic environments). There is a growing interest in biofertilization and reproductive efficiency, areas that may also incorporate genetic engineering approaches for their future development.

The Future of Agricultural Biotechnology Research in South America

Biotechnology owes a great deal to molecular biology. Even so, scientific discoveries are just a partial, although very important, component of modern biotechnology. To generate products and processes, basic research requires integration with other fields such as process technologies, which are essential for large-scale production, or plant and animal breeding for disseminating valuable genes identified through molecular markers. Moreover, several different agents must interact for transformation of science into technological innovation: research centers, companies, government departments, and financing organizations.

Consortia linking public research institutions and private entrepreneurs are mandatory for the development of commercial biotechnology in South American countries. Government expenditure should be moved from subsides to investment in science and technology (in 1993, in Brazil, Chile, and Argentina spending on research and development as 0.6%, 0.8%, and 0.4% of GDP, respectively). Governments should act more firmly in matters like biosafety rules and property rights in order to define the scenario for private and public investments.

Modern biotechnologies are centered in a few technological trajectories: recombinant DNA; genomic genetics; and advanced techniques in cell biology and bioprocessing. These trajectories increase extraordinarily the possibilities of controlling and modifying biological systems. However, according to the economy of technology parameters, these trajectories are not revolutionary, but coevolutionary, because: (1) biotechnologies are mainly applied in traditional industries (e.g., forestry, chemicals, food, and agriculture), being almost excluded from the dynamic segments of the service sector; (2) they usually depend on classical tech-

Table 4. Ranking of Agricultural Biotechnologies for Brazilian Agribusiness*

	Sectors			
Technological Trajectories	Forestry	Grains and Sugarcane	Horticulture	Livestock
Genomic Genetics	+++	+++	+++	+>
rDNA	+>	+>	+>	+/−
Vegetative Propagation and Micropropagation	+++	+	+++	−
Reproduction/ Physiology	+>	+>	+>	++
Microbial Inocula	+>	+>	−	−

* Key: +++ very important; +> growing importance; − not applicable/available. Moreira-Filho et al (1994).

nologies for the obtention of products and processes; and (3) they will not generate a large number of new consumer products (Buttel, 1989).

In spite of these facts, biotechnology is crucially important for global and South American agribusiness because it will become the technical basis of many important industrial sectors (OTA, 1991). This is particularly true for the three largest "Cone Sur" economies, as discussed earlier.

ACKNOWLEDGMENTS

This work was partially supported by FINEP/PADCT/SBIO grant No. 64.91.0288.00.

REFERENCES

Al-Janabi SM, Honeycutt RJ, McLelland M, Sobral B (1993): A genetic linkage map of *Sacharum spontaneum* L. SES 208. *Genetics* 134:1249–1260

ANDIMA (Associação Nacional das Instituições do Mercado Aberto) (1993): *Relatorio Economico Mercosul.* Rio de Janeiro: ANDIMA

ANFPC (Associação Nacional dos Fabricantes de Papel e Celulose) (1994): Avaliação do Setor de Papel e Celulose. *Celulose e Papel* 12(48):10–13

Agroanalysis: (1986/87): vol 10 No. 7 and vol 11 No. 7. Rio de Janeiro: FGV

Barahona P, Quiroz J (1989): Policy reforms and agricultural response: the case of Chile. In: Maunder AH, Valdes A, eds. *Proceedings of the Twentieth International Conference of Agricultural Economists*; Buenos Ayres: Darthmouth Aldershot

Barros JRM, Mello MCB (1988): Proálcool: avaliando o programa. *FIPE Informações* 101:9–10

Brandão LG (1984): The new eucalypt forest. In: *The Marcus Wallenberg Foundation Symposia Proceedings.* Falun Sweden: The Marcus Wallenberg Foundation

Burnquist HL, Bacchi MRP, Figueiredo NMS, Vieira WC (1994) *Liberalização comercial: um fator de desenvolvimento do setor agrícola brasileiro.* Brasília: IPEA

Buttel FH (1989): *Modern Biotechnology: Its Prospective Production and Socioeconomic Impact.* Ithaca NY: Cornell University

Campanário M, Pizysieznig-Filho J (1988): Proálcool: a saída é a competição. *FIPE Informações* 103:12–13

Cavallo D (1989): Agriculture and growth: the experience of Argentina, 1913–84. In: Maunder AH, Valdes A, eds. *Proceedings of the Twentieth International Conference of Agricultural Economists*; Buenos Ayres: Darthmouth Aldershot

Coeymans JE, Mundlak Y (1991): Aggregate agricultural supply response in Chile, 1962–1982. *Food Policy* 16:1–11

Cruz HN, Moreira-Filho CA, Silva ME (1988): *Avaliação do Impacto Economico das Biotecnologias no Setor de Celulose e Papel.* São Paulo: FIPE

Desouzart O (1994): O Brasil e o comércio mundial de carnes. In: *Políticas agrícolas e o comércio mundial*, Fagundes MH, ed. Brasília: IPEA

Dias GLS (1989): The role of agriculture in the structural adjustment process of Brazil. In: Maunder AH, Valdes A, eds. *Proceedings of the Twentieth International Conference of Agricultural Economists*; Buenos Ayres: Darthmouth Aldershot

Dias GLS (1992): Inflation and agriculture: ten years of high inflation and government debt in Brazil. In: Peters GH, Stanton BF, eds. *Proceedings of the Twenty-First International Conference of Agricultural Economists*; Tokyo: Darthmouth Aldershot

FAO (Food and Agriculture Organization; United Nations) (1992): *World Forest Products Demand and Supply 1990 and 2000*. Rome: FAO

FINEP (Financiadora de Estudos e Projetos) (1995): *Relação dos projetos contratados pelo Subprograma de Biotecnologia do PADCT 1990–1994*. Rio de Janeiro: FINEP

Fundacion Chile (1989): *Economic policy and agricultural development: the Chilean case 1974–1988*. Santiago: Fundacion Chile

Garcia JG (1989): The impact of macroeconomic policies on the performance of agriculture in Latin America. In: Maunder AH, Valdes A, eds. *Proceedings of the Twentieth International Conference of Agricultural Economists*; Buenos Ayres: Darthmouth Aldershot

Homem de Mello F (1988): *Gazeta Mercantil*. 31 Dec 1988:16

IBGE (Instituto Brasileiro de Geografia e Estatística) (1994): *Levantamento Sistemático da Produção Agricola*. Rio de Janeiro: IBGE

Jank MS (1994): O Brasil e o comércio mundial de grãos. In: *Políticas Agrícolas e o Comércio Mundial*, Fagundes MH, ed. Brasília: IPEA

Jorge, MM (1994): Competitividade da indústria de celulose. In: *Seminário: Estudo da Competitividade da Indústria Brasileira*. São Paulo: NPGCT-USP

Megido JLT, Xavier C (1993): *Marketing & Agribusiness*. São Paulo: Editoria Atlas

Moreira-Filho CA (1990): The economic impact of modern biotechnologies on animal production. *Rev Bras Reprod Anim Supl* 2:82–87

Moreira-Filho CA (1994): From molecular biology to biotechnology: the role of markets and institutions in the Brazilian scenario. *Ci Cult J Braz Assoc Adv Sci* 45:269–272

Moreira-Filho CA, Silva ME, Dias GLS (1992): Biotecnologia e desenvolvimento agroindustrial. In: *Anais do XVII Simposio Anual da ACIESP* Ribeirão Preto: São Paulo State Academy of Sciences

Moreira-Filho CA, Silva ME, Dias GLS, Xavier APA (1994): *Biotecnologia e Desenvolvimento Agroindustrial*. São Paulo: FIPE

Muchnik ER (1992): Impact of policy reforms on the agriculture sector in Chile. In: Peters GH, Stanton BF, eds. *Proceedings of the Twenty-First International Conference of Agricultural Economists*; Tokyo: Darthmouth Aldershot

OTA (1991): *Biotechnology in a Global Economy* (OTA-BA-494). Washington, DC: US Government Printing Office

Possas ML, Salles-Filho SLM, Mello ALA (1994): *O Processo de Regulamentação na Biotecnologia: as Inovações na Agricultura e na Produção Agroalimentar.* Brasília: IPEA

Reca LG, Garramon CJ (1989): Argentine interactions between the adjustment programme and the agricultural sector. In: Maunder AH, Valdes A, eds. *Proceedings of the Twentieth International Conference of Agricultural Economists*; Buenos Ayres: Darthmouth Aldershot

Sills GR, Bridges W, Al-Janabi SM, Sobral BWS (1995): Genetic analysis of agronomic traits in a cross between sugarcane (*Saccharum officinarum L.*) and its proposed progenitor (*S. robustum* Brandes and Jesw. ex Grassl). *Mol Breed*: in press

Silva JAG, Sorrels ME, Burnquist WL, Tanskley SD (1993): RFLP linkage map and genome analysis of Sacharum spontaneum. *Genome* 36:782–791

Silva JAG, Honeycutt RJ, Burnquist W, Al-Janabi SM, Sorrels ME, Tanksley SD, Sobral BWS (1995): *Saccharum spontaneum* L. 'SES 208' genetic linkage map combining RFLP- and PCR-based markers. *Mol Breed* 1:165-179

Sobral BWS, Honeycutt RJ (1993): High output genetic mapping of polyploids using PCR-generated markers. *Theor Appl Genet* 86:105–112

Tanksley SD, Young ND, Patterson AH, Bonierbale NW (1989): RFLP mapping in plnt breeding: new tools for an old science. *Biotechnology* 7:257–264

The Economist (1994): A survey of Argentina. *The Economist* 333 (7891 suppl):1–19

Welsh J, McClelland M (1990): Fingerprinting genomes using PCR with arbitrary primers. *Nucl Acids Res* 18:7213–7218

Williams JGK, Kubelik AR, Livak KJ, Rafalski JA, Tingley SV (1990): DNA plymorphism amplified by arbitrary primers are useful as genetic markers. *Nucl Acids Res* 18:6531–6535

World Bank (1989): *Public expenditure, subsidy policies and budgetary reform.* Washington DC: World Bank

World Bank (1990): *Brazil agricultural sector review: policies and prospects (Report No. 7798-BR).* Washington DC: World Bank

World Bank (1991): *World Development Report.* New York: Oxford University Press

World Bank (1994): *World Development Report.* New York: Oxford University Press

Zandonadi R (1994): Comportamento das exportacões brasileiras 1980–1993. *Revista de Política Agrícola* 3:5–10

Keyword Index

This index was established according to the keywords supplied by the authors.

GPSR Compliance

The European Union's (EU) General Product Safety Regulation (GPSR) is a set of rules that requires consumer products to be safe and our obligations to ensure this.

If you have any concerns about our products, you can contact us on ProductSafety@springernature.com

In case Publisher is established outside the EU, the EU authorized representative is:

Springer Nature Customer Service Center GmbH
Europaplatz 3
69115 Heidelberg, Germany

Batch number: 10577878

Printed by Printforce, the Netherlands